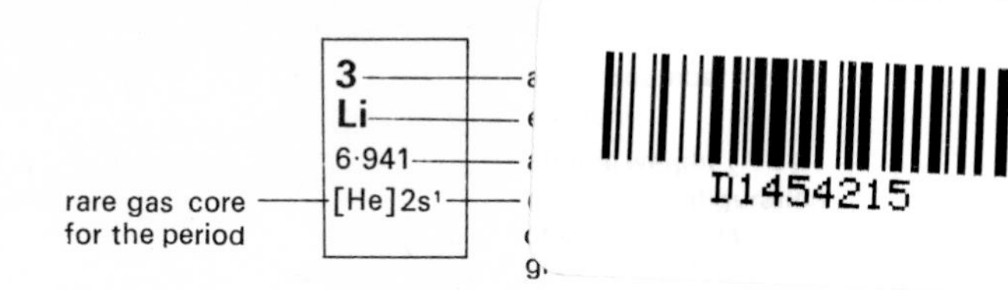

		Group III	Group IV	Group V	Group VI	Group VII	Group 0
		5 B 10·811 $[He] 2s^22p^1$	**6 C** 12·011 $[He] 2s^22p^2$	**7 N** 14·007 $[He] 2s^22p^3$	**8 O** 15·999 $[He] 2s^22p^4$	**9 F** 18·998 $[He] 2s^22p^5$	**10 Ne** 20·179 $[He] 2s^22p^6$
		13 Al 26·98 $[Ne] 3s^23p^1$	**14 Si** 28·086 $[Ne] 3s^23p^2$	**15 P** 30·974 $[Ne] 3s^23p^3$	**16 S** 32·064 $[Ne] 3s^23p^4$	**17 Cl** 35·453 $[Ne] 3s^23p^5$	**18 Ar** 39·948 $[Ne] 3s^23p^6$
29 Cu 63·546 $[Ar] 3d^{10} 4s^1$	**30 Zn** 65·37 $[Ar] 3d^{10} 4s^2$	**31 Ga** 69·72 $[Ar]3d^{10} 4s^24p^1$	**32 Ge** 72·59 $[Ar]3d^{10} 4s^24p^2$	**33 As** 74·922 $[Ar]3d^{10} 4s^24p^3$	**34 Se** 78·96 $[Ar]3d^{10} 4s^24p^4$	**35 Br** 79·909 $[Ar]3d^{10} 4s^24p^5$	**36 Kr** 83·80 $[Ar]3d^{10} 4s^24p^6$
47 Ag 107·87 $[Kr] 4d^{10} 5s^1$	**48 Cd** 112·40 $[Kr] 4d^{10} 5s^2$	**49 In** 114·82 $[Kr]4d^{10} 5s^25p^1$	**50 Sn** 118·69 $[Kr]4d^{10} 5s^25p^2$	**51 Sb** 121·75 $[Kr]4d^{10} 5s^25p^3$	**52 Te** 127·60 $[Kr]4d^{10} 5s^25p^4$	**53 I** 126·90 $[Kr]4d^{10} 5s^25p^5$	**54 Xe** 131·30 $[Kr]4d^{10} 5s^25p^6$
79 Au 196·97 $[Xe]4f^{14} 5d^{10}6s^1$	**80 Hg** 200·59 $[Xe]4f^{14} 5d^{10}6s^2$	**81 Tl** 204·37 $[Xe]4f^{14} 5d^{10}6s^26p^1$	**82 Pb** 207·19 $[Xe]4f^{14} 5d^{10}6s^26p^2$	**83 Bi** 208·98 $[Xe]4f^{14} 5d^{10}6s^26p^3$	**84 Po** (210) $[Xe]4f^{14} 5d^{10}6s^26p^4$	**85 At** (210) $[Xe]4f^{14} 5d^{10}6s^26p^5$	**86 Rn** (222) $[Xe]4f^{14} 5d^{10}6s^26p^6$

65 Tb 158·92	**66 Dy** 162·50	**67 Ho** 164·93	**68 Er** 167·26	**69 Tm** 168·93	**70 Yb** 173·04	**71 Lu** 174·97

97 Bk (247)	**98 Cf** (249)	**99 Es** (254)	**100 Fm** (253)	**101 Md** (256)	**102 No** (253)	**103 Lw**

Penguin Education
Penguin Library of Physical Sciences

The Typical Elements

A. G. Massey

The Typical Elements

A.G. Massey

Penguin Books

Penguin Books Ltd, Harmondsworth, Middlesex, England
Penguin Books Inc, 7110 Ambassador Road, Baltimore, Md 21207, USA
Penguin Books Australia Ltd, Ringwood, Victoria, Australia

First published 1972

Made and printed in Great Britain by
William Clowes & Sons Ltd, London, Beccles and Colchester
Set in Monophoto Times

Contents

Editorial Foreword 9
Preface 11

1 **The Periodic Table** 13

1.1 Quantum numbers 13
1.2 s-Orbitals 16
1.3 p-Orbitals and d-orbitals 18
1.4 The hydrogen atom 21
1.5 The helium atom 22
1.6 The first short period: lithium to neon 24
1.7 The second short period: sodium to argon 29
1.8 The first long period 29
1.9 The second long period 31
1.10 The heavy elements 31
1.11 Possible extensions to the periodic table 32
1.12 The role of outer *n*d orbitals in bonding 32
1.13 p_π–p_π Bonding 35
1.14 d_π–p_π Bonding 38

2 **Hydrogen** 40

2.1 Isotopes of hydrogen 40
2.2 *Ortho* and *para* hydrogen 43
2.3 The position of hydrogen in the periodic table 44
2.4 Atomic hydrogen 46
2.5 The chemistry of hydrogen 46
2.6 The hydrides of boron 54
2.7 Metallic hydrides 59
2.8 The hydrogen bond 60

3 **Group I. The Alkali Metals: Lithium, Sodium, Potassium, Rubidium, Caesium and Francium** 67

3.1 Introduction 67
3.2 Solutions of the alkali metals 78
3.3 Occurrence of the alkali metals 79
3.4 Extraction 80
3.5 Compounds of the alkali metals 81
3.6 A comparison of copper with the alkali metals 90

4 **Group IIa. The Alkaline-Earth Metals: Beryllium, Magnesium, Calcium, Strontium, Barium and Radium** 93

4.1 Introduction 93
4.2 Occurrence of the alkaline-earth metals 99
4.3 Extraction 100
4.4 The elements 101
4.5 Compounds of the alkaline-earth metals 102
4.6 Organometallic compounds 107
4.7 Similarities between beryllium and aluminium 110

5 **Group IIb. Zinc, Cadmium and Mercury** 111

5.1 Introduction 111
5.2 A comparison of zinc with beryllium 115
5.3 Occurrence of zinc, cadmium and mercury 117
5.4 Extraction 118
5.5 The elements 119
5.6 Compounds of zinc, cadmium and mercury 119
5.7 Complexes of zinc, cadmium and mercury 121
5.8 Organometallic compounds 123
5.9 Mercury(I) derivatives 124

6 **Group III. Boron, Aluminium, Gallium, Indium and Thallium** 126

6.1 Introduction 126
6.2 Occurrence of the group III elements 130
6.3 Extraction 131
6.4 The elements 132
6.5 Compounds of the group III elements 132
6.6 Organometallic compounds 141
6.7 Hydroboration 142
6.8 The chemistry of thallium(I) 143

7 **Group IV. Carbon, Silicon, Germanium, Tin and Lead** 144

7.1 Introduction 144
7.2 Occurrence of the group IV elements 152
7.3 Extraction 152
7.4 The elements 153
7.5 Hydrides of the group IV elements 158
7.6 Oxides of carbon 159
7.7 Oxides of silicon, germanium, tin and lead 166
7.8 Silicates 168
7.9 Halides of the group IV elements 169
7.10 Carbides 171
7.11 Organo-derivatives of silicon, germanium, tin and lead 172
7.12 Semiconductor properties of the group IV elements 173

8 **Group V. Nitrogen, Phosphorus, Arsenic, Antimony and Bismuth** 176

8.1 Introduction 176
8.2 Occurrence of the group V elements 184
8.3 Extraction 184
8.4 The elements 185
8.5 Hydrides of the group V elements 187
8.6 Oxides and oxy-acids of the group V elements 191
8.7 Further compounds of the group V elements 204
8.8 Organo-derivatives of the group V elements 210

9 **Group VI. Oxygen, Sulphur, Selenium, Tellurium and Polonium** 211

9.1 Introduction 211
9.2 Occurrence of the group VI elements 218
9.3 Extraction 219
9.4 The elements 220
9.5 Semiconductor properties of selenium 222
9.6 Hydrides of the group VI elements 223
9.7 Peroxides and peroxy-salts 230
9.8 Metallic oxides, sulphides, selenides and tellurides 231
9.9 Oxides of sulphur, selenium and tellurium 232
9.10 Oxy-acids of sulphur, selenium and tellurium 234
9.11 Halides of the group VI elements 237
9.12 Organo-derivatives of the group VI elements 239

10 **Group VII. The Halogens: Fluorine, Chlorine, Bromine, Iodine and Astatine** 241

10.1 Introduction 241
10.2 Mixed halogen compounds 245
10.3 Pseudohalides and pseudohalogens 250
10.4 Occurrence of the halogens 251
10.5 Extraction 252
10.6 The elements 253
10.7 Halogen hydrides 254
10.8 Oxides and oxy-acids of the halogens 255
10.9 Fluorocarbons 263

11 **Group 0. The Rare Gases: Helium, Neon, Argon, Krypton, Xenon and Radon** 265

11.1 Introduction 265
11.2 Occurrence of the rare gases 268
11.3 The elements 269
11.4 Compounds of the rare gases 269

12 **Non-Aqueous Solvents** 279

12.1 Introduction 279
12.2 Solubility 281
12.3 Self-ionization of solvents 282
12.4 Acid–base behaviour in non-aqueous solvents 284
12.5 Levelling action of a protonic solvent 286
12.6 Metal solutions in non-aqueous solvents 287
12.7 Glacial acetic acid as an ionizing solvent 288
12.8 Liquid ammonia as a solvent 292
12.9 Liquid sulphur dioxide as a solvent 296

Formula Index 301

Subject Index 308

Editorial Foreword

The chemistry section of the Penguin Library of Physical Sciences is planned to cover the normal content of honours degree courses of British universities, and of other courses of comparable standard – CNAA, Royal Institute of Chemistry, etc. Most of the optional subjects, which are becoming a prominent feature of present-day courses, will be included.

In inorganic chemistry undergraduate degree courses, it has often been customary to rely upon one large comprehensive textbook, using smaller books only for special topics or for ancillary studies. This situation must now change; no truly comprehensive single book of inorganic chemistry exists and, as courses in chemistry continue to proliferate, ancillary and other topics cover a very wide range. In these circumstances, it is more appropriate to provide the student with a series of smaller texts, covering both 'basic' and 'ancillary' aspects of inorganic chemistry, from which a choice appropriate to a particular course of study can be made. Moreover, recent changes in GCE A-level curricula imply that student entrants to chemistry degree courses will be better equipped in their understanding of the principles of the subject; but these students will also need, and expect, readily accessible sources of information about the more factual side of inorganic chemistry to be available. This need has been kept in mind during the preparation of the present series of books.

A.K.H.

Preface

This book has been written for students of chemistry starting more advanced studies. They may be taking HNC or the graduate examination of the Royal Institute of Chemistry, or they may be doing a university degree course.

Chapter 1 is a brief summary of some of the theoretical ideas in inorganic chemistry. One aspect discussed in this chapter is worthy of comment here: this concerns the role of *n*d orbitals in σ-bonding. For molecules of the non-transition elements in which the central atom has a coordination number greater than four, there are two alternative descriptions, each of which is an approximation to a full molecular-orbital treatment. One alternative is that an adequate description at this level is provided by sp^3d and sp^3d^2 hybridization schemes. The other is to use the three-centre, four-electron bond approach. In molecules such as SF_6, ClF_3 and XeF_4, there is no available experimental technique which can indicate whether d-orbitals play a large or small part in the bonding; thus both descriptions have been given at the relevant places. But in molecules like phosphine, where the bond angles are close to 90°, I have neglected the possibility of small amounts of d-character in the bonds and suggested that the central atoms use pure p-orbitals to bond to the ligands.

Chapters 2 to 11 inclusive cover the chemistry of the non-transition elements taken in order of their groups. The first part of each chapter (except that on hydrogen) discusses the general features of the group and any differences or trends to be expected in the properties of the elements; this is followed by a brief description of the occurrence, extraction and properties of each element, and the properties of their simple compounds. As non-aqueous solvents are derived from main-group elements, it seemed appropriate to summarize such solvents in the final chapter, rather than to describe the properties of, for example, ammonia in Chapter 8 and sulphur dioxide in Chapter 9.

I should like to take this opportunity to thank my former colleagues at Queen Mary College, London, for many stimulating discussions over the past nine years; these have contributed greatly to the content of the book. Peter Thornton, John Hagel and Michael Linehan kindly read the whole manuscript and made many valuable comments. My wife, Sylvia, showed great patience in typing much of the manuscript; in this task she had considerable help from Emma, without which the job would have been finished much earlier.

Chapter 1
The Periodic Table

1.1 Quantum numbers

Although basically the periodic table is an array of the elements in order of their atomic number, it is the Pauli exclusion principle which gives us an understanding as to why elements with similar chemical properties fall into such well-defined groups. One way of stating this principle is that no two electrons within the same atom may have an identical set of quantum numbers, the four quantum numbers themselves arising from solutions of Schrödinger's wave equation. These quantum numbers are a convenient shorthand way of summarizing the energy of an electron or, more commonly but strictly less accurately, the energy of a particular orbital in which the electron is to be found. The principal quantum number n summarizes the predominant part of an orbital's energy, whilst the secondary (azimuthal) quantum number l and the magnetic quantum number m_l essentially describe the particular orientation of the orbital in space. The fourth quantum number m_s describes the 'spin' state of the electron within the orbital, and there are only two allowable values for m_s; these are usually represented by $+\frac{1}{2}$ and $-\frac{1}{2}$. It follows from Pauli's exclusion principle that two electrons may only occupy the same orbital if their spins are $+\frac{1}{2}$ and $-\frac{1}{2}$ (such electrons are said to have their spins 'paired').

The quantum numbers may take any of the following values:

$n = 1, 2, 3, \ldots, \infty$;
$l = 0, 1, 2, \ldots, n - 1$ for a given value of n;
$m_l = l, l - 1, l - 2, \ldots, 2, 1, 0, -1, \ldots, -l$;
$m_s = +\frac{1}{2}, -\frac{1}{2}$;

and the conventional shorthand notation for describing a particular orbital in terms of these numbers is arrived at in the following way.

(a) Orbitals having $l = 0$ are termed s-orbitals; orbitals having $l = 1$ are termed p-orbitals; orbitals having $l = 2$ are termed d-orbitals; orbitals having $l = 3$ are termed f-orbitals. After $l = 3$ the orbital letters follow the alphabetical sequence, so that orbitals with $l = 4$ are g-orbitals, those with $l = 5$ are h-orbitals, etc. However, only s-, p-, d- and f-orbitals are of chemical significance and we may ignore the others.

(b) A numerical prefix in front of s, p, d or f describes the principal quantum number to which the orbital belongs: a 2s orbital is an orbital having $n = 2$, $l = 0$ and a 3d orbital has $n = 3$, $l = 2$, for example.

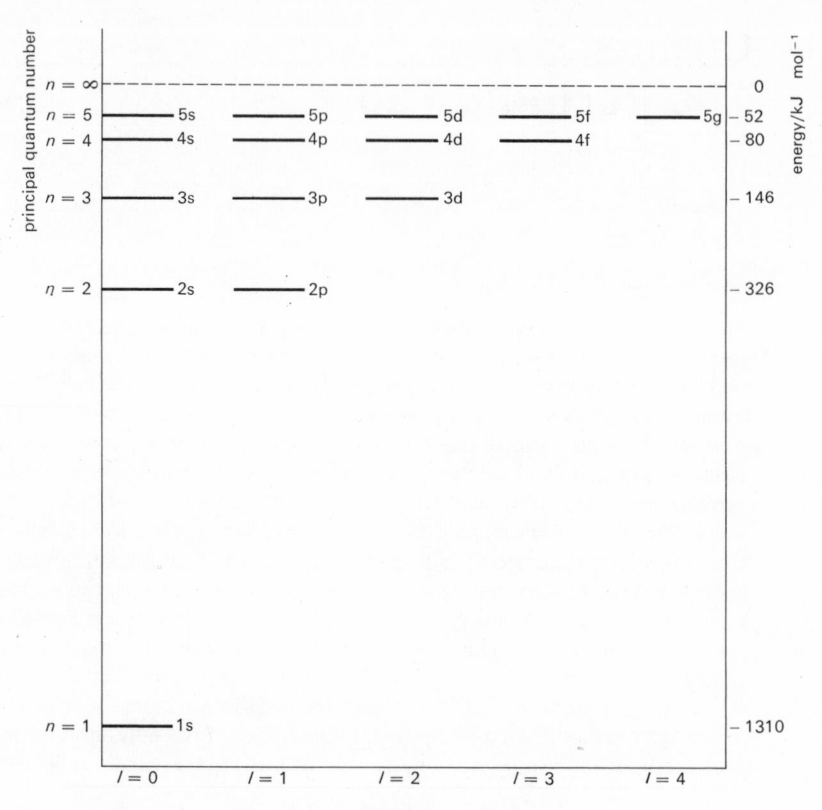

Figure 1 Some electronic energy levels in the hydrogen atom.
(Note: the energy of the nth level is equal to $-1310/n^2$ kJ mol^{-1})

The sequence of orbital energies for the hydrogen atom is shown in Figure 1. It will be noticed that the energies of the orbitals are negative. This implies that the energy of the hydrogen atom, whether the atom is in the ground state ($n = 1$) or an excited state ($n \geqslant 2$), is lower than the energy of the isolated electron and the isolated nucleus ($E = 0$ kJ), a condition which is required for a stable atom.

The magnetic quantum number m_l may take values from $+l$ to $-l$, including zero. For an s-orbital ($l = 0$), m_l can only be zero but, for a p-orbital ($l = 1$), m_l may have the values $+1, 0, -1$, which means that there are always three p-orbitals for $n \geqslant 2$. In the absence of a magnetic field, the three p-orbitals are degenerate (i.e. they have identical energies) but, when a magnetic field is applied to the hydrogen atom, this degeneracy is lifted. This occurs because the orientation in space of the three p-orbitals is different and therefore each orbital interacts with the applied magnetic field in a different manner (Figure 2). Similarly, d-orbitals and f-orbitals are fivefold ($m_l = 2, 1, 0, -1, -2$) and sevenfold ($m_l = 3, 2, 1, 0, -1, -2, -3$) degenerate, respectively.

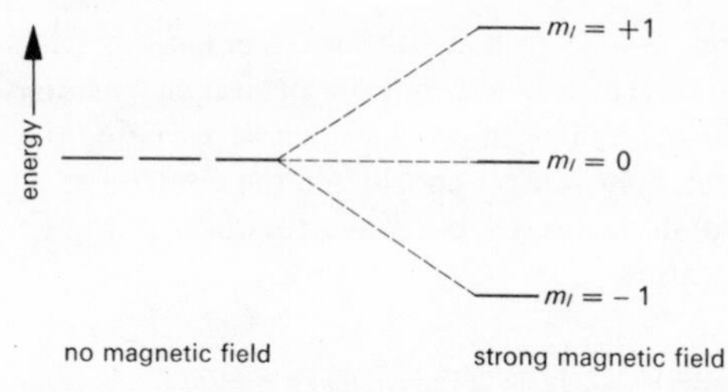

Figure 2 The lifting of the degeneracy of p-orbitals in a strong magnetic field

From Figure 1 it is obvious that, for the hydrogen atom, the s-, p-, d- and f-orbitals of a particular quantum shell are of equal energy – in other words, the energy of the orbitals is defined by the principal quantum number n. In all other atoms the energy of each orbital depends on both n and l; the effect of this on the sequence of orbital energies for the light elements is shown in Figure 3. To understand this partial dependence of the energy on the secondary quantum number l requires a short discussion of orbitals.

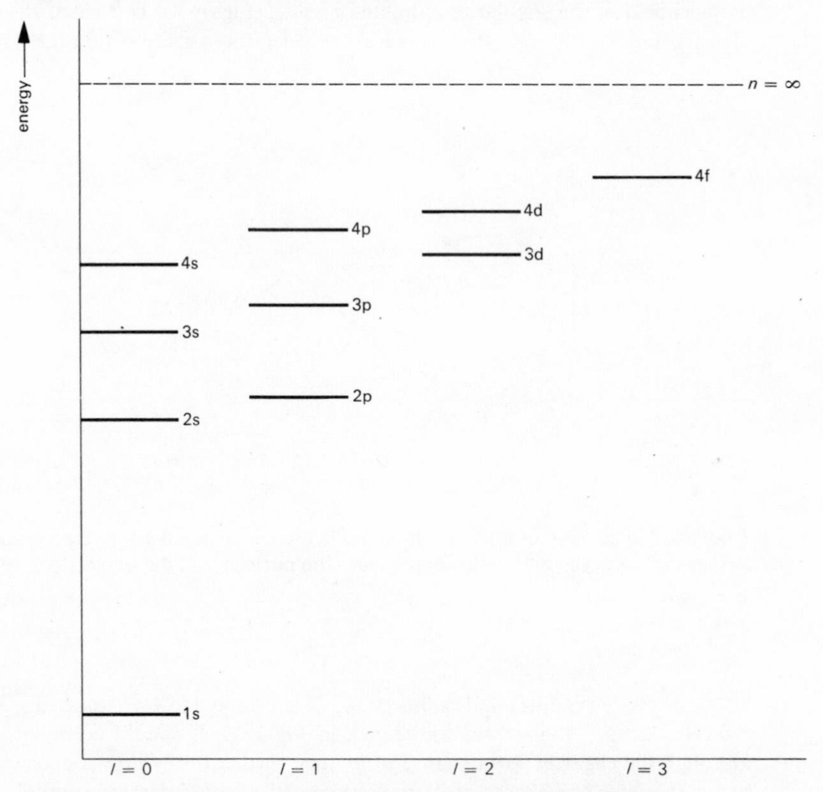

Figure 3 The sequence of orbital energy levels for the lighter elements

At the present time it is impossible to solve the wave equations for multi-electron atoms and it should be remembered that the orbital shapes discussed below are those calculated for the hydrogen atom. When first writing the wave equation of the hydrogen atom, Schrödinger considered the electron as a three-dimensional standing wave and connected the wave function ψ to the total electronic energy E by the equation

$$\frac{1}{r^2}\frac{\partial}{\partial r}\left(r^2\frac{\partial\psi}{\partial r}\right)+\frac{1}{r^2\sin\theta}\frac{\partial}{\partial\theta}\left(\sin\theta\frac{\partial\psi}{\partial\theta}\right)+\frac{1}{r^2\sin^2\theta}\frac{\partial^2\psi}{\partial\phi^2}+$$

$$+\frac{8\pi^2\mu}{h^2}\left[E+\frac{e^2}{r}\right]\psi=0,$$

where μ is the reduced mass of the electron.

1.2 s-Orbitals

These orbitals are spherically symmetrical because their wave functions ψ are independent of the angular coordinates θ and ϕ (Figure 4). The radial electron-density functions $4\pi r^2\psi^2$ (these measure the electron density within the volume

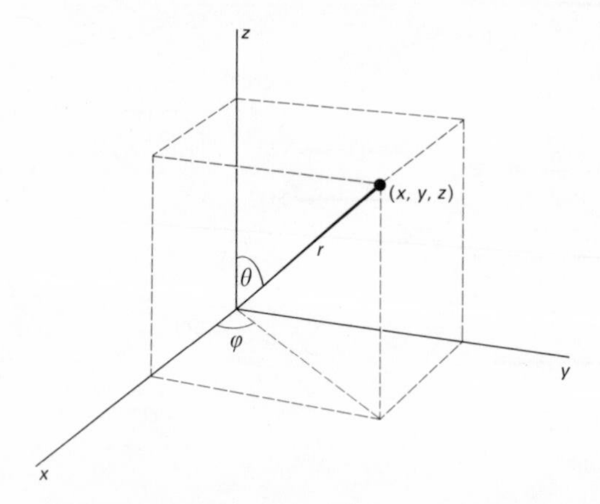

Figure 4 The position of the electron in the hydrogen atom relative to the nucleus, in terms of Cartesian and polar coordinates. The nucleus is at the origin (0, 0, 0)

$z = r\cos\theta$

$y = r\sin\theta\sin\phi$

$x = r\sin\theta\cos\phi$

element between radius r and radius $r+dr$, which is equal to $4\pi r^2$) plotted against r for 1s, 2s and 3s electrons are shown in Figure 5. It should be noted that, although the electron spends more of its time further out from the nucleus the higher the value of n, there is still a finite chance of even a 3s electron being close in

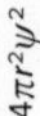

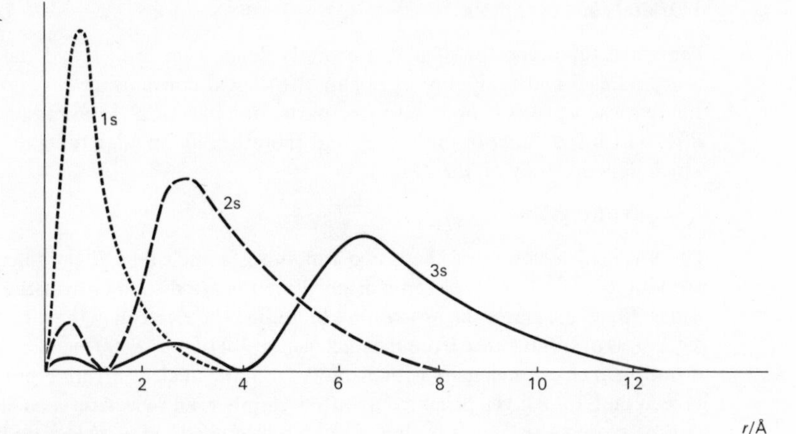

Figure 5 Radial electron-density functions for the 1s, 2s and 3s orbitals of the hydrogen atom

to the nucleus (as evidenced by the small 'humps' in the radial distribution curve). The electrons are said to *penetrate* close in to the nucleus.

The wave functions of s-orbitals having $n > 1$ are zero in certain places and this, owing to the spherical nature of ψ for s-orbitals, gives rise to spherical nodal surfaces in the orbitals – at a node the electron density, which is proportional to ψ^2, is also zero, see Figure 6. The wave function ψ changes sign at a node, being positive on one side and negative on the other. The sign properties of ψ are of primary importance in bonding because only orbitals having wave functions with the same sign may 'overlap' to form a bond.

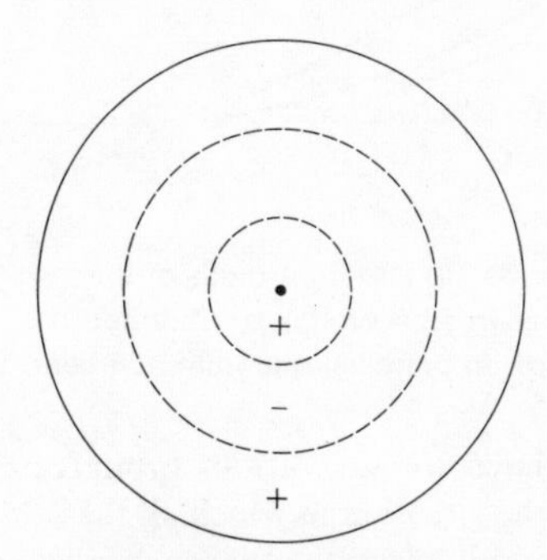

Figure 6 Section through a spherical 3s orbital showing nodes (dotted lines) and change in sign (symmetry) of the wave function ψ. The outer solid line is a section through the spherical surface which encloses about 95 per cent of the electron density

1.3 **p-Orbitals and d-orbitals**

The wave functions for p- and d-orbitals depend on both r and the angular coordinates θ and ϕ, Figure 4. For mathematical convenience it is possible to divide these wave functions into two parts, one of which is the radial function $R(r)$, which is a function only of r, and the other an angular function $A(\theta, \phi)$, which depends only on direction θ, ϕ,

$$\psi = R(r)A(\theta, \phi).$$

The physical meaning of these two functions is such that $R^2(r)$ describes the probability of finding the electron in any direction at a distance r from the nucleus, and $A^2(\theta, \phi)$ measures the probability of finding the electron in the direction θ, ϕ regardless of the distance from the nucleus. A plot of $4\pi r^2 R^2(r)$ against r gives the variation of electron density with distance from the nucleus. From Figure 7 it can be seen that, for a given principal quantum number, an s-electron spends more of its time close to the nucleus than does a p-electron, and a p-electron similarly spends more time near to the nucleus than a d-electron. Thus the penetration effect for electrons in the same quantum shell is in the order s > p > d > f.

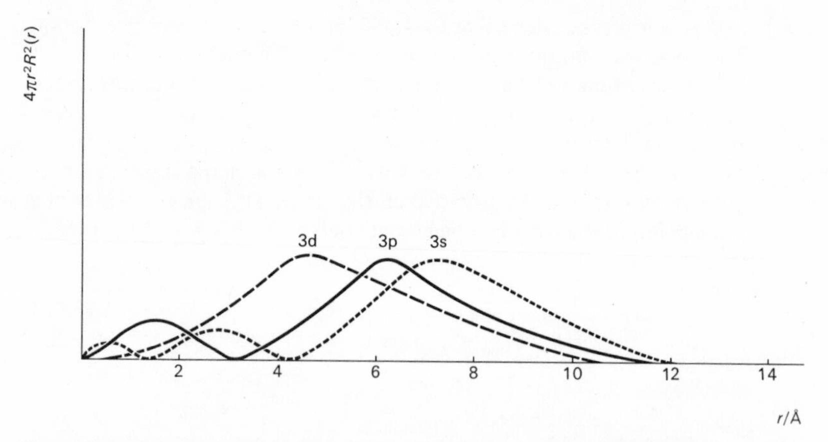

Figure 7 Radial electron-density functions for 3s, 3p and 3d orbitals of the hydrogen atom. Note that there are $n-1$ nodes in an s-orbital, $n-2$ nodes in a p-orbital and $n-3$ nodes in a d-orbital for a given principal quantum number n

The radial distribution functions calculated for the various orbitals of the hydrogen atom necessarily correspond to the situation in which all the orbitals except the one under study are empty. In a multi-electron atom such as sodium, many of the orbitals are occupied; it is interesting to see the effect of orbital penetration on the outer 3s, 3p and 3d orbitals in such an atom (Figure 10). Without the effect of orbital penetration, it would be expected that the electron in sodium's $n = 3$ quantum shell would experience the attraction of the nuclear charge (+11) moderated by the ten electrons in the $n = 1$ and $n = 2$ shells, which are between

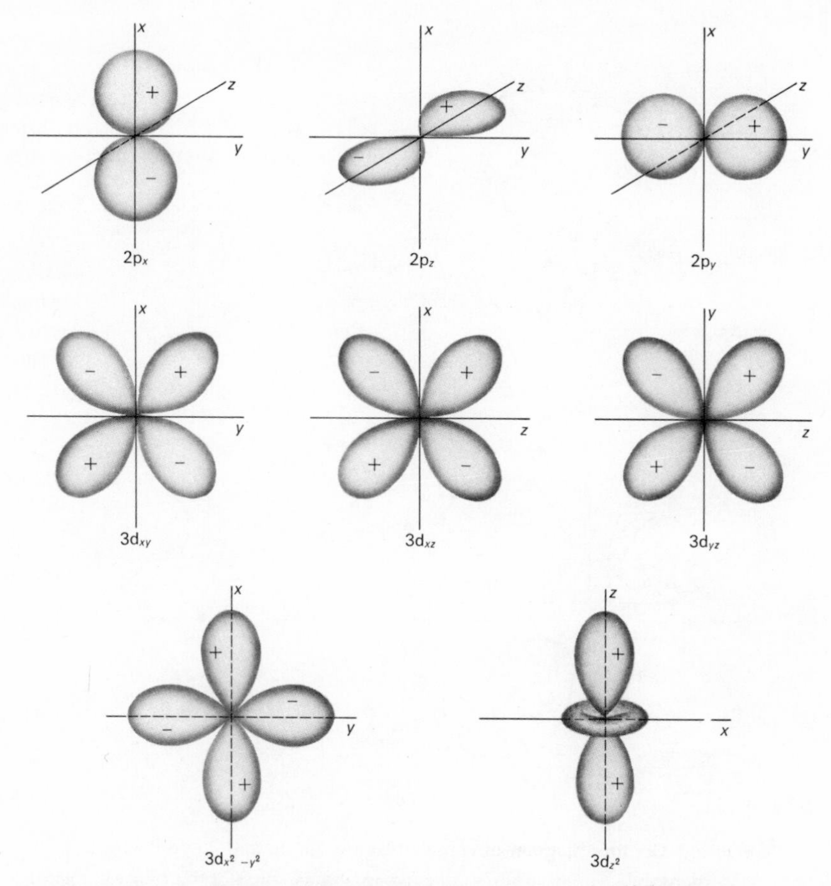

Figure 8 The angular part of the wave function, $A(\theta, \phi)$, for the 2p and 3d orbitals of the hydrogen atom. Note the convention for denoting the different p-orbitals and d-orbitals which differ in their orientation relative to the axes *x*, *y* and *z*

it and the nucleus (remembering that the electrons having the highest value of the principal quantum number are farthest out from the nucleus). The effective nuclear charge felt by the $n = 3$ electron might be expected to equal $+1$ (i.e. $11-10$), in which case the radial distribution functions would appear as in Figure 7, the 3d, 3p and 3s orbitals having their maximum 'humps' at about 4·5, 6 and 6·5 Å from the nucleus, respectively. The 3d orbital certainly has its hump at 4·5 Å from the nucleus, but the 3s and 3p orbitals have their maxima very much closer to the nucleus than expected. This is because the 3s and 3p orbitals penetrate close in to the nucleus through the inner-shell electrons and therefore experience a nuclear charge considerably greater than $+1$; the effect is more

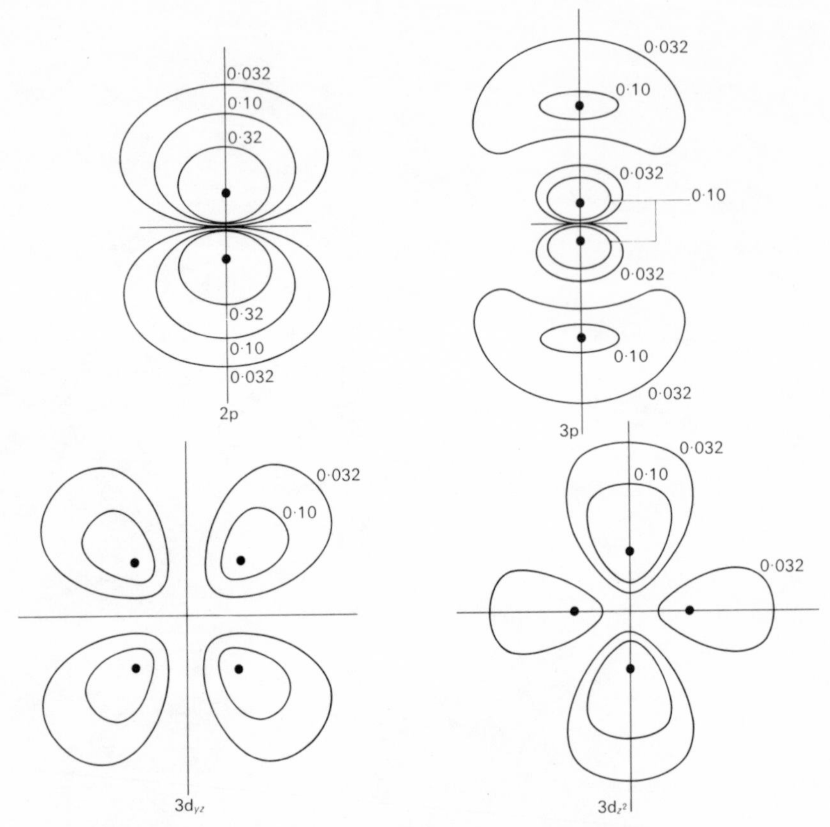

Figure 9 Contour diagram of constant ψ^2 for 2p, 3p and 3d orbitals.
It is impossible to plot in three dimensions the variation of the four variables $\psi(\theta, \phi, r)$, θ, ϕ and r. This difficulty can be overcome to a large degree by holding $\psi(\theta, \phi, r)$ constant and plotting the spatial variation of $\psi^2(\theta, \phi, r)$, which, in two dimensions, results in a contour map of the electron density within the orbital.
It is to be realized that in three dimensions the contours shown above become contour surfaces.
The figures represent the ratio ψ/ψ_{max}; the dots represent the positions of maximum electron density in each part of the orbitals

marked for s-orbitals than p-orbitals, as mentioned above when discussing Figure 7. Hence a 3s electron in the sodium atom would experience a higher effective charge than a 3p electron. The 3d orbital shows no penetration effect and an electron in such an orbital, well outside the inner shell of ten electrons, would be effectively shielded from the nucleus by the inner electrons and would experience an effective nuclear charge of about +1.

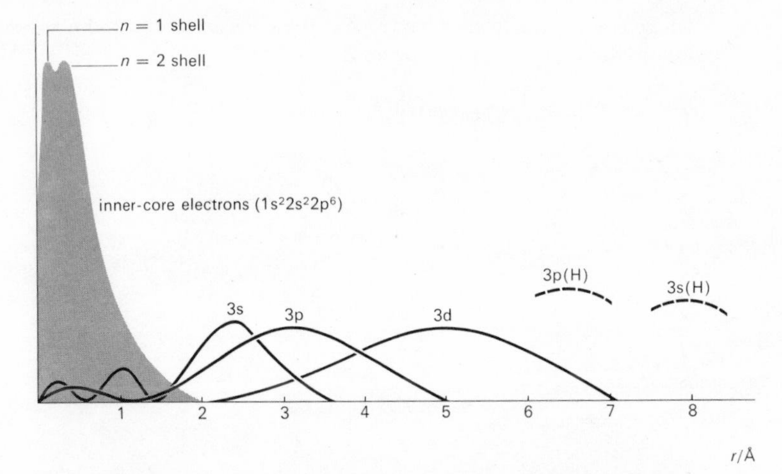

Figure 10 Radial electron-density functions for the sodium atom, showing the contraction effect of orbital penetration on the 3s, 3p and 3d orbitals.
The position of maximum electron density for hydrogen 3s and 3p orbitals is indicated by the partial dotted curves labelled 3s(H) and 3p(H). The hydrogen 3d orbital is similar in position to that shown for sodium

Because of this penetration effect, a 3s electron in the sodium atom is more strongly bound than a 3p electron, which in turn is more strongly bound than a 3d electron. The overall result of penetration on the sequence of orbital energies for the first two dozen or so elements in the periodic table is shown in Figure 3. As is obvious from the radial distribution functions for 3s, 3p and 3d orbitals in Figure 10, the smallest orbitals are those which are the most strongly bound to the nucleus.

Using the sequence of orbital energies given in Figure 3 and Pauli's exclusion principle, we can now begin to build up the periodic table of the elements. We shall use the notation $^{\uparrow}$ to mean a single electron (with, say, spin $+\frac{1}{2}$) and $^{\uparrow\downarrow}$ to mean two electrons occupying the same orbital with their spins paired (i.e. spins $+\frac{1}{2}$ and $-\frac{1}{2}$).

1.4 The hydrogen atom

The solution of the Schrödinger wave equation for a *one-electron* system (e.g. H, He^+, Li^{2+}) gives the energy of the electron as

$$E = -\frac{2\pi^2 me^4 Z^2}{h^2} \times \frac{1}{n^2},$$

where e is the electronic charge; Z is the nuclear charge and n is the principal quantum number. For hydrogen, $Z = 1$ and hence

$$E = -1310 \times \frac{1}{n^2} \text{ kJ (mol H)}^{-1}.$$

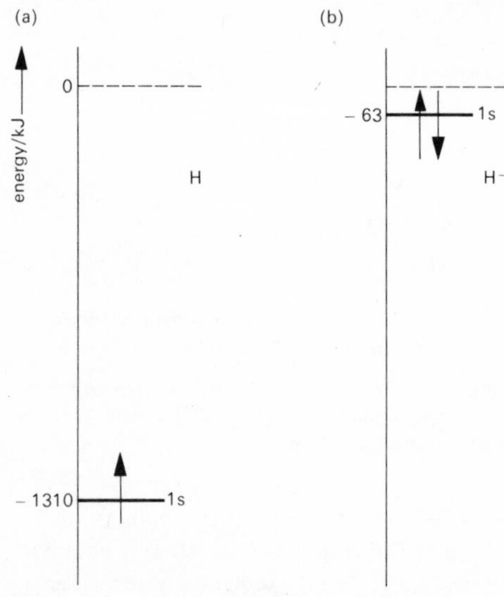

Figure 11 The electronic structures of (a) the hydrogen atom and (b) the hydride ion in their ground states (i.e. $n = 1$)

Figure 11 illustrates the ground-state ($n = 1$) configuration of the hydrogen atom. When two electrons are placed in the 1s orbital, as in the hydride ion H^-, they repel one another so strongly that the binding energy of the electrons falls from -1310 kJ to about -63 kJ. From this it can be recognized that the repulsion energy between the two 1s electrons in H^- must be $1310-63 = 1247$ kJ. Since the electrons are much less strongly bound to the nucleus in H^- than in H, it follows that the hydride ion is considerably larger (radius $\simeq$ 1·5 Å) than the hydrogen atom (radius 0·53 Å).

1.5 The helium atom

The binding energy of the electron in He^+ is four times that in the hydrogen atom because of the increased nuclear charge ($Z = 2$): 1310×2^2 or 5240 kJ (mol He^+)$^{-1}$. This results in the greatly reduced size of the 1s orbital in He^+ compared to the hydrogen 1s orbital. The energy required to remove one electron from the helium *atom* (i.e. the first ionization energy of He) is 2373 kJ, which means that the repulsion energy between the two 1s electrons is $5240-2373$ or

2867 kJ (mol He)$^{-1}$ – an enormous amount of energy when we are used to thinking of electrons which are spin paired as constituting a 'stable' system. Why doesn't the second electron go into another helium orbital so that repulsion effects are minimized? From the solution of the Schrödinger equation we can calculate the energy of the $n = 2$ shell for He^+,

$$E = \frac{-1310 \times 2^2}{4} = -1310 \text{ kJ},$$

hence the $n = 2$ shell of He^+ is less stable than the 1s orbital by

$5240 - 1310 = 3930$ kJ.

It is thus energetically more favourable to place the second electron of helium in the 1s orbital and sacrifice the (much smaller) electron repulsion energy (Figure 12). This forms the basis of the *Aufbau* (or building-up) principle used to decide into which orbitals the available number of electrons are to be placed: for atoms in their ground state, the electrons are placed into orbitals in such a sequence that the energy of the completed atom is the lowest possible.

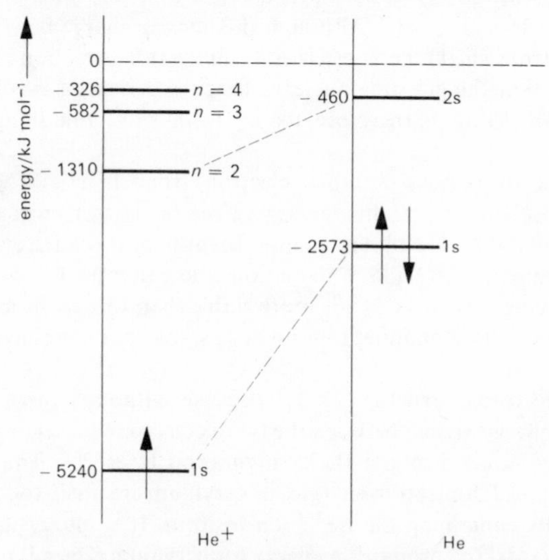

Figure 12 Energy-level diagrams and electron configurations of He^+ and He. Note the destabilizing effect of the second electron on the 1s and 2s orbital energies for the helium atom relative to He^+, due to mutual repulsion between the two electrons

The first ionization energy for helium, 2373 kJ (mol He)$^{-1}$, is much too large an energy term to allow the formation of stable ionic derivatives of He^+. Similarly, the energy required to promote an electron from the 1s orbital to the 2s orbital is $2373 - 460$ or 1913 kJ, far too large to be recouped by the formation of two covalent bonds to helium in a derivative HeX_2, even if X were fluorine.

The most likely compound would be HeF_2 since the helium–fluorine bond would be strong and the dissociation energy of F_2 is small (see Chapter 10). The penetration of a 2s orbital close in to the nucleus suggests that the He^- ion might be capable of existence although no such ion has yet been detected in a stable compound. Without the penetration effect, the 2s electron in He^- would be completely shielded from the nucleus by the 1s electrons and experience no binding energy whatsoever.

1.6 The first short period: lithium to neon

There is only one orbital in the $n = 1$ quantum shell and, from Pauli's exclusion principle, we know that a maximum of two electrons only can be accommodated in it. Hence this quantum shell is completed at helium. The next element, lithium, contains three electrons and from Figure 7 it is obvious that the third electron must occupy the 2s orbital as this is the most stable orbital after the completed $n = 1$ shell. The electron configuration of lithium in its ground state is therefore $1s^2 2s^1$ or, in more detail, $1s^{\uparrow\downarrow}2s^{\uparrow}$. The 2s electron is relatively easily lost, the ionization energy being only 531 kJ mol^{-1}, which is just over one third of that required for the process $H \rightarrow H^+ + e^-$. Although the nuclear charge is +3 for lithium, the two 1s electrons shield the 2s electron from the full attraction of the nucleus with the result that the effective nuclear charge experienced by the 2s electron is only about +1. Lithium, therefore, readily forms salts containing the Li^+ ion.

The energy required to remove another electron from Li^+ is 7291 kJ $(mol\ Li^+)^{-1}$, which represents the binding energy of the 1s electrons when the nuclear charge is increased to +3. Lithium cannot, because of this huge energy term, form salts containing the Li^{2+} ion. Calculations show that the Li^- anion, having the electronic configuration $1s^2 2s^2$, is more stable than $Li + e^-$ by about 50 kJ, which suggests that salts containing Li^- might be stable, but none have yet been prepared.

Beryllium has the electronic structure $1s^2 2s^2$ because, although there is a considerable repulsion energy arising between the two electrons in the 2s orbital, it is less than the energy required to give the configuration $1s^2 2s^1 2p^1$. The first (900 kJ) and second (1756 kJ) ionization energies of beryllium are much too high to allow anhydrous salts containing the Be^{2+} ion to form. It is interesting to compare the $1s^2 2s^2 \rightarrow 1s^2 2s^1 2p^1$ promotion energy for beryllium (264 kJ) to the huge $1s^2 \rightarrow 1s^1 2s^1$ promotion energy for helium (1913 kJ, Figure 12), where the electron has to be promoted to another quantum shell; that is, there is a change in quantum number. Thus, although helium cannot be expected to form divalent compounds HeX_2, the s → p promotion for beryllium (involving only a change in l) is relatively easy and beryllium forms many covalent compounds in which the beryllium atom is sp, sp^2 or sp^3 hybridized.

Boron has five electrons and in the ground state has the configuration $1s^2 2s^2 2p^1$, since the 2p orbital is lowest in energy after the 2s. At carbon, which has six electrons, there is something of a dilemma. From Figure 3 it is evident that the ground-state configuration must be $1s^2 2s^2 2p^2$, but this can represent different spin states such as $1s^{\uparrow\downarrow}2s^{\downarrow\uparrow}2p_x^{\uparrow}2p_y^{\uparrow}$, $1s^{\uparrow\downarrow}2s^{\uparrow\downarrow}2p_x^{\uparrow}2p_y^{\downarrow}$; or $1s^{\downarrow\uparrow}2s^{\uparrow\downarrow}2p^{\uparrow\downarrow}$. It is found that two useful rules apply in such cases.

(a) Electrons avoid being spin paired in the same orbital as much as possible. The reason for this is one of Coulombic repulsion between the similarly charged electrons; the extent of the repulsion varies with the size of the orbital. Two electrons confined to the small 1s orbital of helium repel each other to the extent of 2866 kJ mol^{-1}, but this falls to about 121 kJ mol^{-1} for two paired electrons in a larger carbon 2p orbital.

(b) When unpaired electrons occupy degenerate orbitals, the energy of the system is lowest when their spins are parallel.

From these rules it can be deduced that the carbon atom has the electron configuration $1s^{\uparrow\downarrow}2s^{\uparrow\downarrow}2p_x^{\uparrow}2p_y^{\uparrow}$. The energy required to promote one mole of carbon atoms to the configuration $1s^{\uparrow\downarrow}2s^{\uparrow}2p_x^{\uparrow}2p_y^{\uparrow}2p_z^{\uparrow}$ is reasonably low, being about 403·7 kJ mol^{-1}; hence carbon can be tetravalent. Calculations (see p. 145) show that the formation of CX_4 is energetically more favourable than CX_2, so that carbon is almost always four-covalent in its compounds.

The electronic configurations of the remaining elements in the first short period are built up following the rules used above:

N $1s^{\uparrow\downarrow}2s^{\uparrow\downarrow}2p_x^{\uparrow}2p_y^{\uparrow}2p_z^{\uparrow}$,
O $1s^{\uparrow\downarrow}2s^{\uparrow\downarrow}2p_x^{\uparrow\downarrow}2p_y^{\uparrow}2p_z^{\uparrow}$,
F $1s^{\uparrow\downarrow}2s^{\uparrow\downarrow}2p_x^{\uparrow\downarrow}2p_y^{\uparrow\downarrow}2p_z^{\uparrow}$,
Ne $1s^{\uparrow\downarrow}2s^{\uparrow\downarrow}2p_x^{\uparrow\downarrow}2p_y^{\uparrow\downarrow}2p_z^{\uparrow\downarrow}$.

Although in theory these last four elements might increase their usual valencies via promotion of an electron to the empty 3s orbital, the energy required for this (between 800 and 1600 kJ, depending on the element) is prohibitively large and the covalency maxima remain 3, 2, 1, 0 for nitrogen, oxygen, fluorine and neon, respectively.

The apparently erratic variation in first ionization energies across the $n = 2$ quantum shell from lithium to neon (Figure 13) is explicable in terms of the electronic configurations given above for these elements. The value for beryllium is higher than that for lithium because the increase in nuclear charge more than compensates for the electronic repulsions between the two 2s electrons of beryllium. At boron, configuration $1s^2 2s^2 2p^1$, the least strongly bound electron is now in a 2p orbital and, since 2p orbitals are not stabilized by penetration effects to the same extent as 2s orbitals, the ionization energy of boron is lower than expected by extrapolation from lithium ($1s^2 2s^1$) and beryllium ($1s^2 2s^2$). The ionization energies rise smoothly between boron and nitrogen as one electron is placed successively in each of the three 2p orbitals. As shown in Figure 8, the p-orbitals are orientated at right angles to each other and because of this p-electrons do not shield each other very effectively from the Coulombic attraction

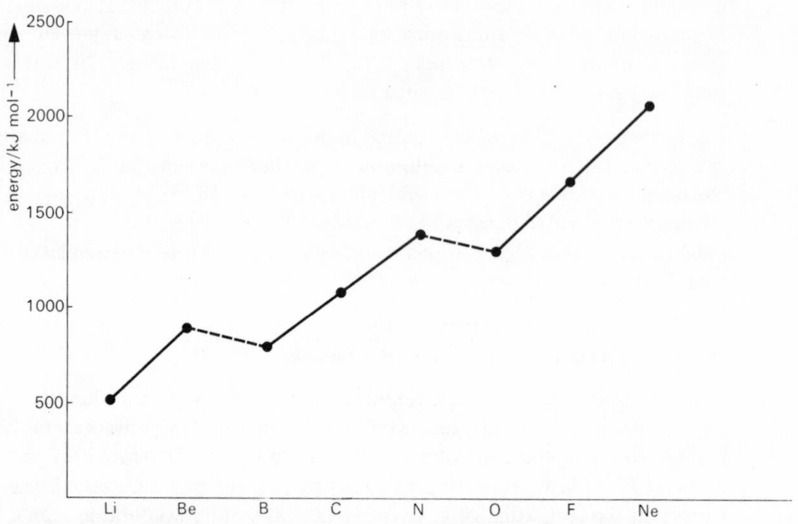

Figure 13 The variation in first ionization energies from lithium to neon

of the nucleus; therefore the increase in nuclear charge from boron to nitrogen means that the 2p electrons become more and more strongly bound and this is reflected in the increasing ionization energies. The drop in ionization energy between nitrogen and oxygen occurs because from oxygen to neon the electrons in the 2p orbitals are being paired up, which results in a repulsion term not present for boron, carbon or nitrogen. The increased nuclear charge, penetrations effects and shielding combine to make the first ionization energy of neon about four times larger than that of lithium, so precluding the formation of ionic salts of neon containing the cation Ne^+. Similarly, derivatives of F^+ are highly unlikely. The above three effects fall off somewhat with increasing atomic number down a particular group; among the halogens, the ionization energy falls steadily, until at iodine it is only 1009 kJ and a few derivatives of I^+ have been isolated.

Figure 14 represents the change in 2s and 2p orbital energies from lithium to neon; note the similarities to Figure 13. The difference between the 2s and 2p energies for a given atom gives the magnitude of the 2s orbital-penetration effect, which even at lithium amounts to 175·7 kJ. As the 2s orbital becomes more strongly bound to the nucleus from lithium to neon, its size decreases and it therefore shields the 2p orbitals more effectively from the full attraction of the nucleus. Hence the divergence of the 2s and 2p orbital energies, especially at oxygen, fluorine and neon, where electronic repulsions cause further destabilization of the 2p orbitals. As the energy separation between the 2s and 2p orbitals increases, the utilization of the 2s orbital in sp^3 hybridization will become less energetically favoured; thus oxygen and (even more so) fluorine do not use their 2s electrons to any significant extent in the bonding of many of their compounds.

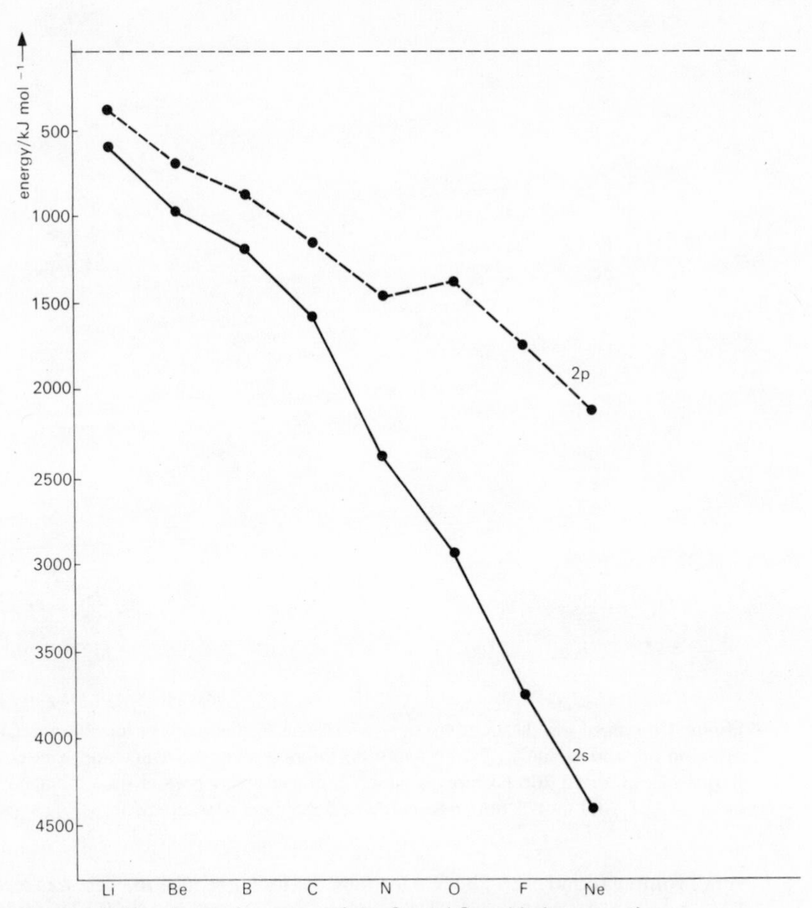

Figure 14 The change in energy of the 2s and 2p orbitals across the first short period

1.6.1 *Effect of atomic size in the first short period*

The size contraction across the first short period from lithium to neon appears to adversely affect the strength of covalent bonds formed by fluorine, the most obvious case being that of the F_2 molecule. When the dissociation energies of I_2, Br_2 and Cl_2 are extrapolated to fluorine (see Figure 15) a dissociation energy of about 266 kJ is obtained for F_2, which is much larger than the actual dissociation energy, 157·7 kJ, found by thermochemical experiments. The unexpectedly low value for F_2 appears to be associated with the exceptionally large electron–electron repulsion terms which occur when eight electrons are packed into the

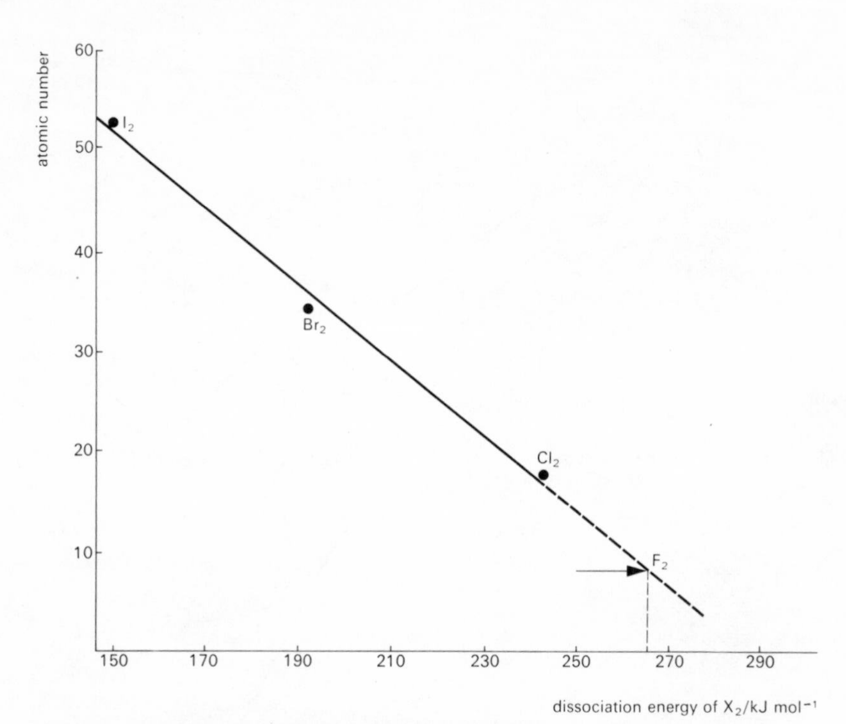

Figure 15 Plot of the dissociation energy against atomic number for chlorine Cl_2, bromine Br_2 and iodine I_2. Extrapolation to fluorine gives a dissociation energy for fluorine F_2 of about 266 kJ mol^{-1}, which is considerably greater than the accepted value of 157·7 kJ mol^{-1} obtained by thermochemical measurements

small volume round each fluorine nucleus. A similar repulsion effect has recently been detected in ionic fluorides, hence earlier suggestions that the low dissociation energy of F_2 was caused by the lone-pair electrons on one atom repelling those on the other fluorine cannot now be considered as wholly correct.

It is possible that this destabilizing effect, due to the outer-electron repulsions, may also be noticeable at oxygen. For example, the dissociation energies of O_2 and S_2 are 494 and 423 kJ respectively; due to the relative weakness of p_π–p_π bonds at sulphur (p. 36), one might have expected the dissociation energy of O_2 to be *much* larger than that of S_2. There is no reason to suppose that either the σ-bonds or the π-bonds in O_2 are weak, hence some other cause must be responsible for the low dissociation energy of O_2; by analogy with F_2 it is suggested that this is due to abnormally high (*intra*-atomic) electronic repulsions.

1.7 The second short period: sodium to argon

At neon, the $n = 2$ quantum shell is completed, so that the next element, sodium, has its last electron placed in the lowest energy orbital of the $n = 3$ shell, the 3s, giving sodium the electronic configuration $1s^2 2s^2 2p^6 3s^1$. The next seven elements complete the 3s and 3p orbitals in a regular manner following the rules adopted for the $n = 2$ shell:

Na $1s^2 2s^2 2p^6 3s^{\uparrow}$,
Mg $1s^2 2s^2 2p^6 3s^{\uparrow\downarrow}$,
Al $1s^2 2s^2 2p^6 3s^{\uparrow\downarrow} 3p_x^{\uparrow}$,
Si $1s^2 2s^2 2p^6 3s^{\uparrow\downarrow} 3p_x^{\uparrow} 3p_y^{\uparrow}$,
P $1s^2 2s^2 2p^6 3s^{\uparrow\downarrow} 3p_x^{\uparrow} 3p_y^{\uparrow} 3p_z^{\uparrow}$,
S $1s^2 2s^2 2p^6 3s^{\uparrow\downarrow} 3p_x^{\uparrow\downarrow} 3p_y^{\uparrow} 3p_z^{\uparrow}$,
Cl $1s^2 2s^2 2p^6 3s^{\uparrow\downarrow} 3p_x^{\uparrow\downarrow} 3p_y^{\uparrow\downarrow} 3p_z^{\uparrow}$,
Ar $1s^2 2s^2 2p^6 3s^{\uparrow\downarrow} 3p_x^{\uparrow\downarrow} 3p_y^{\uparrow\downarrow} 3p_z^{\uparrow\downarrow}$.

1.8 The first long period

Potassium has one more electron than argon and one might predict that this electron would be found in a 3d orbital, giving potassium the configuration $1s^2 2s^2 2p^6 3s^2 3p^6 3d^1$. However, the effect of orbital penetration on the energies of the 3d and 4s orbitals at potassium is such that the 4s orbital is substantially lower in energy (by about 255 kJ) than the five empty 3d orbitals and the true ground-state configuration of potassium is $1s^2 2s^2 2p^6 3s^2 3p^6 4s^1$. A 4s and a 3d orbital are of approximately the same size at this position in the periodic table, so that electrons in 4s orbitals do not shield 3d electrons very effectively from the attraction of the nucleus. At calcium, the increase in nuclear charge and the poor shielding effect of a single 4s electron make the 3d orbitals more tightly bound (i.e. more stable) than they were in potassium; however, their energy still does not fall below that of the 4s orbital, and the twentieth electron of calcium is accommodated in the 4s orbital giving the electron configuration $1s^2 2s^2 2p^6 3s^2 3p^6 4s^2$ for the calcium atom.

Combined effects of increasing nuclear charge and poor shielding by 4s and 4p orbitals on the 3d orbitals (because their sizes are similar) means that at some point in the periodic table the energy of the 3d orbitals must fall below that of the 4p. It so happens that this occurs at the next element after calcium, scandium. Indeed the 3d orbital energy falls so much that it is lower even than the 4s orbital giving scandium the configuration $1s^2 2s^2 2p^6 3s^2 3p^6 3d^1 4s^2$. Although there is room for ten electrons in the 3d orbitals, it will be noticed that two of scandium's electrons are housed in a 4s orbital. This is because the above relative 3d–4s energies are dependent on the 4s orbital being occupied. Had the outer three scandium electrons been in 3d orbitals, it would have been found that, due to penetration effects, the 4s orbital had considerably lower energy than the 3d orbitals, and the configuration $1s^2 2s^2 2p^6 3s^2 3p^6 3d^3$ would not then be the ground state (it is in fact about 400 kJ above the ground state).

The nine elements after scandium are derived from the filling up of the five degenerate 3d orbitals. Slight variations from regularity occur at chromium and copper due both to the closeness in energy of the 4s and 3d orbitals and to a stabilization effect arising from a half-filled, and a filled, set of 3d orbitals. The outer electron configurations of the ten d-transition elements are:

Sc $3d^1 4s^2$,
Ti $3d^2 4s^2$,
V $3d^3 4s^2$,
Cr $3d^5 4s^1$,
Mn $3d^5 4s^2$,
Fe $3d^6 4s^2$,
Co $3d^7 4s^2$,
Ni $3d^8 4s^2$,
Cu $3d^{10} 4s^1$,
Zn $3d^{10} 4s^2$.

The electron configuration of copper suggests that the chemistry of copper might be similar to that of the alkali metal, potassium, which also has a $4s^1$ outer electron configuration. However, the poor mutual shielding effect of 3d and 4s electrons is sufficient to make the first ionization energy of copper almost twice that of potassium. On the other hand, the second ionization energy of copper (where the electron is removed from a 3d orbital) is almost *half* that of potassium (where the second electron has to be removed from a strongly bound 3p orbital). Copper is thus able to have variable valency, commonly occurring as copper(I) or copper(II). In the gas phase, the equilibrium

$$2Cu^+(g) \rightleftharpoons Cu^{2+}(g) + Cu(crystal)$$

is strongly in favour of the cuprous Cu^+ state. However, in crystals and solutions, where the interaction of the copper ions with their surroundings is mainly electrostatic, the interaction energy of the doubly charged Cu^{2+} ion is often sufficiently greater than that of two Cu^+ ions to reverse the copper(I) $\rightleftharpoons$ copper(II) equilibrium in favour of the cupric state. For example, in aqueous solution the hydration energies of Cu^+ and Cu^{2+} are -581 kJ and -2123 kJ respectively; the high value for Cu^{2+} more than compensates for the second ionization energy of copper and the equilibrium constant K of the reaction

$$2Cu^+(aq) \rightleftharpoons Cu^{2+}(aq) + Cu: \qquad K = \frac{[Cu^{2+}]}{[Cu^+]^2}$$

is found to be about 10^6.

Furthermore, although copper is much heavier than potassium, the effect of the poor mutual shielding of the filled d-orbitals in Cu^+ is to make the Cu^+ ion smaller than K^+ ($r_{Cu^+} = 0{\cdot}93$ Å; $r_{K^+} = 1{\cdot}33$ Å; $r_{Na^+} = 0{\cdot}95$ Å). There is, therefore, little to compare in the chemistry of copper and the alkali metals.

When the 3d orbitals are filled at zinc, the next available orbitals are the three degenerate 4p orbitals and gallium thus has the configuration $1s^2 2s^2 2p^6 3s^2 3p^6 3d^{10} 4s^2 4p^1$, which places it in group III below aluminium. The effect of poor shielding by the filled 3d orbitals on the outer 4s and 4p electrons is apparent in the ionization energies (see p. 126) of gallium which are almost equal to those of aluminium, showing that the electrons in gallium are much more tightly bound than would be expected by simple extrapolation of the trends in ionization energies noted in groups I and II (where the energies decrease smoothly down the groups with increasing atomic number of the elements). Continued filling of the 4p orbitals in the normal systematic manner gives the group IV (germanium), V (arsenic), VI (selenium), VII (bromine) and VIII (krypton) elements, which then completes the first long period.

1.9 The second long period

The sequence of orbitals in this period closely resembles that of the first long period: the 5s orbital is below the level of the 4d orbitals at rubidium and strontium, but after strontium ten 4d transition elements occur before the 5p orbitals are finally occupied.

1.10 The heavy elements

In line with all the preceding periods, the 6s orbital now begins to fill at what would have been expected to be the third long period. Thus caesium and barium have the $6s^1$ and $6s^2$ outer electron configurations respectively. At the next element, lanthanum, the expected series of 5d transition elements apparently begins. However, at this point in the periodic table the combination of penetration effects, shielding effects and increasing nuclear charge significantly increases the energy of the 4f orbitals. (At caesium the 4f orbital energy is virtually identical to that of the 4f orbitals in hydrogen, so efficiently are the f-orbitals shielded by the inner electrons.) These effects drop the 4f orbital energy at cerium to a level below that of the 5d and 6p orbitals, so that cerium has the outer configuration $4f^2 5s^2 5p^6 6s^2$; continued filling up of the f-orbitals in a more or less straightforward fashion gives rise to the fourteen rare-earth (or f-transition) elements. When the 4f orbitals are full, the third d-transition series is completed by the filling of the 5d orbitals, which are still lower in energy than the 6p orbitals.

Therefore, the next non-transition element after strontium occurs twenty-five places further on in the periodic table as the 6p orbitals begin to be occupied at thallium. The ionization energies of thallium (see p. 126) are higher than expected by extrapolation from either indium or strontium due to the poor shielding properties of *both* the 5d and 4f electrons (which cause the outer $6s^2 6p^1$ electrons to be strongly bound to the nucleus). The combination of high ionization energies and low Tl—X bond energies (bond energies normally decrease down a group, due to the size and diffuseness of orbitals increasing with atomic number, which leads to poor orbital overlap) make thallium reluctant to form both Tl^{3+} ions and TlX_3 covalent derivatives; more often it exists as the thallous ion Tl^+ which has a $4f^{14}5d^{10}6s^2$ electron core. The chemistry of thallium in the monovalent state is similar to that of the heavier alkali metals (e.g. many of their salts are isomorphous) although salt solubilities (e.g. of the halides) closely resemble those of the silver(I) derivatives. This tendency not to use the tightly bound ns^2 outer electrons is also marked with lead and bismuth and was becoming apparent even at indium and tin in the previous period (where it was partially due to the poor shielding of the 4d electrons).

When the 6p orbitals have been filled, the next to be occupied is the 7s orbital, which contains one electron at francium and two at radium. The rather more complex occupancy of the 6d and 5f orbitals is given in the periodic table inside the front cover.

1.11 Possible extensions to the periodic table

Bismuth is the heaviest element to have a stable isotope, all the elements superseding it being radioactive. Techniques for the production of elements with atomic numbers greater than one hundred consist of bombarding existing heavy nuclei with protons, neutrons or heavier particles in the hope that some of the projectiles might enter a nucleus and create a new element. The experimental difficulties of such research are not made any simpler by the very short half-lives of the new elements: for example, a recently synthesized isotope of element 104 has a half-life of only 0·3 second. However, one theory of the nucleus suggests that the protons and neutrons are confined to 'shells' in a somewhat similar way to the extra-nuclear electrons and that certain numbers of nucleons constitute closed-shell configurations, which confer on those nuclei a special stability towards radioactive decay or fission. It is probable that element 114 could have such a relatively stable nucleus giving it, and perhaps its near neighbours, half-lives as long as 10^6 years; element 114 would follow lead in group IV of the periodic table. Although attempts both to detect such long-lived, very heavy isotopes in nature and to prepare them in the laboratory have failed†, it is probable that the heavier cosmic rays are nuclei of such elements (the heaviest cosmic ray yet found appears to have an atomic number between 99 and 107). The fact that cosmic rays travel in space for millions of years means, of course, that the heavy nuclei constituting them would need to have very long half-lives.

1.12 The role of outer *n*d orbitals in bonding

The s-orbitals and p-orbitals are used extensively in bonding either singly or in the form of sp, sp^2 and sp^3 hybrids. However, there is little or no direct evidence for the participation of d-orbitals in bonding, although sp^3d and sp^3d^2 hybridization schemes are often invoked in discussions of trigonal-bipyramidal and octahedral compounds. It has been known for some years that in those configurations con-

†Good evidence has been presented very recently for the production of element 112 (eka-mercury), but the claim has yet to be confirmed in other laboratories.

taining occupied d-orbitals (for example, the excited configurations $3s^1 3p^3 3d^1$ of phosphorus and $3s^1 3p^3 3d^2$ of sulphur) the d-orbitals are very much larger than the corresponding s- and p-orbitals (see Figure 16). For orbitals to mix effectively with each other in hybrids and to form strong bonds they must be of similar size; such is not the case with the outer d-orbitals in the free atoms. For the s-, p- and d-orbitals to be compatible with each other the d-orbitals have to contract in some way when the central element forms either a five- or a six-coordinate derivative (e.g. PF_5 or SF_6).

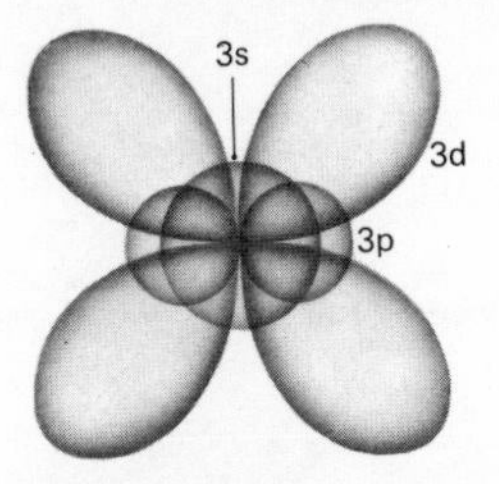

Figure 16 Approximate relative sizes, neglecting nodes, of the outer orbitals of an isolated phosphorus atom in which one of the 3p electrons has been promoted to a 3d orbital

Recent elaborate calculations show that in actual compounds like phosphorus pentafluoride, sulphur hexafluoride and chlorine trifluoride the 3d orbitals of the central atom contract markedly and become much more compatible in size and energy with the 3s and 3p orbitals

Theoretical calculations show that d-orbitals contract markedly when (a) more than one d-electron is promoted from lower-energy orbitals and (b) when the formal charge on the central atom is greatest. Therefore, the 3d orbitals of sp^3d^2 sulphur will contract more than those of sp^3d phosphorus whereas the d-orbitals of both sulphur and phosphorus will contract most of all when highly electronegative ligands surround the central sulphur or phosphorus atom and draw electronic charge away from it: $M^{\delta+}\text{—}\overset{\curvearrowright}{F}^{\delta-}$. This could explain why SH_6 and PH_5 do not exist, whereas SF_6, PF_5 and PCl_5 are well known.

The bonding in molecules such as PF_5, SF_6, ClF_3 and XeF_2 can be described in terms of only s- and p-orbitals; an unusual feature of this bonding approach is the suggestion that linear three-centre molecular orbitals are formed by the interaction of three p-orbitals, one from the central atom and one from each of two ligands lying diametrically opposite to each other in the molecule (Figure

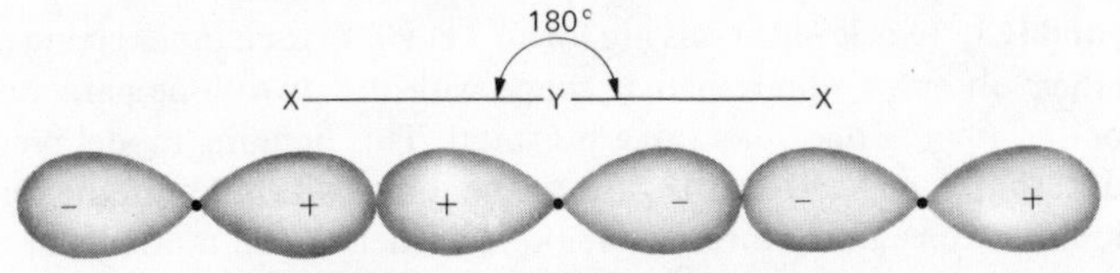

Figure 17 Linear arrangement of three atoms X, Y, X to allow interaction of their p-orbitals

17). Since three atomic orbitals are interacting, they will form three molecular orbitals. (In this case, σ-type molecular orbitals are formed because the p-orbitals are interacting along the X—Y—X axis.) To a first approximation, one of these orbitals will be bonding, one will be nonbonding and the third will be anti-bonding (Figure 18).

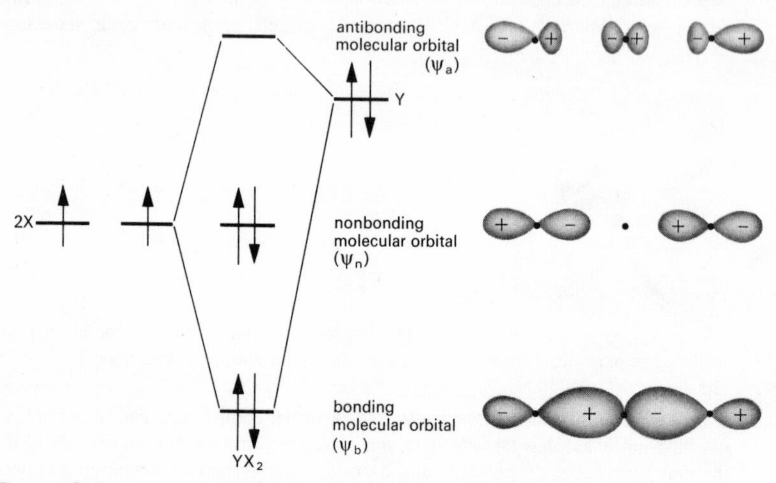

Figure 18 Energy-level diagram for the formation of a three-centre, four-electron bond from three p-orbitals. For simplicity, no nodes are shown in the p-orbitals of either atom

In the specific case of XeF_2, before interaction the xenon 5p orbital contains two electrons and the two fluorine 2p orbitals contain one electron each. There are therefore four electrons available for bonding and these are accommodated $(\psi_b)^2(\psi_n)^2$, giving an overall bonding situation, although *both* fluorines are held to the xenon by only the two electrons in ψ_b; this type of bond is called a three-centre, four-electron (3c, 4e) bond. The two nonbonding electrons in the ψ_n orbital are found by calculation to be localized on the fluorine atoms, which gives a marked polarity to the xenon–fluorine bonds, although the molecule as a whole is non-polar because the bond polarities act in diametrically opposite directions:

$$\overset{\longleftarrow\!\!\!+}{\text{F—Xe}}\overset{+\!\!\!\longrightarrow}{\text{—F}}.$$

In XeF_4 and ICl_4^- two 3c, 4e bonds are formed at 90° to each other giving these molecules their observed square-planar shape with the (two) lone pairs on the central atom occupying one s- and one p-orbital. This bonding model predicts an octahedral shape for XeF_6 and IF_6^-, with the three *n*p orbitals on the central xenon or iodine atoms giving three 3c, 4e bonds which are all mutually at right angles; the lone-pair electrons in such a model would be contained in the spherically symmetrical *n*s orbital. However, it appears that XeF_6 and IF_6^- are probably somewhat distorted from regular octahedral symmetry, and refinements to this simple molecular orbital treatment of the bonding are necessary. The iso-electronic MX_6^{2-} anions, where M = Se or Te, are octahedral in shape and could also be described in terms of three-centre bonds. Both the 3c, 4e bond approach and that requiring the participation of d-orbitals normally give the same predicted geometry for a molecule (Figure 19), and at the present time there

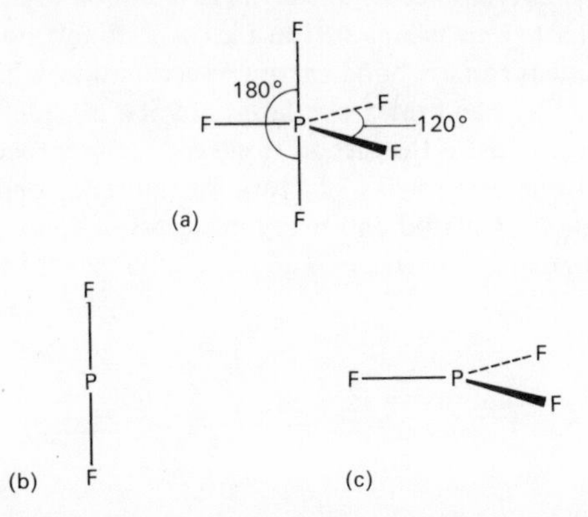

Figure 19 Two ways of explaining the bonding in phosphorus pentafluoride PF_5. (a) Trigonal-bipyramidal structure. (b) Assuming no d-orbital participation, the axial fluorine atoms would be held by a 3c, 4e bond, giving a linear structure; alternatively, if participation of the phosphorus $3d_{z^2}$ orbital is assumed, the molecular orbitals bonding the axial fluorines to the phosphorus would be derived from overlap of two $d_{z^2}p$ hybrid phosphorus orbitals and fluorine 2p orbitals, giving the same linear structure. (c) Whether or not d-orbital participation occurs, the phosphorus atom is sp^2 hybridized and forms normal σ-bonds to the three equatorial fluorine atoms

is no way of deciding which description is nearer to reality. One reason why elements of the first short period do not increase their coordination numbers above four is probably that the atoms are simply not large enough to accommodate the extra groups. (The only apparent violation of a coordination maximum of four for the elements beryllium to neon is the ion CH_5^+ sometimes encountered in mass-spectral studies on organic compounds. The structure of CH_5^+ is unknown, although it is probable that the carbon atom is trigonal-bipyrimidally coordinated to the five hydrogens; significantly, hydrogen is the smallest atom which can bond to carbon.)

1.13 p_π–p_π Bonding

Multiple p_π–p_π bonding occurs relatively frequently in the $n = 2$ quantum shell and especially so in compounds of carbon. In any bond the interaction is strongest when the orbitals involved 'overlap' to the maximum possible extent, the extent being determined by the balancing effect of the inner electron–inner electron and nucleus–nucleus repulsions which occur as the two atoms are brought close together. Taking ethylene as an example, the σ-bond energy between the carbon atoms is at a maximum when the carbon–carbon distance is close to the single-bond value of 1·54 Å (because of repulsion terms the *overlap* is not at a maximum when the carbon atoms are 1·54 Å apart). The sideways overlap of the two carbon p-orbitals involved in the π-bond would be greatest when the two carbon atoms fully coincide (Figure 20) but, because of internuclear and inner-electron repulsions, the maximum allowable overlap occurs when the two carbon atoms are somewhat less than 1·33 Å apart. This means that in the σ + π double bond of ethylene a compromise is reached: some σ-bond energy is sacrificed to achieve a greater π-overlap, but on the other hand the π-overlap is not at a maximum, so that not too much σ-bond energy is lost as the carbon atoms move closer together. Such a compromise appears to be perfectly satisfactory for multiple bonds between the elements in the first short period and many instances of compounds containing, for example, C=C, C≡C, C=O, C≡N and N=N bonds are known.

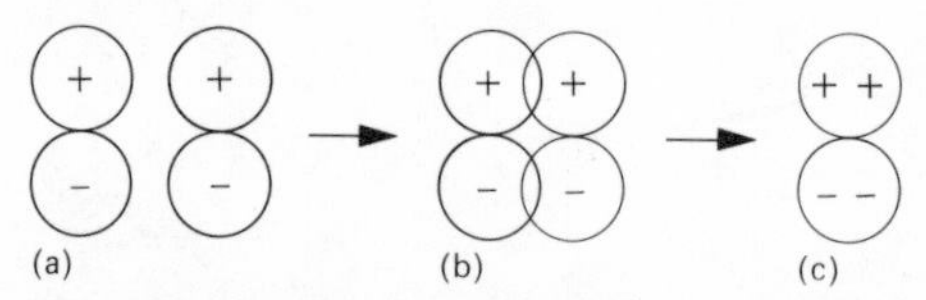

Figure 20 The sideways overlap of two carbon p-orbitals. Overlap would be greatest when the two atoms are fully coincident (c)

Outside the first short period, p_π–p_π multiple bonds between identical atoms are weak relative to σ-bonds, and elements can almost always regain more energy by forming two σ-bonds with the available electrons than one σ-bond and one π-bond. There appear to be two main contributing factors for this:

(a) the electron density in *n*p orbitals for $n > 2$ is more spread out in space (because the orbitals are larger) than in 2p orbitals and hence, to achieve the same amount of overlap, the bonded atoms have to approach each other rather closely; this results in strong inner-electron repulsions and internuclear repulsions;

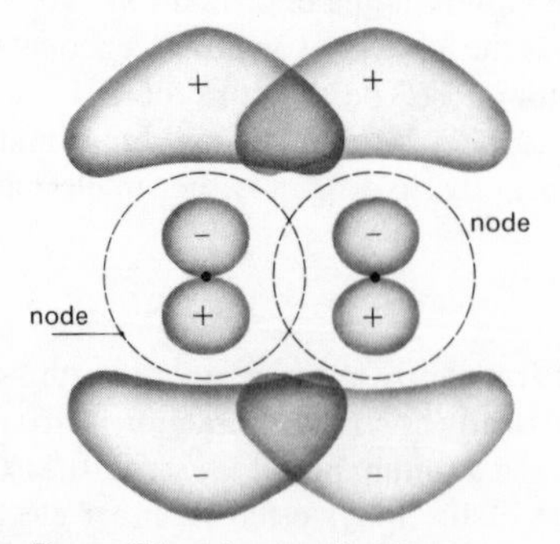

Figure 21 π-Interaction between two 3p orbitals

(b) there are nodes present in the np orbitals when $n > 2$; these nodes, representing regions of zero electron density, cause the overlap of the bonding orbitals to be even lower than when (a) is taken into account (Figure 21).

The p_π–p_π bonds formed between carbon, nitrogen or oxygen and those elements which are not in the first short period (i.e. those elements having $n \geqslant 3$) are also found to be weak relative to σ-bonds. This occurs mainly because the nodes present in np orbitals having $n > 2$ reduce the effective overlap with the 2p orbitals of carbon, nitrogen or oxygen, the overlap already having been adversely affected by the disparity in size between the 2p and np orbitals; see Figure 22. The result of weak π-bonding between 2p and np orbitals is that the

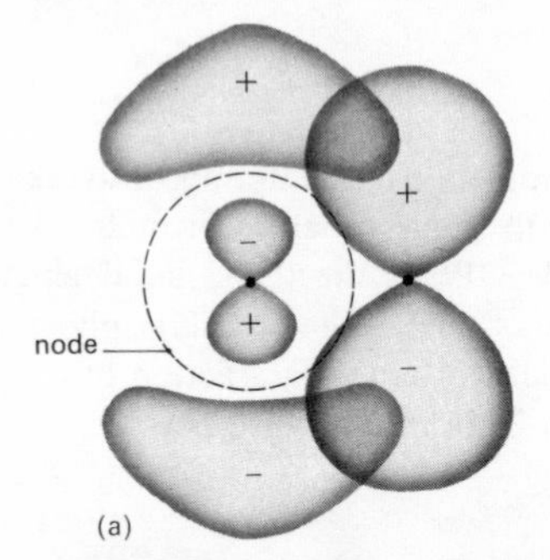

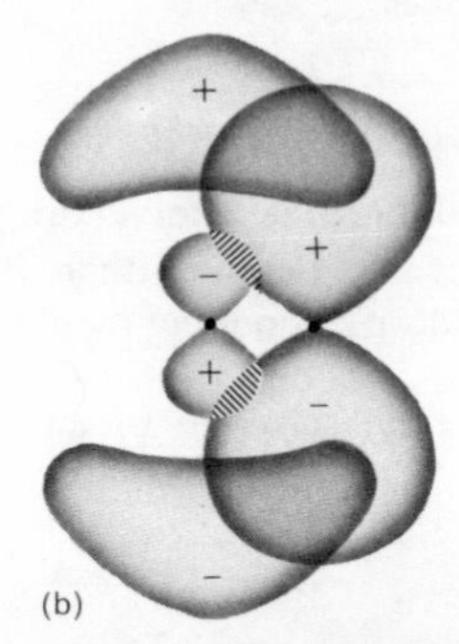

Figure 22 π-Interaction between a 3p and a 2p orbital. (a) The presence of the node and the large size of the 3p orbital reduce the possibility of effective overlap. (b) If the atoms approach more closely, effective overlap is not greatly increased (as are repulsion forces), because the symmetry of ψ in the 2p and 3p orbitals does not match in the hatched regions. Such a mismatch of symmetry constitutes an antibonding situation at that point, which counterbalances any increase in bonding overlap (shaded regions)

elements involved form two σ-bonds in preference to a σ + π double bond and this results in the formation of a polymer. For example, the elements in group IV of the periodic table form dioxides which have the empirical formula MO_2; when M is carbon, strong p_π–p_π bonding occurs between carbon and oxygen to give CO_2 molecules, but when M is silicon the π-bonding between silicon and oxygen is weak relative to σ-bonding and a giant molecule, silica, results in which all four oxygen atoms surrounding each silicon atom are held by σ-bonds (Figure 23).

The polymeric nature of silica is responsible for the high melting point (1710 °C) because the process of melting involves rupture of strong Si—O σ-bonds. In sharp contrast, carbon dioxide sublimes at −78 °C because the molecules of CO_2 are held in the crystal lattice only by weak van der Waals forces. The whole of the silicone industry is similarly based on the fact that silicon and oxygen form very much stronger σ-bonds than they do π-bonds; if this were not so, the main silicone polymer building unit —$OSiR_2$— would resemble the monomeric ketones $R_2C{=}O$ of organic chemistry and have little or no commercial value.

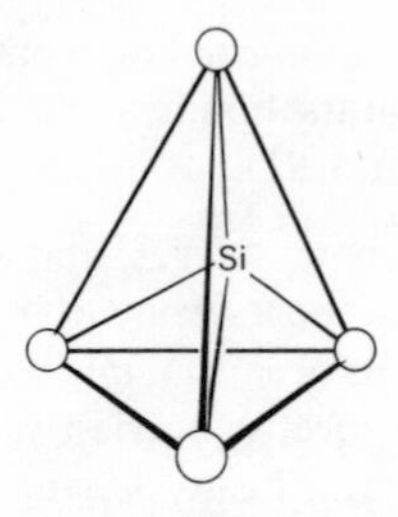

Figure 23 The SiO_4 tetrahedron, which is the basic unit in the structure of silica SiO_2. The oxygen atoms are shared between two silicon atoms, giving Si—O—Si bridges

1.14 d_π–p_π Bonding

Although the outer *n*d orbitals of the non-transition elements appear to take little part in σ-bonding, there is more evidence to suggest that *n*d orbitals take part in d_π–p_π bonding. In phosphoryl chloride $OPCl_3$, the P—O bond length of 1·45 Å is considerably shorter than the sum of the covalent radii for phosphorus and oxygen (1·76 Å), which has been explained by the interaction of a vacant phosphorus 3d orbital with a filled oxygen 2p orbital (Figure 24).

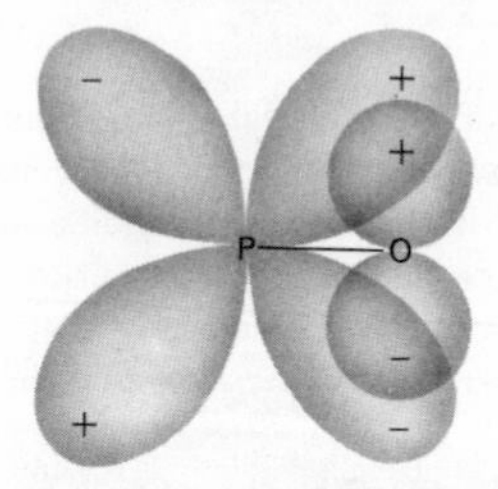

Figure 24 d_π–p_π Bonding between the phosphorus and oxygen atoms of, for example, phosphoryl chloride $OPCl_3$, resulting from the interaction of a vacant phosphorus 3d orbital and a filled oxygen 2p orbital

In moving from left to right across any period of the periodic table, the outer orbitals experience an increasing effective nuclear charge due to the shielding characteristics of the p-orbitals; this causes a contraction of orbital – and hence atomic – size, the greatest contraction being experienced by the *n*d orbitals. (This appears to be a general property of d-orbitals; their size is much more profoundly influenced by changes in effective nuclear charge than either s- or p-orbitals.) The size of d-orbitals is important in determining the strength of d_π–p_π bonds because large, diffuse orbitals cannot overlap effectively with orbitals on other atoms. In the oxyanions MO_4^{n-} of phosphorus, sulphur and chlorine, the M—O bond length is shorter than expected, owing to d_π–p_π bonding between M and O, which augments the σ-bonding (see Figure 25); because of the d-orbital contraction across the second short period, the strength of the π-bonding in the MO_4^{n-} anions would be expected to be in the order Cl > S > P > Si. It has been suggested that weak d_π–p_π bonding between silicon and oxygen is responsible for the occurrence of so many polymeric silicates, whereas the π-bonding in ClO_4^- is sufficiently strong to stabilize the anion against polymerization. The behaviour of phosphorus and sulphur is intermediate; several polyphosphates are known, together with the pyrosulphate $S_2O_7^{2-}$ and trisulphate $S_3O_{10}^{2-}$ anions.

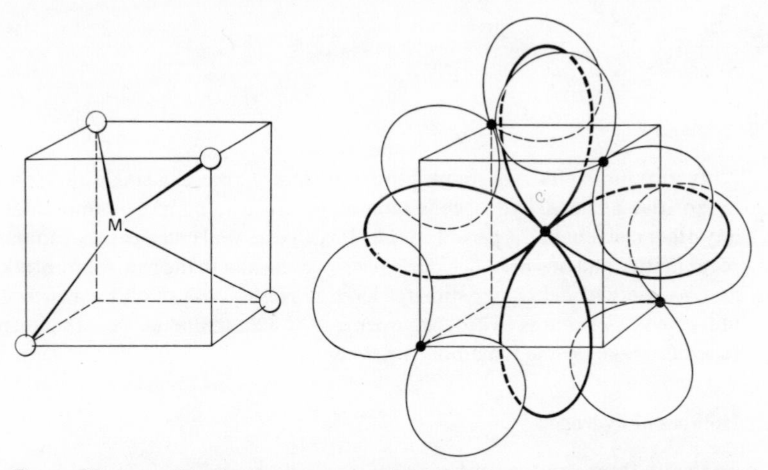

Figure 25 d_π–p_π Bonding in the tetrahedral MO_4^{n-} ion. (a) The tetrahedral MO_4^{n-} ion. (b) π-Orbital overlap in the MO_4^{n-} ion

Chapter 2
Hydrogen

Notwithstanding its very simple atomic structure, that of a single electron and a unipositive nucleus, hydrogen is remarkable in forming more compounds than any other element in the periodic table. It is an essential constituent of animal and plant matter and has been detected spectroscopically in the sun, many other stars and a variety of nebulae. Estimates have suggested that of all the matter in the universe 92 per cent is hydrogen, 7 per cent is helium and all the other elements together make up the remaining 1 per cent.

2.1 Isotopes of hydrogen

There are three isotopes of hydrogen:

(a) Protium (hydrogen-1) or, simply, hydrogen, ^{1}H: single proton in the nucleus; 99·04 per cent.

(b) Deuterium (hydrogen-2), ^{2}H or D: one proton, one neutron in the nucleus; 0·016 per cent.

(c) Tritium (hydrogen-3), ^{3}H or T: one proton, two neutrons in the nucleus.

Tritium is radioactive (half-life 12·26 years) and decays into helium-3 with the expulsion of a beta particle

$$^{3}_{1}H \longrightarrow {}^{3}_{2}He + \beta^{-}.$$

It does not occur naturally in meaningful amounts (minute quantities are produced by the bombardment of nitrogen-14 by cosmic rays in the upper atmosphere) and is made artificially; for example, by bombarding lithium with low-energy neutrons in a nuclear reactor

$$^{6}_{3}Li + {}^{1}_{0}n \longrightarrow {}^{3}_{1}H + {}^{4}_{2}He.$$

Both deuterium and tritium make useful tracers with which to follow reactions involving hydrogen. Although the hydrogen isotopes are chemically identical, quite wide variations in the rates of their reactions often occur because the zero-point vibration energy (which, like all vibration energies, is dependent on mass) of an X to hydrogen bond is lower for deuterium and tritium than for protium.

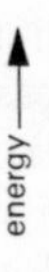
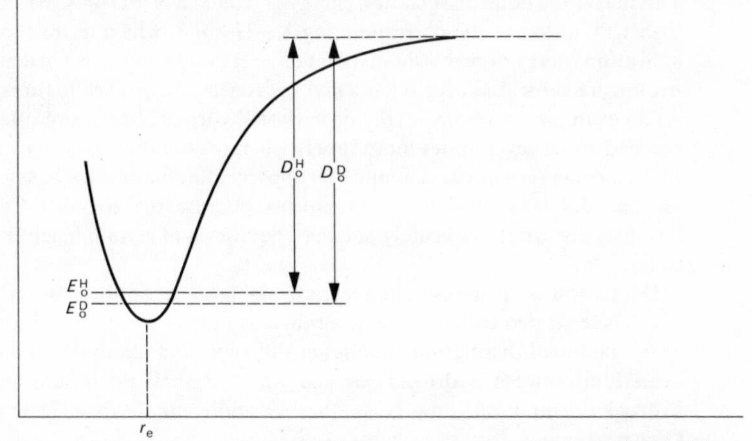

Figure 26 Morse curves for X—H and X—D bonds relating potential energy and interatomic distance. The curves are essentially identical for hydrogen and deuterium, but the zero-point energies E_0^H and E_0^D differ, due to the difference in mass between the hydrogen and deuterium atoms.
The chemical heats of dissociation for the two bonds are given by D_0^H and D_0^D. The difference in the zero-point energies makes D_0^D larger than D_0^H by about 5 or 6 kJ mol^{-1}

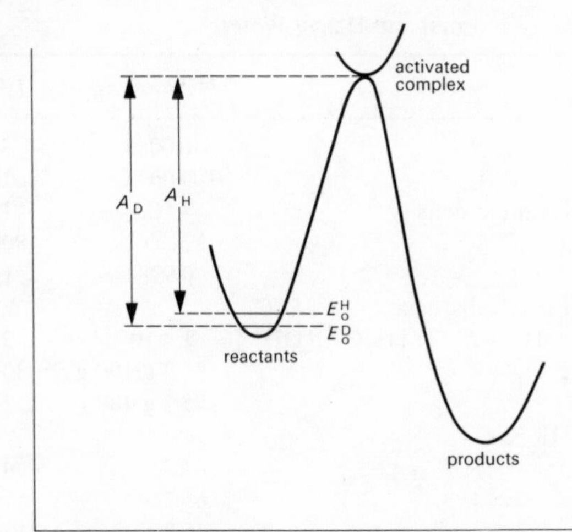

Figure 27 Potential-energy profile for a reaction involving the cleavage of a bond to hydrogen. The effect of the differing zero-point energies is to make the activation energy A_D of the deuterium compound larger than A_H, resulting in a slower rate of cleavage for the bond to deuterium

This makes the bond dissociation energy of either an X—D or X—T bond slightly higher than that of the corresponding X—H bond, which in turn increases the activation energy of reactions involving X—H bond rupture when deuterium and tritium are substituted for the normal hydrogen isotope; see Figures 26 and 27. As an example, reactions involving carbon–hydrogen bonds are often found to proceed about seven times more slowly when protium is replaced by deuterium. In the biosphere, it has been found that appreciable quantities of heavy water D_2O are harmful to several living organisms because the stronger O—D bonds seriously disturb the delicately balanced equilibria of certain reactions involving water.

Deuterium is available commercially in large quantities, usually as D_2O, which is separated from normal water by a variety of methods including electrolysis, fractional distillation, fractional diffusion and catalytic exchange of the deuterium between hydrogen gas and water. Several deuterium analogues of hydrogen compounds may be synthesized quite simply from D_2O using well-known reactions, for example,

$CaC_2 + D_2O \rightarrow DC{\equiv}CD$,
$Mg_3N_2 + D_2O \rightarrow ND_3$,
$Na + D_2O \rightarrow D_2$,
$LiC_6H_5 + D_2O \rightarrow C_6H_5D$ (singly labelled benzene),
$SO_3 + D_2O \rightarrow D_2SO_4$.

Table 1 Properties of Normal and Heavy Water

	H_2O	D_2O
Melting point	0·00 °C	3·82 °C
Boiling point	100·00 °C	101·42 °C
Temperature of maximum density	4 °C	11·6 °C
Dielectric constant	82	80·5
Specific gravity at 20 °C	0·9982	1·1059
Ionic product K_w for self-dissociation at 25 °C		
$2H_2O \rightleftharpoons H_3O^+ + OH^-$ $K_w = [H_3O^+][OH^-]$	1×10^{-14}	3×10^{-15}
Solubility of NaCl	35·9 g/100 g	30·5 g/100 g
Solubility of $BaCl_2$	35·7 g/100 g	28·9 g/100 g
Ionic mobilities at 18 °C:		
K^+	64·2	54·5
Cl^-	65·2	55·3
H^+ (or D^+)	315·2	213·7

Table 1 compares the physical properties of H_2O and D_2O and illustrates the changes which occur on substitution of hydrogen by deuterium. The dielectric constant of D_2O is somewhat less than that of H_2O which accounts, at least in part, for the lower solubility of salts in D_2O relative to H_2O; the dielectric constant and stronger O—D bonds are responsible for the lower self-ionization of D_2O.

Industrially, hydrogen gas is made by any one of several reactions: the electrolysis of water; the steam–water-gas reaction; the thermal cracking of hydrocarbons in the petroleum industry. When high-purity hydrogen ($\sim$99·99+ per cent) is required, it is produced by diffusing hydrogen through heated films of palladium, when the impurities remain behind. Large quantities of liquid hydrogen are at present being used for rocket propulsion because hydrogen represents the ultimate in high-energy fuels.

2.2 *Ortho* and *para* hydrogen

The proton, which forms the nucleus of protium, has a spin $I = \frac{1}{2}$ (measured in units of $h/2\pi$). When two hydrogen atoms are coupled together in the H_2 molecule, the two nuclei can have their spins either parallel or antiparallel to each other (Figure 28).

Figure 28 (a) *Ortho* hydrogen; nuclear spins parallel; resultant spin = 1. (b) *Para* hydrogen; nuclear spins antiparallel; resultant spin = 0

Transitions between the *ortho* and *para* states are theoretically forbidden, but in practice they occur very slowly and as a result the pure gaseous species have half-lives of about three years at 20 °C. At the temperature of liquid air ($\sim$80 K), the molecules are in their lowest rotational energy state and *para* hydrogen is the form thermodynamically favoured; at higher temperatures increasing amounts of *ortho* hydrogen are present in the equilibrium mixture to the limiting ratio of 3 *ortho*:1 *para* (which is essentially the composition at 20 °C). To obtain pure *para* hydrogen the change

$$3\ o\text{-}H_2 + 1\ p\text{-}H_2 \rightarrow \text{pure } p\text{-}H_2$$

is catalysed by adsorbing ordinary hydrogen on oxygenated charcoal at liquid air temperatures and then pumping off the pure *para* hydrogen from the charcoal. It is also possible to separate the two forms of hydrogen by gas chromatography. The change from the *ortho* to the *para* form is exothermic to the extent of 703 J g^{-1}, the heat liberated being sufficient to evaporate over 60 per cent of liquid hydrogen stored as a 3 *ortho*: 1 *para* mixture; liquid hydrogen must therefore be catalytically converted into the pure *para* form before storage. The deuterium nucleus has a

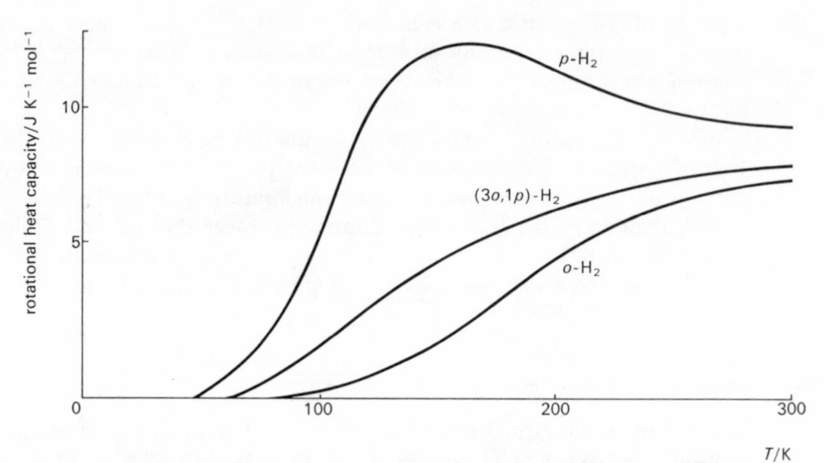

Figure 29 The heat capacities of *ortho*, *para* and normal (3*o*, 1*p*) hydrogen. Mixtures of *ortho* and *para* hydrogen can be analysed by measuring their thermal conductivities, which are proportional to the heat capacities. The boiling points of *ortho* and *para* hydrogen are too similar to be of use in analysis

spin of one and hence a somewhat similar situation occurs in the D_2 molecule, of which the *ortho* and *para* forms have been isolated.

2.3 The position of hydrogen in the periodic table

The electronic configuration of the hydrogen atom is $1s^1$ and over the years various chemists have considered hydrogen as being the first member of no less than three groups in the periodic table. It has been classed with the alkali metals in group I by virtue of the single s-electron, with the halogens in group VII because hydrogen requires only one electron to complete the quantum shell, and with carbon in group IV because the outer quantum shell (1s) is, like that of carbon, only half full.

While it may appear extreme to consider hydrogen with *any* other group, it should be remembered that the elements within groups III to VII have quite similar properties even though their electron 'cores', which govern such things as ionization energies and atomic size, are quite different (see Chapter 1). On these grounds alone a fairly strong case can be made for considering hydrogen with group I, remembering that diverse ionization energies and (because of atomic and ionic sizes) hydration energies are to be expected. Although hydrogen is a diatomic gas under normal conditions and the other group I elements are metallic solids, the situation is not too different from the cases of nitrogen and oxygen relative to the other members of group V and group VI. We are at fault, of course, for expecting any of the elements to fall nicely into groups; a short study of the electronic configurations within the periodic table shows that it is indeed fortuitous that we are able to divide the table up into groups at all.

The ionization energy of hydrogen is very high compared to that of an alkali metal (Table 2). When hydrogen loses its single electron the exceedingly small proton H^+ (radius $\simeq 10^{-5}$ Å) is formed, which cannot exist except in the isolation of high vacuum; certainly it is not present in solid HCl, HF and H_2SO_4. On the other hand, the alkali metals exist as ions in the salts MCl, MF and M_2SO_4. The heat of hydration of the proton is exceptionally high but, because of the large ionization energy term contained in the process

$$\tfrac{1}{2}H_2(g) \xrightarrow{+217{\cdot}5\text{ kJ}} H(g) \xrightarrow{+1310\text{ kJ}} H^+(g) \xrightarrow{-1172\text{ kJ}} H^+(aq),$$

the enthalpy of the reaction $\tfrac{1}{2}H_2 \rightarrow H^+(aq)$ ($\Delta H = +355{\cdot}5$ kJ (mol H)$^{-1}$) is more endothermic than that of the alkali metals,

$$Li(s) \xrightarrow{+159\text{ kJ}} Li(g) \xrightarrow{+518{\cdot}8\text{ kJ}} Li^+(g) \xrightarrow{-514{\cdot}6\text{ kJ}} Li^+(aq)$$

$$Li(s) \longrightarrow Li^+(aq), \quad \Delta H = +163{\cdot}2\text{ kJ mol}^{-1}.$$

So again hydrated H^+ ions are less likely to occur than hydrated alkali-metal cations.

Table 2 First Ionization Energies of Several Elements Compared to Hydrogen

$H(g) \longrightarrow H^+(g)+e;$	1310 kJ mol^{-1}
$Li(g) \longrightarrow Li^+(g)+e;$	518·8 kJ mol^{-1}
$Na(g) \longrightarrow Na^+(g)+e;$	493·8 kJ mol^{-1}
$F(g) \longrightarrow F^+(g)+e;$	1682 kJ mol^{-1}
$Cl(g) \longrightarrow Cl^+(g)+e;$	1255 kJ mol^{-1}
$C(g) \longrightarrow C^+(g)+e;$	1088 kJ mol^{-1}

By gaining an electron the hydrogen atom can form the hydride ion H^-, which might be considered analogous to halide ions formed by the halogens. One great difference is that the hydride ion is unstable in water,

$$H^- + H_2O \longrightarrow H_2 + OH^-.$$

Energetically, it is also found that the heat of formation of gaseous H^- is positive whilst that of the halide ions is highly negative,

$$\text{e.g.}\quad \tfrac{1}{2}H_2 \longrightarrow H^-(g) \qquad +146{\cdot}4\text{ kJ},$$

$$\tfrac{1}{2}F_2 \longrightarrow F^-(g) \qquad -351{\cdot}5\text{ kJ},$$

$$\tfrac{1}{2}Cl_2 \longrightarrow Cl^-(g) \qquad -251{\cdot}0\text{ kJ}.$$

Although the ionization energies of hydrogen and chlorine atoms are very similar (Table 2), the Cl^+ ion is very much larger than H^+ and hence the solvation energy of Cl^+ is much too small to stabilize it in any known solvent. It is of considerable interest to note that the heat of formation of gaseous Na^- ions is actually lower than that of H^-,

$$Na(s) \longrightarrow Na^-(g) \qquad +38\text{ kJ (mol Na)}^{-1}.$$

It is not very profitable to pursue the comparison of carbon with hydrogen except to mention that both elements form covalent compounds and have some thermodynamic properties (e.g. ionization energies and electron affinity) which are rather similar.

In view of the data presented above which show, for example, that it is easier to form Cl^{+}(g) or Na^{-}(g) than H^{+}(g) or H^{-}(g), hydrogen should not be placed with any group in the periodic table but warrants discussion in its own right.

2.4 Atomic hydrogen

As expected from the high heat of dissociation of the H_2 molecule (431 kJ mol^{-1}), hydrogen atoms are formed only at high temperature: at one atmosphere pressure, the degree of dissociation is about 0·001 at 2000 °C and 0·95 at 5000 °C. Hydrogen atoms can be made more conveniently either by passing an electrical discharge through hydrogen gas at low pressure or by exposing mixtures of hydrogen and mercury to the ultraviolet radiation produced by a mercury arc. At low pressure the atoms recombine only relatively slowly with a half-life of about one second at 0·2 mm pressure because a collision between two hydrogen atoms cannot give a molecule unless a third body (e.g. a vessel wall or another hydrogen atom) is present to remove the energy of reaction.

When hydrogen molecules enter into a reaction one of the reaction steps is the breaking of the strong hydrogen–hydrogen bond. Because of the high energy required for this dissociation, the activation energy of molecular hydrogen reactions is also high and hence many take place only very slowly. Atomic hydrogen is much more reactive because dissociation has occurred prior to reaction and it is found that hydrogen atoms attack a wide variety of non-transition elements such as germanium, tin, arsenic, antimony and tellurium,

e.g. $As + 3H \longrightarrow AsH_3$.

As might be expected elements which form very unstable hydrides (e.g. lead and bismuth) do not appear to react with atomic hydrogen.

2.5 The chemistry of hydrogen

Basically, there are three main options open to the hydrogen atom when it is undergoing chemical reaction:

(a) It can lose an electron to form H^{+}.

(b) It can gain an electron to form H^{-}.

(c) It can form normal covalent bonds by sharing two electrons with another atom.

Although the vast majority of hydrogen compounds are covered by these options, hydrogen will participate in three-centre bonds, as in the hydrides of boron, aluminium and beryllium, form interstitial hydrides with transition metals, and when coupled to a highly electronegative atom (fluorine, oxygen or nitrogen) will also exhibit an unusual attractive force called 'hydrogen bonding'. Each of these various types of hydrogen derivative will now be discussed more fully.

2.5.1 *Hydrogen present as* H^+

A large amount of energy is required to form H^+ from the H_2 molecule,

$$\tfrac{1}{2}H_2 \xrightarrow{217{\cdot}5\ \text{kJ}} H(g) \xrightarrow{1310\ \text{kJ}} H^+(g) + e,$$

$$\text{therefore } \tfrac{1}{2}H_2 \longrightarrow H^+(g) \qquad +1527\ \text{kJ (mol H)}^{-1}.$$

The product of this ionization is the bare proton, which cannot exist except in the isolation of high vacuum because of its extremely small size (its radius is only about $\frac{1}{50\,000}$ that of the Li^+ ion). When generated in the presence of solvents having lone-pair electrons, the proton can be stabilized by solvation as, for example, in water,

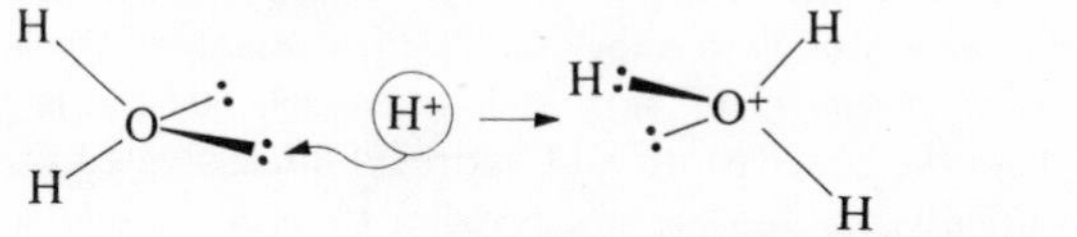

the oxonium ion H_3O^+

In the oxonium ion the three hydrogen atoms are more positively charged than hydrogen atoms in the bulk of the surrounding water due to partial electron migration towards the central (charged) oxygen atom. This will lead to the binding, via the phenomenon of hydrogen bonding (see section 2.8) of at least three more molecules of water to H_3O^+ (2.1).

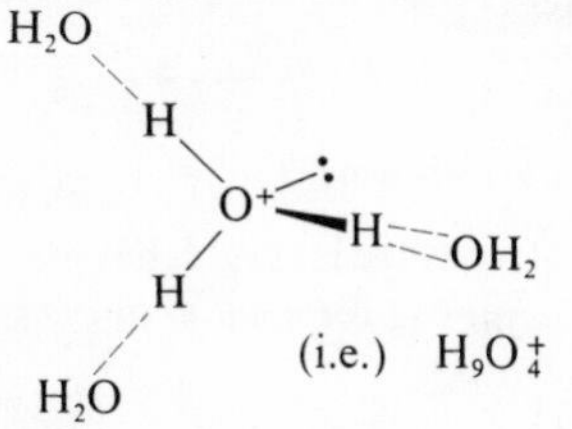

(i.e.) $H_9O_4^+$

(2.1)

Due to thermal agitation, the various water molecules in contact with the proton are continually being changed. Mass-spectroscopic studies on water vapour have shown, by the presence of the ion $H(H_2O)_{10}^+$, that up to ten water molecules can be associated with a proton even under conditions of high vacuum; in liquid water the number of 'coordinated' water molecules must be considerably greater than this. Bagster and Cooling, in a classical proof of the existence of oxonium ions, mixed H_2O and HBr in anhydrous liquid sulphur dioxide and found that the two compounds dissolved in equimolar proportions, thus indicating that a 1:1 interaction had occurred. On electrolysis of the solution, water and hydrogen were formed at the cathode and bromine at the anode in proportions expected from Faraday's law if the 1 : 1 product was $H_3O^+Br^-$. Since that time the existence of H_3O^+ in solid hydrates of a variety of acids has been demonstrated by both X-ray structural analysis and spectroscopy. Some typical examples are

$$\begin{aligned} HClO_4.H_2O &: H_3O^+ClO_4^-, \\ HNO_3.H_2O &: H_3O^+NO_3^-, \\ H_2SO_4.2H_2O &: (H_3O^+)_2SO_4^{2-}, \\ H_2PtCl_6.2H_2O &: (H_3O^+)_2PtCl_6^{2-}. \end{aligned}$$

Pure water exhibits a minute, but definite, electrical conductivity ($\sim 4 \times 10^{-6}\ \Omega^{-1}\ m^{-1}$) due to a slight self-ionization

$$2H_2O \rightleftharpoons H_3O^+ + OH^-,$$
$$K_w = [H_3O^+][OH^-] \simeq 1 \times 10^{-14}\ mol^2\ l^{-2} \text{ at } 25\ ^\circ C,$$

which forms the basis of the familiar pH scale of 'hydrogen ion' concentration. By definition, aqueous acids are those substances which dissolve in water to increase the concentration of the H_3O^+ ion and, conversely, bases are those substances which increase the concentration of hydroxyl ion. Solvated protons can exist in a variety of solvents and are not peculiar to aqueous media. For example, in liquid ammonia the proton is solvated to give the NH_4^+ ion, widely known in numerous ammonium salts. Liquid ammonia, like water, possesses a minute electrical conductivity due to the self-ionization

$$2NH_3 \rightleftharpoons NH_4^+ + NH_2^-.$$

It is therefore possible to define acids in liquid ammonia as those solutes which increase the concentration of solvated protons:

$$HCl(g) + NH_3(l) \longrightarrow NH_4^+ + Cl^- \text{ (by reaction with solvent)},$$
$$NH_4Cl(s) \xrightarrow{NH_3(l)} NH_4^+ + Cl^- \text{ (by direct ionization)}.$$

Bases in this solvent are those substances which dissolve to increase the concentration of amide ions (NH_2^-) in the solvent (see Chapter 12 for a wider discussion of acids and bases in non-aqueous solvents).

2.5.2 *Hydrogen present as* H^-

The alkali and alkaline-earth metals form colourless, salt-like hydrides on being heated in a hydrogen atmosphere. Only lithium hydride is sufficiently stable to be fused (m.p. 688 °C) and electrolysis of the melt produces hydrogen at the *anode* in amounts required by Faraday's law if H^- ions are present. Similar electrolytic experiments have been carried out on solutions of the other hydrides in molten salts or salt eutectics (e.g. the LiCl–KCl eutectic melting at about 360 °C). The electrolyses cannot be carried out in aqueous media because the hydrides are hydrolytically unstable.

The alkali-metal hydrides have the sodium chloride crystal structure. The ionic radius of H^- determined from the crystallographic data is found to be somewhat dependent on the alkali metal, varying from 1·44 Å in lithium hydride to 1·54 Å in caesium hydride (cf. the radius of F^-, which is 1·36 Å). Thus the hydride ion is larger than the iso-electronic helium atom, radius 1·2 Å; this is due to hydrogen having the smaller nuclear charge resulting in a weaker nuclear attraction of the two electrons in H^- compared to those in the helium atom. On the other hand, Pauling calculated that the ionic radius of the free hydride ion should be 2·08 Å, and the reason that the crystallographic radius is smaller than this is probably due to the high compressibility of the rather tenuous electron cloud in the hydride ion. The densities of the saline hydrides are *higher* than those of the metals (the increase in density is almost 50 per cent for the alkali-metal hydrides) due to the strong Coulombic attraction between the metal and hydride ions, which draws them close together and results in a relatively dense crystal.

2.5.3 *Covalent hydrides*

The derivatives discussed in this section are those in which hydrogen is coupled to another element by normal (i.e. two-electron, two-centre) covalent bonds; the elements forming such hydrides are to be found in groups IV, V, VI and VII. Hydrogen atoms are small and their bonds tend to become weaker with the increasing atomic number of the other element because the disparity in orbital size makes overlap poor for the heavier elements in each group – for example, methane is thermally very stable, whereas plumbane PbH_4 has only a very fleeting existence. This decreasing stability of the M—H bond with increasing atomic number of the element M means that catenation in the hydrides also decreases down a given group of the periodic table. This is particularly marked in group IV, where carbon, and to lesser extents silicon and germanium, form long-chain hydrides, but the maximum chain length for tin is two (and even Sn_2H_6 is highly unstable) and for lead is one. This is to be compared to the case where Pb—Pb bonds are stabilized with groups other than hydrogen, when it is possible to couple five lead atoms together as in the tetrahedral compound $Pb[Pb(C_6H_5)_3]_4$ (2.2).

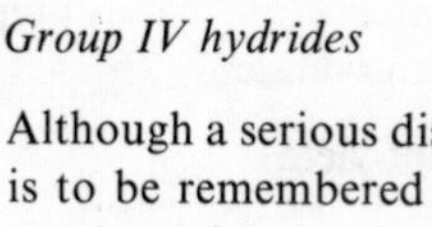

Ph_3Pb
Pb $PbPh_3$ $PbPh_3$
$PbPh_3$

(2.2)

2.5.4 *Group IV hydrides*

Although a serious discussion of the hydrocarbons will not be attempted here, it is to be remembered that the saturated hydrocarbons C_nH_{2n+2} can exist in a number of isomeric forms when n is greater than three; for example, when

n = 20 there are over 360 000 structural isomers possible. Not unexpectedly it has been shown that a variety of structural isomers exist for the higher hydrides of silicon and germanium; for example, three pentagermanes Ge_5H_{12} are known (2.3–5).

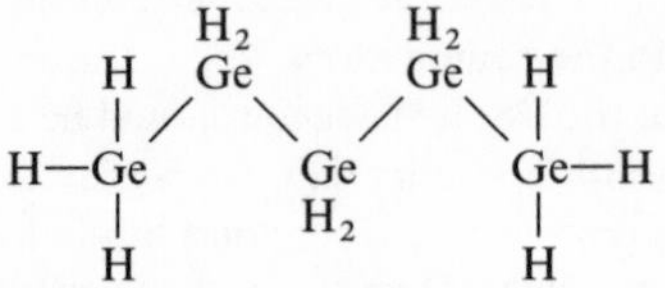

```
        H2         H2                       H2
    H   Ge         Ge   H              H    Ge         GeH3
    |  /  \       /  \  |              |   /  \       /
H—Ge       Ge         Ge—H         H—Ge        Ge
    |      H2           |              |        H \
    H                   H              H           GeH3
```

(2.3) n-pentagermane

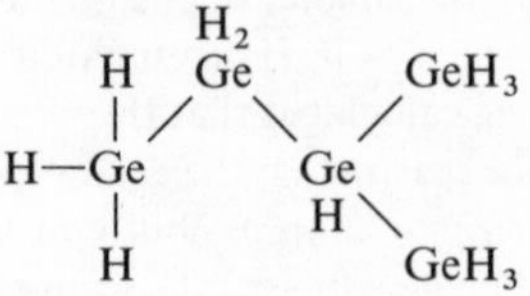

(2.4) isopentagermane

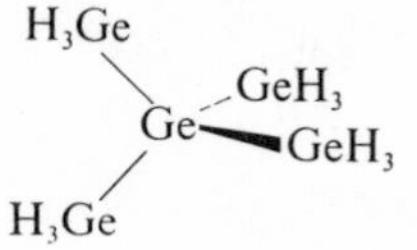

```
H3Ge
    \      GeH3
     Ge----GeH3
    /
H3Ge
```

(2.5) neopentagermane

The heats of formation of the simple group IV hydrides are

CH_4 -75 kJ mol^{-1},
SiH_4 $+33{\cdot}5$ kJ mol^{-1},
GeH_4 $+92$ kJ mol^{-1},
SnH_4 $+163$ kJ mol^{-1},

from which it is apparent that silane, germane and stannane are metastable compounds at room temperature. However, silane only shows noticeable signs of decomposition at temperatures around 300 °C. In the complete absence of oxygen, stannane deposits tin very readily at room temperature

$$SnH_4 \longrightarrow Sn + 2H_2,$$

but a trace of oxygen markedly retards the decomposition rate, probably by oxidizing the metallic tin surface upon which the stannane decomposition occurs.

Carbon forms a number of unsaturated hydrocarbons (e.g. olefins and acetylenes), in which some of the carbon atoms are linked together by p_π–p_π multiple bonds. The other elements in group IV do not form strong p_π–p_π bonds, both because there are nodes present in the np orbitals and also inner-shell electron repulsions are thought to be large (see p. 36). Due to this weak π-bonding, no silicon or germanium analogues of ethylene or acetylene (or compounds containing Si=C and Ge=C bonds) have yet been isolated.

The simple MH_4 derivatives of silicon, germanium and tin are readily prepared by reducing the tetrahalides with lithium tetrahydroaluminate

$$SiCl_4 + LiAlH_4 \longrightarrow SiH_4 \quad \text{(monosilane; 100 per cent yield).}$$

The method may be applied to the higher hydrides when the corresponding halides are available,

$$Si_2Cl_6 + LiAlH_4 \longrightarrow Si_2H_6 \quad \text{(disilane)},$$

$$Si_3Cl_6 + LiAlH_4 \longrightarrow Si_3H_8 \quad \text{(trisilane)}.$$

The acid hydrolysis of either magnesium silicide or magnesium germanide yields a complex mixture of hydrides which can be separated by vapour-phase chromatography; in the case of silicon, twenty-one silanes have been detected, the heaviest being n-octasilane Si_8H_{18}. When treated with phosphoric acid, a finely ground mixture of magnesium silicide and magnesium germanide generates 'mixed' hydrides (e.g. $H_3Si{-}GeH_3$), as well as silanes and germanes. The highly unstable plumbane PbH_4 can only be prepared in trace quantities via the acid hydrolysis of magnesium–lead alloy. It has been characterized by mass spectrometry but has proved to be far too unstable for its chemistry to be studied.

An interesting property of the higher silanes is that, like the hydrocarbons, they can be 'cracked' into other silicon hydrides on heating,

$$\text{e.g.} \quad Si_5H_{12} \longrightarrow 2(SiH)_x + Si_2H_6 + SiH_4.$$

Conversely, monosilane and monogermane can be 'polymerized' into higher species by subjecting them to electrical discharges, the products again being separated by vapour-phase chromatography.

Alkali-metal derivatives of the MH_4 hydrides are known for silicon, germanium and tin, and these are the analogues of the metal methyls; for example, potassium germyl is formed when potassium metal reacts with monogermane in liquid ammonia

$$K + GeH_4 \longrightarrow KGeH_3 + \tfrac{1}{2}H_2.$$

It reacts in a typical manner with methyl chloride to give methylgermane

$$KGeH_3 + CH_3Cl \longrightarrow CH_3GeH_3 + KCl,$$

but the products of reaction between $KGeH_3$ and metal halides are almost always found to be unstable; a rare exception is germylmanganese pentacarbonyl,

$$KGeH_3 + BrMn(CO)_5 \longrightarrow H_3GeMn(CO)_5 + KBr.$$

2.5.5 *Group V hydrides*

As found for the group IV hydrides, the stability of the hydrides in this group decreases with increasing atomic weight of the element, so that bismuthine BiH_3 decomposes rapidly at room temperature. Except for that of bismuth, it is possible to synthesize all the $H_2M{-}MH_2$ hydrides and these probably adopt the gauche configuration in the liquid and gaseous phases, i.e. (2.6a and b).

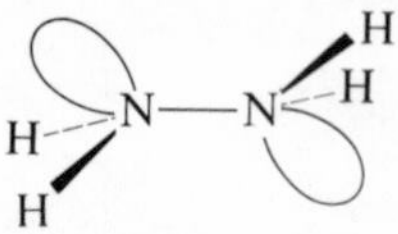

(2.6a) hydrazine

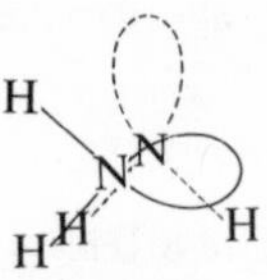

(2.6b) end-on view of N_2H_4

Although there is good evidence for the fleeting existence of diazine N_2H_2, the other elements in the group are not expected to form such a hydride, because the nitrogen atoms in diazine are held by p_π–p_π multiple bonds, (2.7).

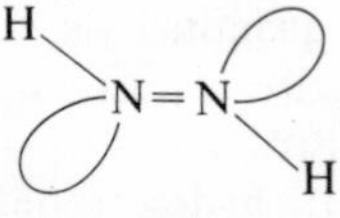

(2.7a) *trans*-diazine (di-imide)

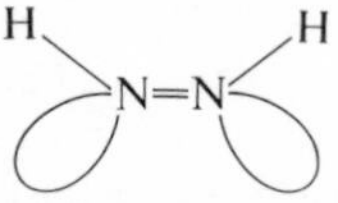

(2.7b) *cis*-diazine (di-imide)

Table 3 Bond Angles and Bond Lengths in the MH_3 Hydrides

	HMH *bond angle*	M—H *distance*/Å
NH_3	107°	1·015
PH_3	93·5°	1·42
AsH_3	92°	1·52
SbH_3	91·5°	1·71

The HMH bond angles in phosphine, arsine and stibine suggest that the central atoms are using their p-orbitals to bond to hydrogen. The variations from the expected 90° angle can be explained by assuming that hydrogen–hydrogen steric repulsions are partially relieved by a slight opening up of the HMH angle, less opening being required as the M—H bond length increases. The lone-pair electrons on the central atoms are in a spherical s-orbital which will make phosphine, arsenic and stibine less potentially useful as donor molecules than ammonia, which has the nitrogen lone-pair electrons in a highly directional sp^3 hybrid orbital. The steric requirements of the lone-pair electrons in ammonia are considered responsible for the slight closing of the HNH angle from the expected tetrahedral (sp^3) value of 109° 28′.

2.5.6 *Group VI hydrides*

A normal hydride, MH_2, is formed by all the group VI elements although H_2Po was only detected by tracer experiments using about 10^{-10} g of the very radioactive polonium. Except for water, where the oxygen is thought to be sp^3 hybridized, the HMH bond angle in these hydrides suggests that the central atoms bind to hydrogen via their outer p-orbitals. H_2S, H_2Se and H_2Te are conveniently prepared by the action of acids on metallic sulphides, selenides and tellurides; they are all very poisonous gases which have offensive smells. Although the heats of formation of H_2Se and H_2Te are positive, the compounds decompose only slowly at room temperature.

Oxygen, sulphur and probably selenium form an M_2H_2 hydride; H_2O_2 and H_2S_2 are nonplanar molecules (2.8).

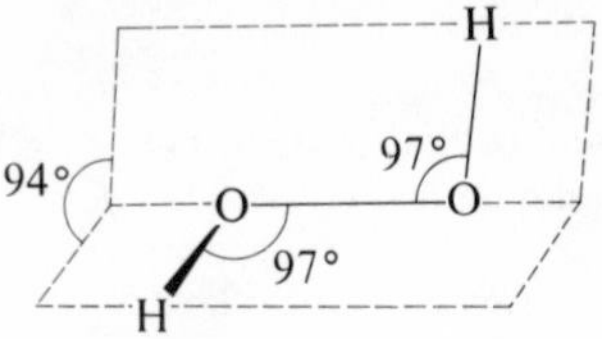

(2.8a)

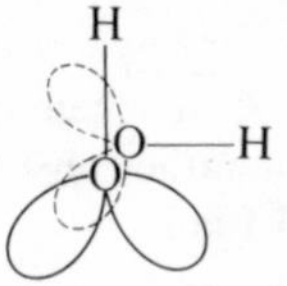

(2.8b) end-on view showing lone pairs

Table 4 Physical Data on the Group VI Hydrides MH_2

	HMH *bond angle*	M—H *bond length*/Å	K_1†	K_2‡
H_2O	104·5°	0·96	—	—
H_2S	92°	1·32	10^{-7}	10^{-15}
H_2Se	91°	1·47	$1{\cdot}3\times10^{-4}$	10^{-11}
H_2Te	89·5°	1·69	$2{\cdot}3\times10^{-3}$	2×10^{-11}

†K_1 corresponds to the reaction $MH_2 \overset{aq}{\rightleftharpoons} MH^- + H^+(aq)$.
‡K_2 corresponds to the reaction $MH^- \overset{aq}{\rightleftharpoons} M^{2-} + H^+(aq)$.

Sulphur readily forms chains with itself (cf. the behaviour of elemental sulphur, p. 214) and it is the only member of group VI to form hydrides containing chains of more than two M atoms. These higher sulphanes are all thermodynamically unstable with respect to H_2S and sulphur, and owe their metastable existence at room temperature to a rather high activation energy for the decomposition (the activation energy in the case of H_2S_2 is about 105 kJ mol^{-1}). The components of a sulphane mixture have to be separated by distillation, which causes difficulties due to simultaneous decomposition, but it has proved possible to isolate sulphanes from H_2S_3 to H_2S_8 in a pure state, whilst others up to H_2S_{17} or H_2S_{18} have been obtained in an impure form. Kilogramme quantities of the sulphanes containing up to six sulphur atoms may be made via the acid hydrolysis of sodium polysulphides,

$$Na + S \longrightarrow Na_2S_x,$$

$$Na_2S_x \xrightarrow{HCl} \underset{n\,=\,4,\,5,\,6}{H_2S_n} \xrightarrow[\text{thermal cracking}]{\text{distillation and}} \underset{n\,=\,2,\,3,\,4,\,5,\,6}{H_2S_n}.$$

2.5.7 *Group VII hydrides*

The hydrides are known for all the members of this group. Only hydrogen fluoride is strongly associated in the solid state, a zig-zag polymer being formed by hydrogen bonding (2.9). The liquid is also associated and the vapour contains

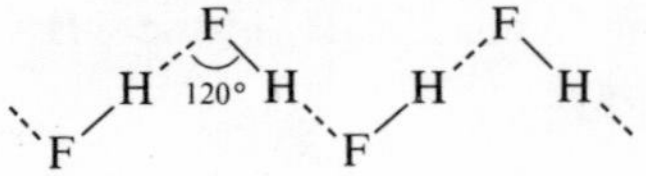

(2.9)

polymers up to 60 °C, the main species present probably being a cyclic hexamer. Hydrogen iodide is rather unstable thermodynamically, the heat of formation being about $+25\ \mathrm{kJ\ mol^{-1}}$.

2.6 The hydrides of boron

2.6.1 *Diborane* B_2H_6

The reduction of a boron halide with lithium tetrahydroaluminate results in the formation of B_2H_6 (diborane, 2.10) and not the expected monomer BH_3 (borane),

$$BF_3 + LiAlH_4 \xrightarrow{\text{ether}} B_2H_6;$$

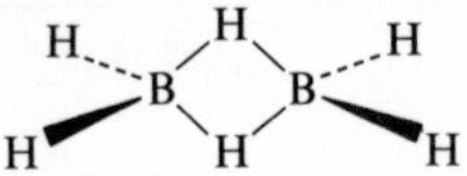

(2.10)

When diborane is heated under reduced pressure it is possible to detect dissociation of B_2H_6 into the highly unstable BH_3 by mass spectrometry; the borane unit can be stabilized by suitable donor molecules, and treatment of diborane with such molecules readily results in the symmetrical cleavage of the B—H—B bridges.

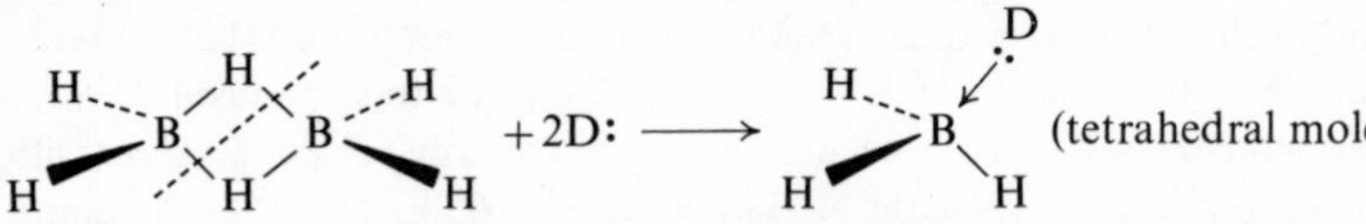

$$B_2H_6 + 2D\!: \longrightarrow 2\,H_3B{\leftarrow}D \quad \text{(tetrahedral molecule)}$$

e.g. D = $(CH_3)_3N$; $(CH_3)_3P$.

The question as to why BH_3 dimerizes so readily is probably the wrong one to ask; the question ought to be, 'Why are the halides and alkyls of boron monomeric and not dimeric?' (in line with aluminium and the other members of group III). A monomeric derivative MX_3 of a group III element is planar and the central atom M has a vacant p-orbital above and below the plane of the molecule (2.11).

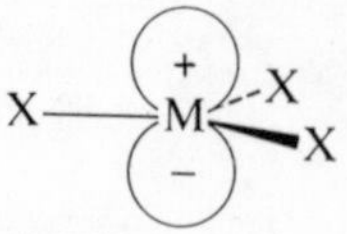

(2.11)

Such a vacant orbital endows the M atom with very strong electron-acceptor properties which are satisfied, for example, in the case of the aluminium halides and lower alkyls, by Al—X—Al bridging (2.12). Calculations show that boron is

X X X
Al Al
X X X

(2.12)

too small for a stable four-atom

X
B B
X

ring system to form for all X-groups except hydrogen. The size effect is rather delicately balanced, because at beryllium, which is somewhat larger than its neighbour boron (atoms become smaller on moving left to right across the periodic table; see Chapter 1), the steric repulsions in the bridge are relieved sufficiently to allow the formation of bridged species among the halides and lower alkyls of beryllium.

The eight atoms in diborane are held together by only twelve electrons so clearly there are insufficient electrons to form normal two-centre, two-electron bonds between the atoms; for this reason diborane is often said to be an 'electron-deficient molecule'. An approximate description of the bonding in B_2H_6 can be obtained by assuming that both boron atoms are sp^3 hybridized and make

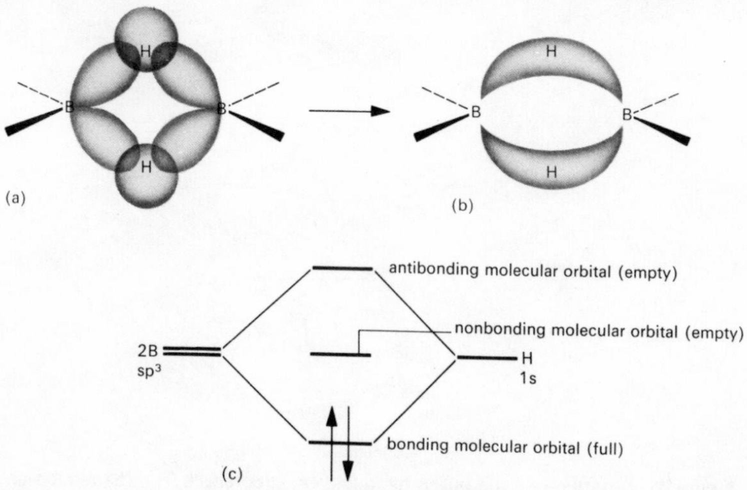

Figure 30 The formation of three-centre bonds in diborane B_2H_6. (a) The atomic orbitals used. (b) The bonding molecular orbital formed. (c) Energy-level diagram

normal σ-bonds to the four terminal hydrogen atoms by overlap of their sp^3 orbitals with the hydrogen 1s orbitals. In each half of the BHB bridge, each boron contributes an sp^3 orbital; these interact with the 1s orbital of the bridge hydrogen atom to give a three-centre molecular orbital containing two electrons (Figure 30).

2.6.2 *Higher boron hydrides*

When diborane is heated it decomposes into a number of other boron hydrides. By subtle variation of the conditions employed during heating, the thermal cracking process can be used as a preparative technique which is essentially unique for a particular hydride, although sometimes a small amount of another hydride may be formed simultaneously; fractional distillation under conditions of high vacuum usually enables a ready separation of the two products. For example, when diborane is passed through a tube heated in a furnace to 180–220 °C

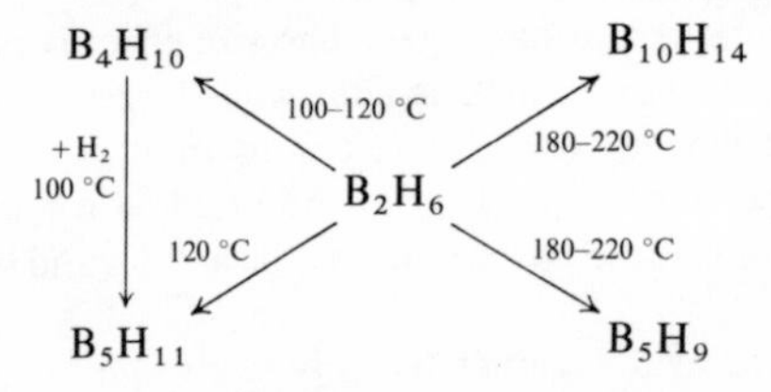

B_5H_9 (60 per cent yield) and $B_{10}H_{14}$ (6 per cent yield) are formed; when the heating is carried out under static conditions, pure $B_{10}H_{14}$ results.

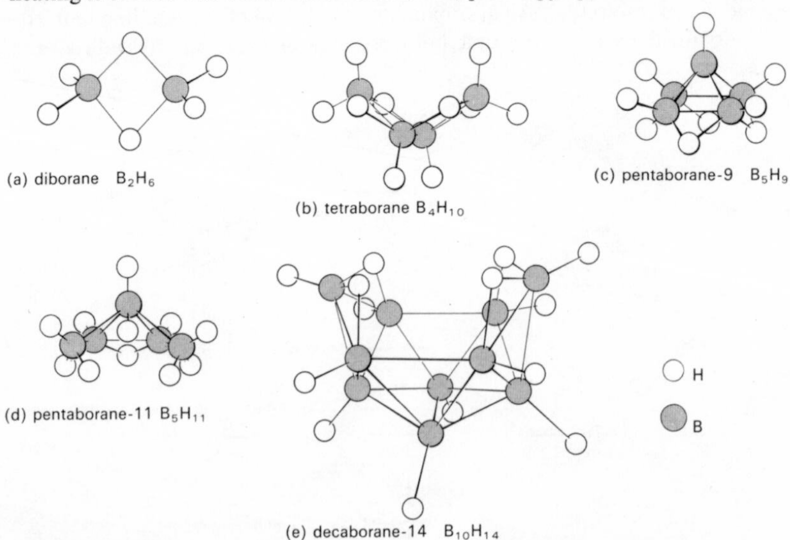

Figure 31 Structures of the boron hydrides: (a) diborane B_2H_6, (b) tetraborane B_4H_{10}, (c) pentaborane-9 B_5H_9, (d) pentaborane-11 B_5H_{11}, (e) decaborane-14 $B_{10}H_{14}$

Almost all of the boron hydrides which have been prepared so far fall into two main formula-groups.

B_nH_{n+4} *hydrides*		B_nH_{n+6} *hydrides*	
B_2H_6	diborane; b.p. −92·5 °C	B_4H_{10}	tetraborane; b.p. 18 °C
B_5H_9	pentaborane-9; b.p. 48 °C	B_5H_{11}	pentaborane-11; b.p. 63 °C
$B_{10}H_{14}$	decaborane-14	B_9H_{15}	nonaborane; m.p. −20 °C
		$B_{10}H_{16}$	decaborane-16

(When the nomenclature is ambiguous, as for B_5H_9 and B_5H_{11}, the number of hydrogen atoms present is added to the name using a hyphen: pentaborane-9 and pentaborane-11.)

The structures of the boron hydrides (boranes) have been determined by X-ray diffraction techniques and involve open three-dimensional cages of boron atoms (Figure 31). The bonding in these molecules is complex and will not be discussed in detail, but involves multicentre bonds between boron atoms as well as B—H—B bridges.

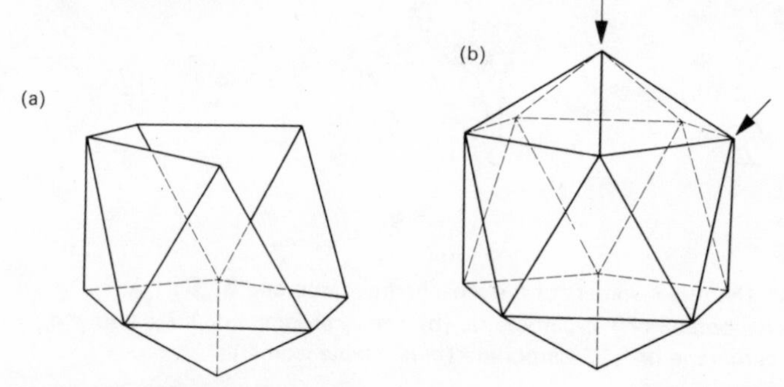

Figure 32 (a) The boron skeleton of decaborane-14 $B_{10}H_{14}$. (b) The icosahedron, a figure having twenty equilateral triangular faces and a total of twelve vertices. Arrows indicate the two atoms added to the decaborane-14 skeleton to complete the cage

By treating certain of the boranes with suitable molecules it has proved possible to form compounds having 'closed' cages of boron atoms. Most of the work in this direction has been carried out using decaborane-14; the boron skeleton in this molecule lacks only two atoms which would otherwise complete a closed icosahedral cage (Figure 32). These two atoms can be added to decaborane-14 relatively simply by treating it with either an acetylene, or an amine complex of BH_3:

$$B_{10}H_{14} + RC{\equiv}CR \longrightarrow B_{10}C_2R_2H_{10} + 2H_2;$$
(2.13) a 'carborane'

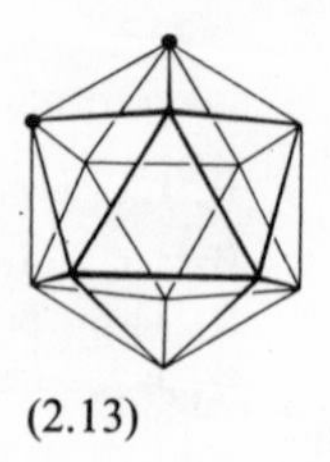

(2.13)

$$B_{10}H_{14} + 2(C_2H_5)_3NBH_3 \longrightarrow ((C_2H_5)_3NH^+)_2[B_{12}H_{12}]^{2-}.$$

These cages are exceptionally stable as demonstrated by the thermal isomerization of the carborane synthesized above (Figure 33).

The bonding in the $B_{n-2}C_2H_n$ carboranes and $B_nH_n^{2-}$ anions is very complex but it appears from molecular-orbital calculations that a B_nH_n cage lacks just two electrons needed to fill the last of several bonding molecular orbitals; hence the

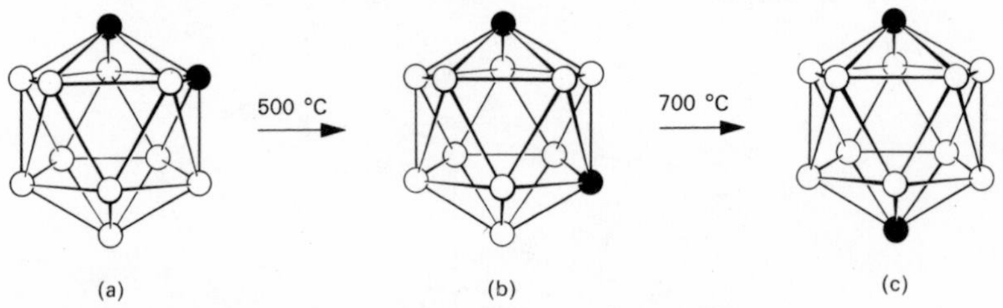

Figure 33 The three isomers of the icosahedral carborane $B_{10}C_2H_{12}$:
(a) *ortho*-carborane or 1,2-carborane, (b) *meta*-carborane or 1,7-carborane,
(c) *para*-carborane or 1,12-carborane (most stable isomer)

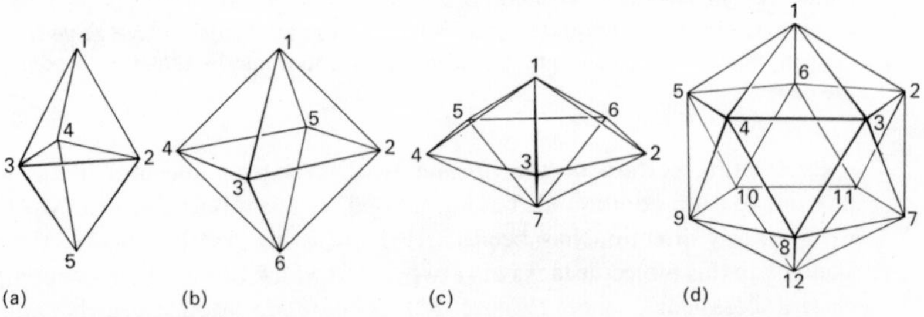

Figure 34 The cages present in the structures of some $B_nH_n^{2-}$ anions and $B_{n-2}C_2H_n$ carboranes. (a) Trigonal bipyramid – $B_3C_2H_5$ carborane; carbon atoms can be at positions 1,2; 2,3; 1,5, giving three possible isomers. (b) Octahedron – $B_6H_6^{2-}$ anion; $B_4C_2H_6$ carborane; carbon atoms can be at positions 1,2; 1,6, giving two possible isomers. (c) Pentagonal bipyramid – $B_5C_2H_7$ carborane; carbon atoms can be at positions 1,2; 2,3; 2,4; 1,7, giving four possible isomers.
(d) Icosahedron – $B_{12}H_{12}^{2-}$ anion; $B_{10}C_2H_{12}$ carborane; carbon atoms can be at positions 1,2; 1,7; 1,12, giving three possible isomers

formation and stability of the $B_nH_n^{2-}$ ions (molecular-orbital calculations predicted the stability of $B_{12}H_{12}^{2-}$ several years before it was synthesized). A carbon atom is iso-electronic to B^-, hence two carbon atoms substituted for borons in a B_nH_n cage would supply the two extra electrons required for a carborane cage's stability. A few of the anions and carboranes which have been prepared in recent years are shown in Figure 34. The presence of two carbon atoms in a carborane cage gives rise to the possibility of isomers (Figure 34) and it has been found empirically that the most stable carborane isomers are those in which the carbon atoms are (a) as far from each other as possible, and (b) bonded to the fewest neighbour atoms.

2.7 Metallic hydrides

Several of the d- and f-transition elements absorb hydrogen to a widely variable degree forming hydrides which exhibit many of the physical characteristics of metals: high thermal conductivity, high electrical conductivity, hardness and

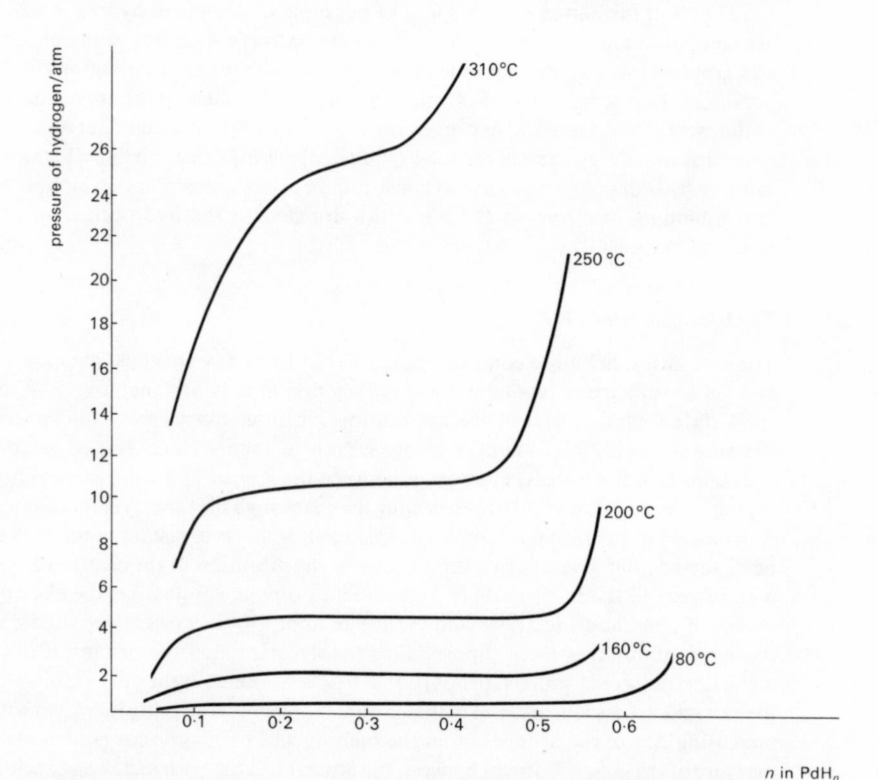

Figure 35 Pressure–composition isotherms for the palladium–hydrogen system

lustre. However, they are usually more brittle than the parent metals. An unusual feature of these hydrides is that they are non-stoichiometric, the composition varying with temperature and external pressure of hydrogen gas (see Figure 35). A considerable expansion of the metal crystal structure occurs during the process of hydrogen absorption, so that the hydrides are less dense than the parent metals; in many cases the lattice expansion is sufficient to cause a rearrangement of the original metal structure into completely new structures. Typically, the metals titanium, zirconium and hafnium form hydrides of composition between $MH_{1\cdot6}$ and $MH_{1\cdot8}$ which have the fluorite structure (Figure 50, p. 105) in which certain of the 'anion' lattice sites are vacant; with the absorption of more hydrogen the fluorite structure becomes unstable and a change to a face-centred tetragonal structure occurs having the composition $MH_{1\cdot8}$–$MH_{1\cdot98}$.

No unifying bonding scheme has been developed for these metallic hydrides. They are certainly not to be considered just as solid solutions with the hydrogen atoms occupying interstices in the metal lattice; the fact that the metal lattices either increase markedly in size or the solid adopts a completely different structure indicates a much stronger interaction between metal and hydrogen (indeed many have heats of formation of about 170 kJ per mole of combined hydrogen which are comparable to the values obtained for the salt-like hydrides of the group I and group II metals). Perhaps one of the most reasonable approximations to the bonding is that some of the electrons which normally make up the crystal lattice in the metal are donated to hydrogen (giving H^-) whilst the remainder continue to occupy the energy band of the (now expanded) metal lattice. On this scheme the salt-like hydrides of groups I and II constitute an extreme example in which all the lattice-binding electrons in the metal are donated to the hydrogen atoms so resulting in typically ionic lattices.

2.8 The hydrogen bond

The rare gases, helium, neon and argon, do not form any chemical compounds and yet at sufficiently low temperatures they first liquefy and then freeze to the solid state (helium requires the application of about twenty-five atmospheres pressure to enable it to solidify). Hence weakly attractive forces (called van der Waals, or London, forces) must occur between the atoms. This cohesion was first explained by London in 1930 by assuming that, although the time-average electron distribution in the rare-gas atoms was spherical, at any one instant in time it will be distorted, and give rise to a dipole, due to the vibration of the electron cloud with respect to the nucleus. This instantaneous dipole will polarize the electron clouds of neighbouring atoms and induce an appropriately orientated dipole in these neighbours; as these dipoles are suitably orientated the net result is an attraction the size of which is proportional to α^2/r^6, where α is the polarizability of the rare-gas atoms which are at distance r apart. The polarizability increases with increasing size of the atoms so that the melting and boiling points (which are a measure of the cohesive forces between the atoms) will rise with increasing atomic

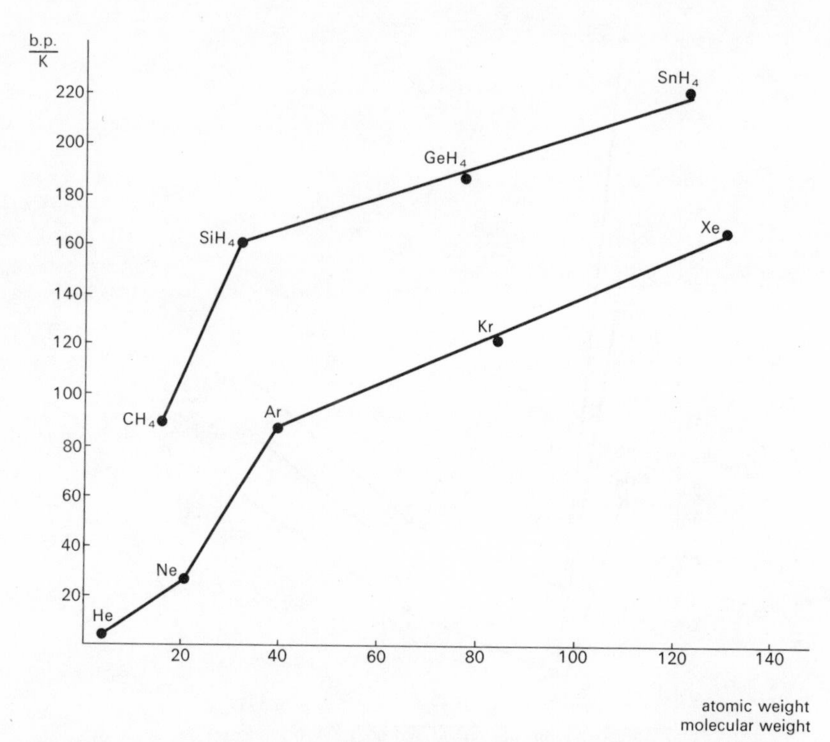

Figure 36 Plot of boiling point against atomic weight or molecular weight for the rare gases and the hydrides of the group IV elements

size. As seen from Figure 36, the boiling points of the rare gases increase in a fairly regular manner with increasing atomic weight of the elements. The same phenomenon is observed for the boiling points (and melting points) of the non-polar hydrides of carbon, silicon, germanium and tin.

However, when the boiling (or melting) points of the MH_3 hydrides of group V, the MH_2 hydrides of group VI or the MH hydrides of group VII are plotted against their molecular weights it is seen from Figure 37 that NH_3, H_2O and HF have boiling (or melting) points which are very much higher than the values predicted from the trends within their group. In other words, the cohesive forces between the NH_3, H_2O or HF molecules are bigger than those arising from just the normal van der Waals and dipole–dipole attractions. This extra cohesive force between these molecules is attributed to 'hydrogen bonding' and such bonding only occurs when a hydrogen atom is bonded to highly electronegative atoms such as nitrogen, oxygen and fluorine. The hydrogen atom taking part in hydrogen bonding is almost always found to lie on, or very near to, a straight line joining the two other atoms involved in the interaction. This line usually points in the direction of regions of high electron density (e.g. lone pairs) on the

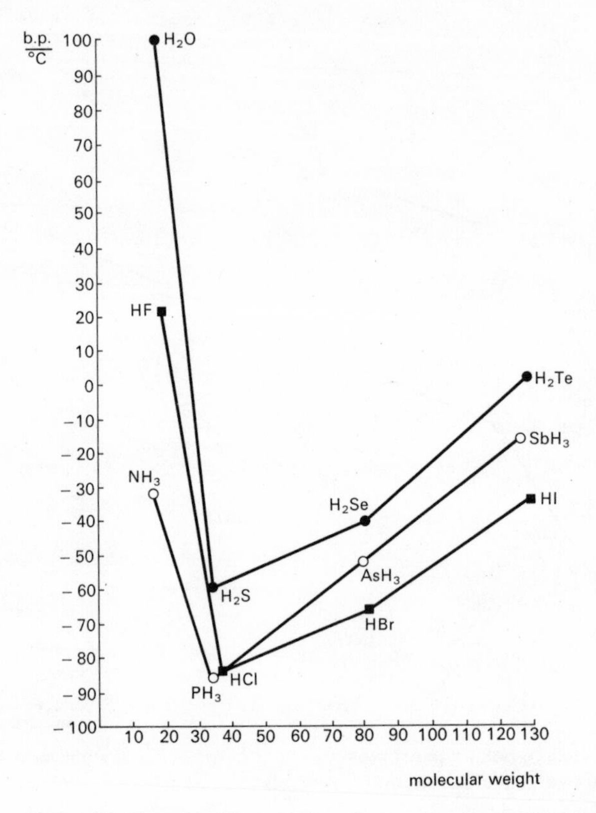

Figure 37 Plot of boiling point against molecular weight for the hydrides of the elements of groups V, VI and VII

atom to which the hydrogen is not covalently bonded; (2.14) is an example for oxygen.

$$\overset{\delta-}{O^1}—\overset{\delta+}{H}\cdots\cdots O^2$$

(2.14)

Due to the electronegative nature of oxygen there will be an electron drift towards O^1 from the O^1—H covalent bond, which will leave the hydrogen atom positively charged; hence a crude description of hydrogen bonding could be given in terms of a purely electrostatic attraction between the positive hydrogen atom and the electrons on O^2. However, the interaction is considerably more complex than this, so much so that a completely satisfactory interpretation of all the factors involved has yet to be given. Coulson suggests that four effects probably contribute to hydrogen bonding:

(a) electrostatic attraction: $\overset{\delta-}{O^1}$—$\overset{\delta+}{H}\cdots\overset{\delta-}{O^2}$

(b) Delocalization (or covalent bonding) effect; the electrostatic attraction in (a) will polarize (i.e. pull) the electron clouds in O^2 towards the H—O^1 region of space.

(c) Electron-cloud repulsion; the sum of the van der Waals radii for hydrogen and oxygen is 2·6 Å, and yet in O···H hydrogen bonding the two atoms often approach to within 1·6 Å. Thus the normal electron–electron repulsive forces will occur when the charge clouds of the O^1—H bond and O^2 atom begin to overlap.

(d) London (or van der Waals) forces; as in all intermolecular interactions these forces contribute to the binding, but their effect is relatively small.

Table 5 Estimates of the Energy Contributions to Each Separate Hydrogen Bond in Ice

Type of energy contribution	*Energy* / kJ mol^{-1}
(a) Electrostatic	25
(b) Delocalization	34
(c) Repulsion	−34
(d) London forces	13
Total	38
(cf. Experimental value, 25·5 kJ mol^{-1})	

From Table 5 it is seen that the electron–electron repulsion term is, numerically, one of the largest terms in the hydrogen bond interaction and, if it became larger by only a factor of two, it would cancel out the other bonding contributions entirely. Other elements cannot bond in a similar manner to hydrogen because the presence of their inner electron shells would increase the repulsion term to such an extent that the overall interaction would be antibonding.

2.8.1 *Detection of hydrogen bonding*

(a) Abnormal melting and boiling points can be indicative of hydrogen bonding as shown in Figure 37.

(b) Unexpectedly high molecular weights of hydrogen-containing species are often due to hydrogen bonding. For example, many carboxylic acids are dimeric in hydrocarbon solutions and in the vapour state (2.15). The dimeric structures

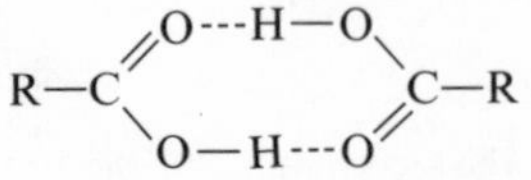

(2.15)

of gaseous formic and acetic acids (R = H, CH_3) have been confirmed by electron-diffraction measurements.

(c) The infrared stretching frequency of an O—H or N—H group taking part in hydrogen-bond formation will be lowered because the 'freedom' of the hydrogen atom is reduced by the hydrogen-bond interaction. (In a simple analogy it is as if the weight of the hydrogen atom has been increased by attaching the second atom to it, thus making it vibrate more slowly.) For example, the O—H stretching frequency of H_2O is 3756 cm^{-1} in H_2O gas, 3453 cm^{-1} in liquid water and 3256 cm^{-1} in ice.

(d) X-ray and neutron-diffraction studies on single crystals give the precise positions of the atoms within the crystal; the abnormally close approach of hydrogen atoms to electronegative groups can be interpreted in terms of hydrogen bonding. X-rays are diffracted by regions of high electron densities and, since hydrogen atoms possess relatively few electrons, X-ray diffraction techniques cannot always detect hydrogen positions accurately in crystals. Neutrons are diffracted by nuclei and, although the experimental difficulties are formidable, the positions of hydrogen atoms (or, more easily, deuterium atoms, since these have a larger nucleus) can be accurately determined.

2.8.2 *Examples and importance of hydrogen bonding*

Water. It is evident from Figure 37 that, without the effects of hydrogen bonding, water vapour would only begin to liquefy at about −70 to −80 °C, or almost 200 °C lower than the actual boiling point. Under such conditions life as we know it, with its absolute dependence on liquid water, could not have evolved. Also hydrogen bonding plays a major role in the chemistry of a wide variety of reactions basic to life.

The isolated water molecule has the central oxygen atom sp^3 hybridized with two sets of lone electron pairs occupying two of the sp^3 orbitals (2.16). In ice,

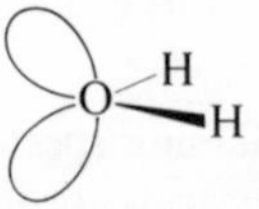

(2.16)

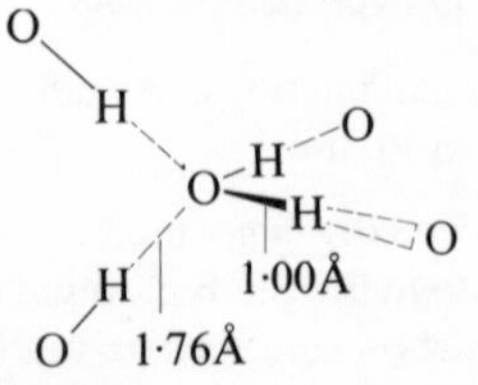

1·00Å 1·76Å

(2.17)

the water molecules are found to be aligned so that hydrogen bonds are formed along the direction of the oxygen lone pairs, resulting in a tetrahedral coordination for each oxygen atom, the two covalently bonded hydrogens being closer to the oxygen than the hydrogen-bonded atoms (2.17).

In a normal solid with no strong *directional* forces between the molecules the latter pack as close together as space and molecular shape allow. However, in ice the highly directional hydrogen bonds are stronger than other intermolecular forces and the water molecules have to adopt a much more open structure than would otherwise be the case. On warming, hydrogen bonds in ice begin to rupture but sufficient still remain at 0 °C for ice to retain an open structure and have a density lower than liquid water at the same temperature; about half the maximum number of hydrogen bonds are still present in water at 20 °C.

Solid ammonium fluoride NH_4F. Ammonium fluoride has, like water, a very open structure in the solid state. This is thought to be due to hydrogen bonding between the ammonium NH_4^+ and fluoride F^- ions, which results in tetrahedral coordination of both ions (2.18, 2.19). In contrast ammonium chloride adopts the caesium chloride structure, in which the ions have eight coordination.

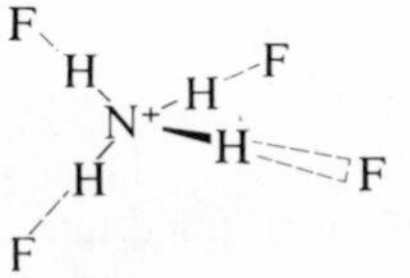

F H F H N⁺ H F H F

(2.18)

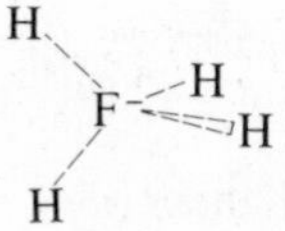

H H F H H

(2.19)

Hydrogen fluoride. The structure of solid hydrogen fluoride (2.9) is that of a zig-zag polymeric chain. Although the reason for the HFH bond angle of 120° is not clear, the angle is very different from that expected (180°) if the structure were dominated purely by electrostatic alignment of the highly polar HF molecules (2.20). Hydrogen bonds persist in liquid hydrogen fluoride and the vapour is even associated up to 60 °C, probably as $(HF)_4$ and $(HF)_6$ polymers.

$$\overset{\delta+}{H}-\overset{\delta-}{F}\cdots\overset{\delta+}{H}-\overset{\delta-}{F}\cdots\overset{\delta+}{H}-\overset{\delta-}{F}.$$

180°

(2.20)

'*Hydrogen bifluoride' ion* HF_2^-. Fluoride ions readily attach themselves to free hydrogen fluoride molecules to give the 'hydrogen bifluoride' ion

$$F^- + HF \longrightarrow [F-H-F]^-.$$

Unlike most other cases of hydrogen bonding, when one bond to hydrogen is shorter than the other, the two H—F bond lengths in HF_2^- are identical. Careful

X-ray measurements show that there is a substantial amount of electron density around the hydrogen nucleus so that the interaction cannot be a purely electrostatic one between two fluoride ions and a proton, i.e. $F^-H^+F^-$. The energy of the dissociation

$$[F—H—F]^- \longrightarrow HF + F^-$$

is very much larger ($\sim$243 kJ mol^{-1}) than the other hydrogen-bond energies shown in Table 5. For this reason it has been suggested that the bonding in HF_2^- may be unique and perhaps involve three-centre bonding, the hydrogen 1s orbital interacting with two fluorine 2p orbitals.

Bicarbonates. In solid bicarbonates, such as $NaHCO_3$, hydrogen bonds link the planar CO_3^{2-} ions into polymeric chains, the crystal as a whole being bound together by electrostatic interaction between the metal cations and the negatively charged bicarbonate chains (2.21).

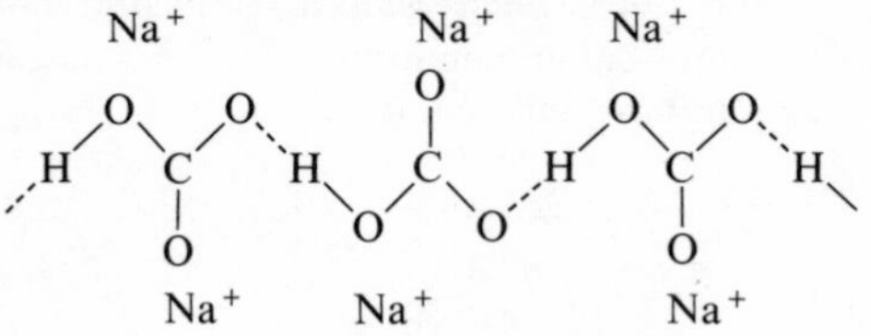

(2.21)

Nitrophenols. Of the three isomeric mononitrophenols, it is found that *ortho*-nitrophenol (2.22) has the lowest melting point, the lowest boiling point and the highest solubility in organic solvents. This is considered to be due mainly to *intra*-molecular hydrogen bonding in this isomer. The *meta* and *para* isomers form *inter*-molecular hydrogen bonds resulting in relatively high intermolecular forces in the solid and liquid states which affect physical properties such as melting and boiling points and solubility in non-polar solvents. Several other *ortho*-substituted phenols are also internally hydrogen bonded, e.g. (2.23, 2.24).

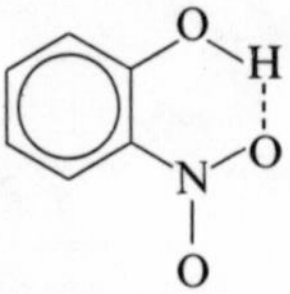

(2.22)

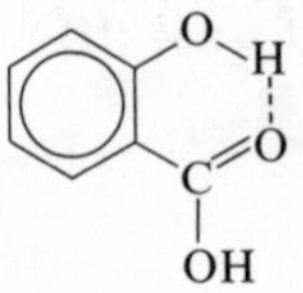

(2.23) salicylic acid

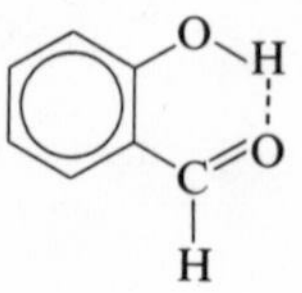

(2.24) salicylaldehyde

Chapter 3
Group I. The Alkali Metals: Lithium, Sodium, Potassium, Rubidium, Caesium and Francium

3.1 Introduction

The electronic configuration of the alkali metals indicates that the simplest process by which the metals may achieve a rare-gas electronic structure is to form M^+ via the loss of the outer s-electron. Hence in the great majority of their compounds the alkali metals exist as unipositive ions: often as 'bare ions' in the solid state, for example Na^+Cl^-, or as solvated ions when dissolved in suitable solvents, for example $Na(H_2O)_x^+$ for sodium ions in water. However, even the formation of M^+ ions requires the absorption of 400–500 kJ mol^{-1} and it is pertinent to inquire where the energy for the ionization process arises.

Table 6 The Group I Elements

Element		*1st IE* kJ mol^{-1}	*2nd IE* kJ mol^{-1}	*3rd IE* kJ mol^{-1}	*Radius of* M^+/Å
Lithium Li	$1s^22s^1$	520·0	7297	11 810	0·60
Sodium Na	$[Ne]3s^1$	496·0	4564	6911	0·95
Potassium K	$[Ar]4s^1$	418·8	2932	4602	1·33
Rubidium Rb	$[Kr]5s^1$	412·3	2655	3850	1·48
Caesium Cs	$[Xe]6s^1$	375·9	2423	—	1·69
Francium Fr	$[Rn]7s^1$	—	—	—	—

IE = ionization energy at 0 K; add ~6 kJ mol^{-1} for values at 25 °C.

The reaction between sodium and chlorine takes place with the evolution of 395·8 kJ for each mole of crystalline sodium chloride formed. The reaction may be broken down, theoretically, into a cycle of fundamental energy processes such that the summation of the total energy in the cycle is equal to the heat of formation of sodium chloride (the Born–Haber cycle).

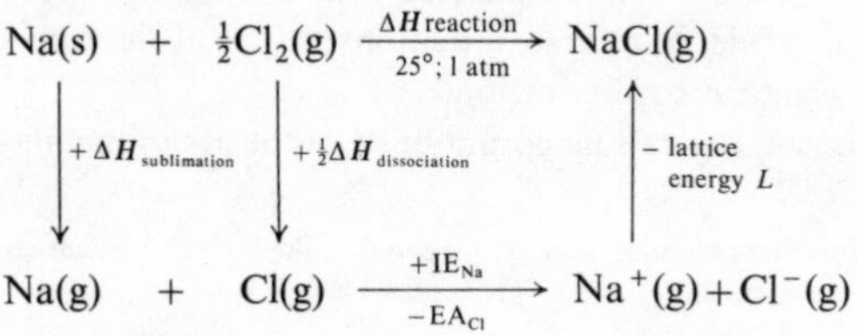

$$Na(s) + \tfrac{1}{2}Cl_2(g) \xrightarrow[25°;\ 1\ atm]{\Delta H\ \text{reaction}} NaCl(g)$$

$$Na(s) \xrightarrow{+\Delta H_{\text{sublimation}}} Na(g) \qquad \tfrac{1}{2}Cl_2(g) \xrightarrow{+\frac{1}{2}\Delta H_{\text{dissociation}}} Cl(g)$$

$$Na(g) + Cl(g) \xrightarrow[-EA_{Cl}]{+IE_{Na}} Na^+(g) + Cl^-(g) \xrightarrow{-\text{lattice energy } L} NaCl$$

By Hess's law of heat summation,

$$\begin{aligned}\Delta H_{\text{reaction}} &= \Delta H^{\circ}_{\text{sublimation}} + \tfrac{1}{2}\Delta H^{\circ}_{\text{dissociation}} + \text{IE}_{\text{Na}} - \text{EA}_{\text{Cl}} - L \text{ kJ mol}^{-1}\\ &= 108{\cdot}4 + \frac{242{\cdot}7}{2} + 502{\cdot}0 - 348{\cdot}5 - 779{\cdot}1\\ &= -395{\cdot}8 \text{ kJ (mol NaCl)}^{-1}\\ &= \text{heat of formation of one mole of crystalline sodium chloride at } 25\,^{\circ}\text{C}.\end{aligned}$$

Thus the main source of energy which compensates for the formation of Na^+ ions is the lattice energy of the crystalline product, sodium chloride. If the reaction products are formed in aqueous solution, the ionization energy of sodium is offset by the high heat of hydration of the sodium and chloride ions,

i.e. $Na^+(g) + Cl^-(g) \xrightarrow[\text{water}]{\text{dissolve in}} Na^+(aq) + Cl^-(aq) - 759{\cdot}8 \text{ kJ mol}^{-1}$.

Fluorine has the highest reactivity of all the elements which attack the alkali metals and would therefore stand the best chance of oxidizing these metals to the M^{2+} state. The feasibility of such a reaction can be tested theoretically using a suitable Born–Haber cycle to sum all the energy terms involved.†

$$\begin{array}{ccccc} NaF(s) & + & \tfrac{1}{2}F_2(g) & \xrightarrow[25^{\circ}C;\ 1\ atm]{\Delta H} & NaF_2(s)\\ \downarrow L_{NaF} & & \downarrow \tfrac{1}{2}\Delta H^{\circ}_{diss} & & \uparrow -L_{NaF_2}\\ Na^+(g) + F^-(g) & + & F(g) & \xrightarrow[IE_{Na^+}]{-EA_F} & Na^{2+}(g) + 2F^-(g)\end{array}$$

By Hess's law of heat summation,

$$\begin{aligned}\Delta H &= L_{\text{NaF}} + \tfrac{1}{2}\Delta H^{\circ}_{\text{diss}} - \text{EA}_{\text{F}} + \text{IE}_{\text{Na}} - L_{\text{NaF}_2} \text{ kJ mol}^{-1}\\ &= 899{\cdot}4 + \frac{157{\cdot}7}{2} - 349{\cdot}5 + 4564 - 2908\\ &= 2284{\cdot}7 \text{ kJ mol}^{-1}.\end{aligned}$$

(It is assumed in the above calculation that the lattice energy of NaF_2 would be approximately equal to that of MgF_2.) Hence this reaction is energetically highly unfavourable (as are all other possible routes to NaF_2) and from the Born–Haber cycle it is seen that the term which prevents the realization of preparing M^{2+} compounds of these elements is their high second ionization energy. The formation of M^{2+} ions would involve the removal of an electron from an inner quantum shell where the electrons are closer to the nucleus and thus more strongly held than the outer s-electron; this is, of course, a demonstration of the well-known 'stability of the rare-gas electron configuration'.

The presence of M^+ ions in crystalline compounds of the alkali metals can be

†This discussion neglects entropy effects because, strictly, we should calculate the free energy change $\Delta G = \Delta H - T\,\Delta S$, not the change in enthalpy.

demonstrated by accurate X-ray diffraction measurements on single crystals of, for example, sodium chloride. Such experiments enable the number of electrons around the sodium and chloride nuclei to be calculated, since the extent of diffraction of the X-rays by the ions making up the crystal is a function of the electron density at the ion sites. In solution the presence of ions can be demonstrated by electrolytic techniques.

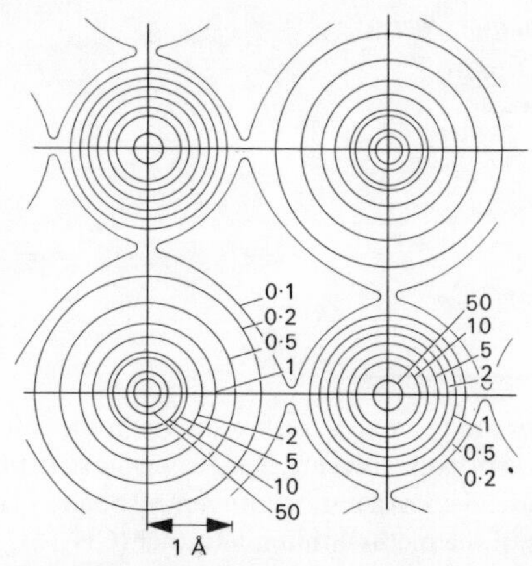

Figure 38 Contour map of the electron density round the nuclei in crystalline sodium chloride
The figures give the electron densities in electrons per Å^3, round each nucleus. By integration, the number of electrons was calculated to be 10·05 round the sodium nucleus and 27·75 round the chlorine nucleus; the close agreement with the theoretical values of 10 electrons for Na^+ and 28 for Cl^- demonstrates that the crystal is made up of ions (Reproduced from H. Witte and E. Wölfel, *Zeitschrift für Physikalische Chemie*, vol. 3, 1955, p. 296, with permission)

The high sublimation and ionization energies, the key energy terms involved in the initial stages of a reaction, are usually said to make lithium the least reactive alkali metal; conversely caesium is the most reactive. However, this must be a purely kinetic effect since the reaction products for lithium have the highest lattice and hydration energies, making the lithium reactions thermodynamically favoured.

Not all compounds of the alkali metals are purely ionic in nature. The vapour above the liquid metals, though predominantly monatomic, contains a small proportion of dimer, and probably tetramer, molecules. The bond between the atoms in the dimers can be assumed to arise by overlap of the outer s-orbitals,† but it will be much weaker than the corresponding bond in H_2 because of (a) the strong repulsion forces arising between the inner electron shells of the two

†This is a slight oversimplification because spectroscopic studies show the σ-bond to contain a small amount of p-character.

metal atoms and (b) the diffuse nature of the outer *ns* orbitals, which results in their poor overlap. The strength of the bond in these dimers will progressively decrease from lithium to caesium because both these factors increase as the size of the alkali metal increases; see Table 7.

Table 7 Bond Lengths and Dissociation Energies of the Alkali-Metal Dimers at 25 °C.

Dimer	*Bond length*/Å	*Dissociation energy*/kJ mol^{-1}
Li_2	2·673	113·8
Na_2	3·079	73·2
K_2	3·923	49·4
Rb_2		47·3
Cs_2		43·5
cf. H_2	0·74	435·9

Many organolithium derivatives are tetrameric or hexameric in the solid and gaseous states and when dissolved in organic solvents. The electronic structures of these polymers are thought to involve covalent, multi-centre bonds between lithium and carbon, as in the case of the methyllithium tetramer $(CH_3Li)_4$. This compound has the four lithium atoms arranged in the form of a tetrahedron, the four methyl groups being placed symmetrically above the midpoint of each tetrahedral face (Figure 39). If each lithium atom is approximately sp^3 hybridized, the three lithium atoms in each tetrahedral face can form a four-centre molecular orbital, containing two electrons, by combining their sp^3 hybrids with a carbon sp^3 hybrid atomic orbital. In this way only four electron pairs are required to hold the tetramer together (Figure 40).

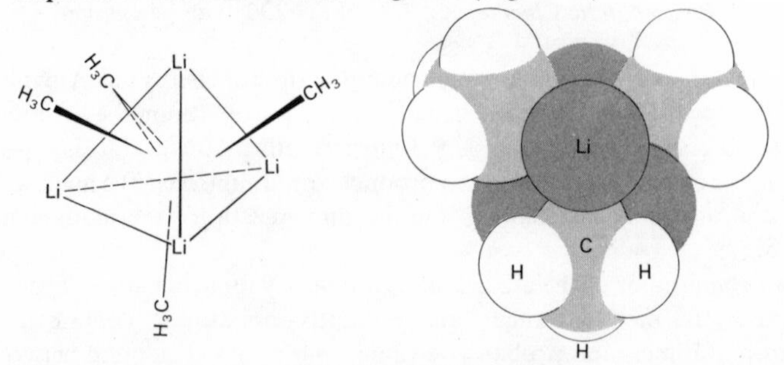

Figure 39 The structure of the methyllithium tetramer

On strong heating in a vacuum the alkali-metal halides vaporize to the highly polar diatomic molecules MX. Their high dipole moments and nuclear quadrupole resonance spectra indicate that very little covalent bonding occurs between metal and halogen, the ion pair being held mainly by Coulombic attraction of the

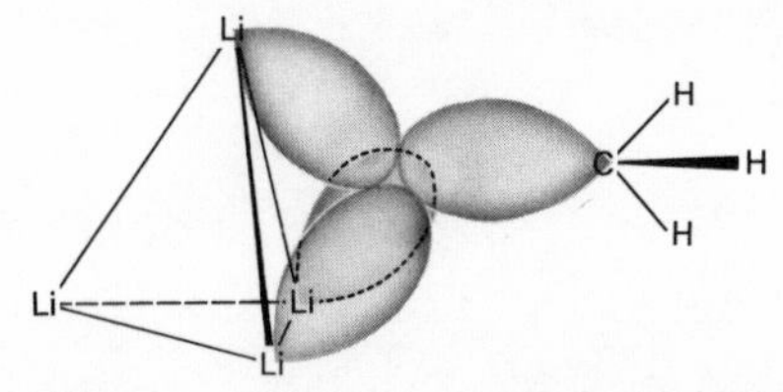

Figure 40 The four sp^3 atomic orbitals used in forming a molecular orbital in one face of the methyllithium tetramer

oppositely charged ions.† The normally spherical electron clouds of the ions will be distorted by the unsymmetrical electric field in the ion pair which will make the dipole moments smaller than that calculated assuming two spheres of charge +1 and −1 separated by the internuclear distance. The distortion will be greatest when M^+ is small (lithium) and X^- is large (iodine) and will lead to the most favourable chance of covalent bond formation in the lithium iodide molecule. The vapours of alkali-metal halides also contain higher aggregates of ions such as M_2X_2, M_3X_3 and M_4X_4. Electron-diffraction measurements show Li_2Cl_2 and Li_2F_2 to be planar, diamond-shaped molecules (Figure 41). The alkali-metal hydroxides, and many other salts, also vaporize to give ion clusters, see Table 8.

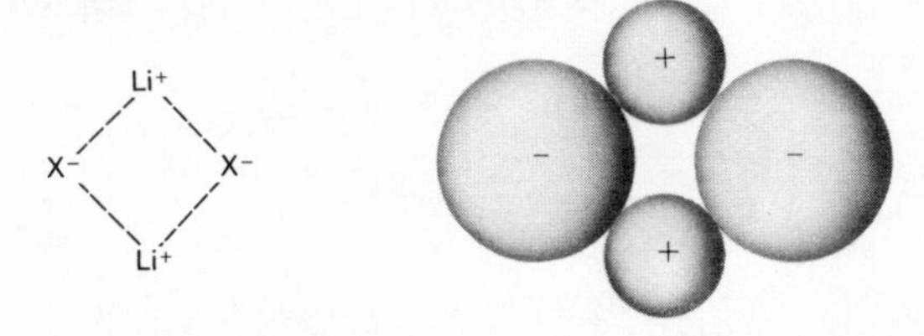

Figure 41 The diamond-shaped structure of lithium fluoride and lithium chloride dimers

The water molecules in the hydration shell of alkali-metal ions dissolved in water are held only by relatively weak ion–dipole forces with the result that many salts of alkali metals separate from solution as anhydrous crystals. As the ion–dipole interaction is inversely dependent on the size of the metal ion it is no surprise to find that lithium forms the largest number of stable solid hydrates; for example, many simple lithium salts crystallize from solution as trihydrates $LiX.3H_2O$, where X = Cl^-, Br^-, I^-, ClO_3^-, ClO_4^-, MnO_4^-, NO_3^- and BF_4^-. In the perchlorate (and presumably many of the other salt) trihydrates, the lithium ion has been shown to be octahedrally surrounded by six water molecules, which it shares with two neighbouring lithium ions to form a linear polymer as the basic unit in the crystal (3.1).

† From nuclear quadrupole coupling data, a maximum of 4 per cent covalent character has been assigned to the sodium chloride molecule.

$$\begin{array}{c} \diagdown \quad \mathrm{OH_2} \quad\quad \mathrm{OH_2} \quad \diagup \\ -\mathrm{Li^+}-\mathrm{OH_2}-\mathrm{Li^+}-\mathrm{OH_2}-\mathrm{Li^+}- \\ \diagup \quad \mathrm{OH_2} \quad\quad \mathrm{OH_2} \quad \diagdown \end{array}$$

(3.1)

Table 8 Typical Gaseous Molecules of the Alkali Metals

Molecules†	*Dipole moment of monomer/debye*	*Bond length of diatomic molecule/Å*
LiF, Li_2F_2, Li_3F_3	6·28	1·56
LiCl, Li_2Cl_2, Li_3Cl_3, Li_4Cl_4		
$LiNaF_2$ (i.e. 'mixed' dimer)		
NaF, Na_2F_2, Na_3F_3, Na_4F_4		
KCl, K_2Cl_2, K_3Cl_3	10·5	2·67
CsF, Cs_2F_2	7·9	2·35
CsCl, Cs_2Cl_2	10·4	2·91
CsI	12·1	3·315
NaOH, $Na_2(OH)_2$		
CsOH, $Cs_2(OH)_2$	7·1 (dimer is non-polar)	Cs—O (in monomer) = 2·40
$NaK(OH)_2$ (i.e. 'mixed' dimer)		
$NaNO_3$		
$NaNO_2$		
$RbNO_3$		
Cs_2O_2	0·0	
Cs_2SO_4	0·0	

†For a given halide, the polymers are less readily formed at a particular temperature the larger the alkali-metal ion. For example, lithium chloride is predominantly a dimer in the gas phase at 820 °C whereas gaseous caesium chloride is monomeric at this temperature. This is mainly due to a decrease in the Coulombic attraction between the ions as the metal-ion radius increases.

The alums $M^IM^{III}(SO_4)_2.12H_2O$, M^I = alkali metal and M^{III} = a trivalent ion, represent one of the few cases where the alkali metals coordinate to six separate water molecules. Lithium does not form alums, presumably because the ion $Li(H_2O)_6^+$ is too small to be accommodated satisfactorily into the alum crystal.

The single positive charge on the alkali-metal ions results in their having very few well-characterized complexes because, as in the hydrates, the ion–dipole interactions between a potential ligand and the cations are weak. However, under strictly anhydrous conditions, isonitrosoacetophenone HA (3.2) forms solid complexes of the alkali metals, the stoichiometry of which can be changed by altering the reaction conditions: for example KA (red), KA.HA (yellow-orange), RbA (red), CsA.HA (yellow). From spectroscopic studies it appears probable that these compounds have chelated structures (3.3 and 3.4). They dissolve in a variety of dry organic solvents with retention of the ligand on the metal but in water the complexes break up to give the ions $M^+(aq)$ and $A^-(aq)$. 8-Hydroxyquinoline gives similar compounds but the complexes formed by salicylaldehyde are much less well defined.

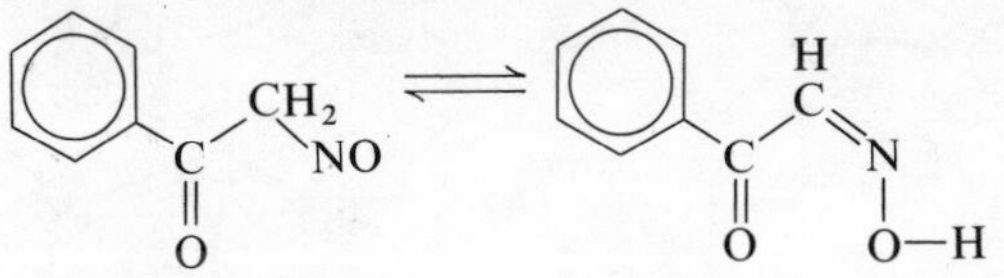

(3.2) HA

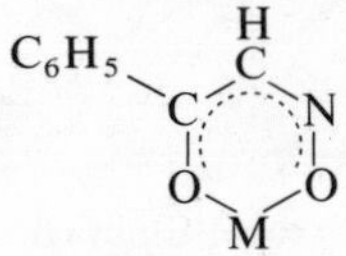

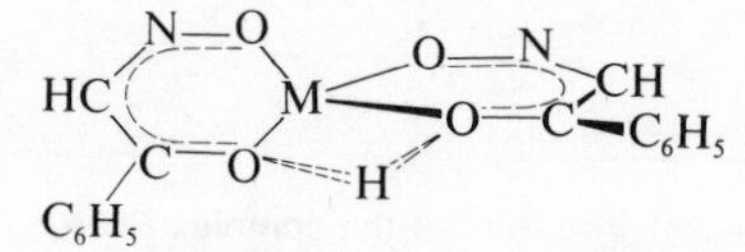

(3.3) MA

(3.4) tetrahedral complex MA.HA

Although most alkali-metal salts are highly insoluble in non-polar organic solvents, their solubility can be increased dramatically by the addition of cyclic polyethers to the solvent, and in many cases crystalline complexes formed between the metal salts and the polyethers can be isolated. Considerable interest is being shown in these polyether complexes because similar naturally occurring cyclic compounds may act as ion carriers in the transport of alkali-metal ions across cell membranes. Typical of these cyclic polyethers is (3.5), which is often referred to

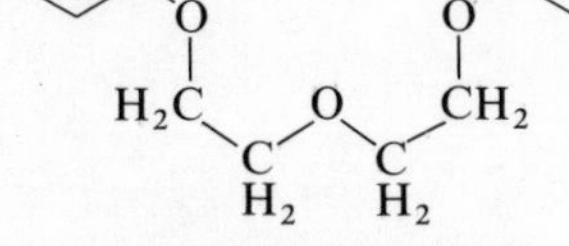

(3.5) dibenzo-18-crown-6

by its shortened name dibenzo-18-crown-6 (i.e. there are eighteen atoms in the crown-shaped ring, six of which are oxygen atoms). When this ligand forms a complex the alkali-metal ion is situated centrally in the ring with all the oxygen atoms coplanar. By making the polyether ring larger, as in dibenzo-30-crown-10, it is possible for the ligand to wrap completely round the metal so that all ten oxygen atoms coordinate to the metal (Figure 42).

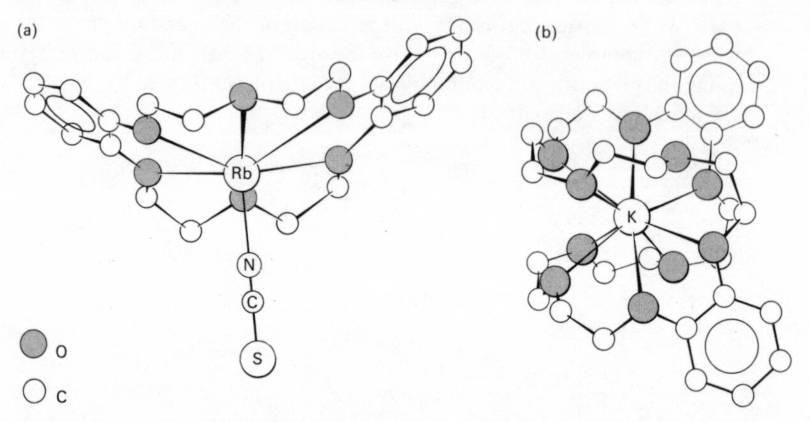

Figure 42 (a) Structure of the complex RbNCS(dibenzo-18-crown-6), in which the six oxygen atoms are coplanar and approximately equidistant from the rubidium atom. (b) Cation in the complex KI(dibenzo-30-crown-10). The potassium ion is approximately equidistant from all ten oxygen atoms

In the solid state the alkali metals normally adopt a body-centred cubic structure (Figure 43), in which the metal atoms are surrounded by eight equidistant neighbours, although at low temperatures lithium forms the hexagonal close-packed arrangement (Figure 44). The lithium atoms in each layer have six neighbours in the same plane arranged at the corners of a regular hexagon; alternate layers are staggered relative to each other so that any one lithium atom also has three near neighbours in both the adjacent layers giving a total coordination of 12. On the current theory of bonding, only the outer *n*s electrons of the

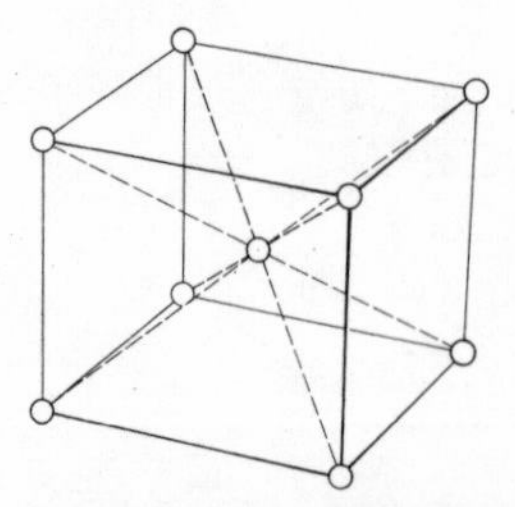

Figure 43 The body-centred cubic structure adopted by the alkali metals. Each atom has eight nearest neighbours at the corners of a cube. The structure is related to that of caesium chloride (Figure 46), in which both the caesium and chloride ions are replaced by metal atoms

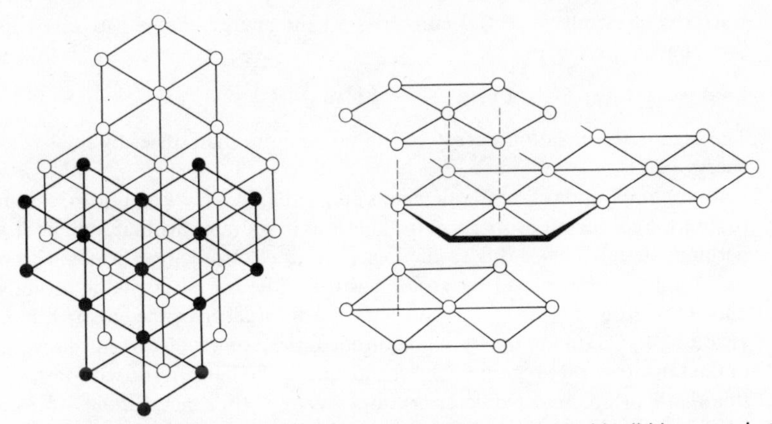

Figure 44 The hexagonal close-packed arrangement adopted by lithium metal at low temperatures

metal atoms participate in bonding and these are delocalized into a large number of molecular orbitals extending over the whole crystal. In most other metals each atom contributes two, or even more, electrons to the binding of the lattice, so that the alkali metals are relatively soft metals and have comparatively low melting and boiling points. As the metal in the lattice is changed from lithium through to caesium, repulsion forces due to nonbonding inner-shell electrons increase resulting in lower cohesion in the case of the heavier metals; this in turn will lower the melting point and the hardness down the group. By extrapolation of the data in Table 9 we can predict that should francium ever be isolated in quantity it would be found to be a very soft metal with a melting point slightly below room temperature.

From Table 9 it can be seen that lithium has the largest standard electrode potential, which is opposite to the general trend in the group as a whole. This is probably linked to the high solvation energy of the relatively small lithium ion

Table 9 Some Physical Properties of the Alkali Metals and their Ions

Property	Li	Na	K	Rb	Cs
Melting point /°C	180	98	64	39	29
Boiling point /°C	1330	892	759	700	690
Heat of sublimation /(kJ mol^{-1})	218·0	108·5	89·9	85·8	—
Approximate hydration energy of M^+ /(kJ mol^{-1})	515	395	310	285	259
Electrode potential /V	−3·05	−2·71	−2·93	−2·92	−2·92

since the electrode potential embodies all the energy changes involved in the hypothetical processes

$$Li(s) \longrightarrow Li(g) \longrightarrow Li^+(g) \longrightarrow Li^+(aq).$$

The high first ionization energy of lithium is more than offset by the hydration energy of Li^+.

Ionic compounds of the alkali metals are, with only a few exceptions, colourless (unless the colour is a property of the anion, as in sodium chromate and potassium permanganate). This is due to the fact that no electronic transitions within the metal ions can be excited by visible light; i.e. the lowest available orbitals are above the ground state by a factor very much larger than the energy quanta of visible light. Quanta of the right magnitude for exciting such transitions would be in the far ultraviolet region of the spectrum. In the flame test used for detecting the alkali metals, electronic transitions occur within metal *atoms* transiently formed in the flame. Some of these transitions (e.g. $ns \rightarrow np$ and $ns \rightarrow (n+1)p$ orbitals) happen to be of the correct energy for their associated spectral frequencies to fall in the visible region of the spectrum and give the characteristic colour of the flame: lithium – crimson; sodium – golden yellow; potassium – red-violet; rubidium – blue-violet; caesium – blue-violet. In the flame photometer a given alkali-metal salt solution is injected into an oxygen–gas flame and the intensity of the metal's emission spectrum measured using colour filters to remove light at other wavelengths. By comparing this intensity to that of a given standard solution of the same metal, the concentration of the unknown sample can be determined.

The alkali-metal salts are completely ionized in aqueous solution and, because water of hydration round the metal ions is not held by strong covalent bonds, hydrolyses of the type found in metals having strongly coordinated water (e.g. Al^{3+}, Fe^{3+}),

$$M^{m+}(H_2O)_{n+1} \rightleftharpoons [(H_2O)_nM(OH)]^{(m-1)+} + H^+(aq)$$

do not occur. The hydroxides are strong bases (a strong base being defined as that which ionizes completely in aqueous solution to give hydroxyl ions); salts of all but the very weakest acids may be made simply by mixing equivalent amounts of alkali hydroxide with the acid in question and crystallizing the solution

$$NaOH + HA \longrightarrow H_2O + NaA.$$

Virtually all alkali-metal salts are soluble in water. As the process of solution involves separation of the ions by water of solvation it suggests that the hydration energy and lattice energy of each of their salts must be almost equal (a fact verified for the halides in Table 10). The least soluble halide is that containing the two smallest ions – only 0·13 g of lithium fluoride dissolve in 100 g of water at 20 °C. Therefore there are few salts available as reagents for the gravimetric determination of the alkali metals in aqueous solution and, indeed, lithium and sodium cannot be estimated *accurately* in this manner (sodium zinc uranyl acetate

$NaZn(UO_2)_2(CH_3COO)_9H_2O$ is probably the most insoluble sodium salt and can be used in all but the most accurate work for the gravimetric determination of sodium). However, for the three heaviest metals it is found that their tetraphenylborates, $MB(C_6H_5)_4$, have solubilities less than that of silver chloride and hence the addition of sodium tetraphenylborate to the solution results in their essentially quantitative precipitation,

$$NaB(C_6H_5)_4 + M^+ \longrightarrow Na^+ + MB(C_6H_5)_4\downarrow \quad (M = K, Rb, Cs).$$

The precipitates can be filtered off, dried at 100 °C and weighed in the usual way. The low solubilities are probably connected with the very low hydration energy of the large tetraphenylborate ion. Several other large anions (e.g. perchlorate, cobaltinitrite, hexachloroplatinate, hexafluorosilicate) behave in the same way and precipitate the potassium, rubidium or caesium salts on the addition of the cation.

The solution chemistry of the alkali metals is not very extensive because of their salts being soluble and ionized. Many of their 'reactions' are not truly properties of the cations but represent reactions of the anion as illustrated by the neutralization of a hydroxide and the precipitation of a silver halide

$$(M^+) + OH^- + H^+(aq) \longrightarrow (M^+) + H_2O,$$
$$(M^+) + Cl^- + Ag^+ \longrightarrow (M^+) + AgCl.$$

Only when the metal ions are directly involved, as in the precipitation of the salts mentioned above, are the reactions really associated with the metal. It is mainly in the solid state that the difference in physical and chemical properties, due to changes in size and mass, become apparent within the group. The greatest percentage size change occurs between lithium and sodium so that in many ways lithium appears to be somewhat out of place among the alkali metals; lithium and

Table 10 Lattice Energies and Hydration Energies of Some Alkali-Metal Halides at 25 °C

Halide	*Lattice energy*/kJ	*Hydration energy*/kJ
LiF	1034	1004
LiCl	845	883
LiBr	799	849
LiI	716	808
NaCl	778	762
KCl	699	682
RbCl	673	657
RbBr	644	624
RbI	607	577
CsCl	648	632

magnesium seem to be more similar in chemical behaviour. Some of the typical properties which *increase* from caesium to lithium (i.e. have the order Li > Na > K > Rb > Cs) are (a) hardness, melting points and sublimation energies of the metals, (b) ease of thermal decomposition of salts containing polyatomic anions (e.g. carbonate, nitrate, perchlorate, hydroxide, peroxide and polyhalides) and (c) lattice energies of the salts (although there are irregularities when the salts contain a small ion due to inefficient packing in the lattice).

The structure of the halides changes from the face-centred cubic sodium chloride lattice to that of the caesium chloride type to give a more favourable lattice energy as the ions increase in size; that is, a change in coordination number from six to eight occurs.

3.2 Solutions of the alkali metals

A particularly unusual property of the alkali metals is their high solubility in anhydrous liquefied ammonia (Table 11). The dilute solutions are coloured a very

Table 11 Solubilities of the Alkali Metals in Liquid Ammonia

Metal	*Solubility* / g/(kg NH_3)	*Temperature*/°C
Li	109·0	−33·4
Li	113·1	0
Na	250·7	−33·4
Na	230·0	0
K	471·1	−33·4
K	485·0	0

(−33·4 °C is the boiling point of ammonia at one atmosphere pressure)

deep blue but at high concentration the colour changes to bronze. There is still much conjecture as to the exact nature of these solutions, but it is generally supposed that at low concentrations the metals dissociate into solvated M^+ ions and solvated electrons:

$$M \longrightarrow M^+(am) + e^-(am).$$

The electron, being very small and therefore strongly polarizing, is able to orientate the solvent ammonia molecules round itself so that their protons point towards the electron (3.6). Essentially the electron has made itself a cavity in the body of the solvent. These dilute solutions, containing unpaired electrons, are paramagnetic and exhibit electron spin resonance spectra. As the concentration is

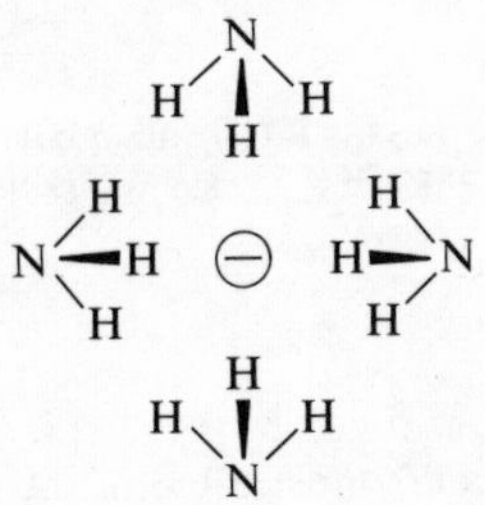

N
H H
H
H H
N—H ⊖ H—N
H H
H
H H
N
(3.6)

increased the electron is thought to be more and more closely associated with the protons of those ammonia molecules which are solvating the metal ions,

$$M^+(am) + e^-(am) \longrightarrow M(NH_3)_n e.$$

The entity thus formed (called an expanded monomer) is essentially the solvated *metal*; in agreement with this some metals, notably the alkaline earths, can be isolated from their liquid-ammonia solutions as solvates; for example, $Ca(NH_3)_n$, $n \simeq 6$. These monomers will also be paramagnetic but do not constitute more than a small percentage of the dissolved metal even under optimum conditions; the dimerization (or formation of even higher polymers) of the monomers by electron pairing causes decreasing paramagnetism as the solutions are made more concentrated.

In summary, the process of dissolution can be represented

$$\underset{\text{'dilute'}}{M^+(am) + e^-(am)} \longrightarrow M(NH_3)_n \longrightarrow M_2(am) \longrightarrow \underset{\text{'concentrated'}}{M_x(am)}.$$

As would be expected these solutions have very strong reducing powers and have found extensive use in preparative chemistry, see Chapter 12. Similar, but much more dilute, blue solutions can be made by dissolving the alkali metals in certain ethers and amines; blue metal solutions have even been prepared in ice but of course such aqueous solutions are highly unstable.

3.3 Occurrence of the alkali metals

3.3.1 *Lithium*

65 p.p.m. total abundance; mainly in spodumene $LiAl(SiO_3)_2$ and lepidolite $Li_2(F, OH)_2Al_2(SiO_3)_3$; ^6Li 7·3 per cent, ^7Li 92·7 per cent.

3.3.2 *Sodium*

28 300 p.p.m. total abundance; largest deposits are as rock salt NaCl.

3.3.3 *Potassium*

25 900 p.p.m. total abundance; mainly in natural brines (as KCl) and as carnallite $KCl.MgCl_2.6H_2O$; of the three natural isotopes ^{39}K, ^{40}K, ^{41}K, one (^{40}K 0·01 per cent) is radioactive (β^-) with a half-life of $1{\cdot}5 \times 10^{13}$ years.

3.3.4 *Rubidium*

310 p.p.m. total abundance; no single source, occurs most often in the mica, lepidolite (0–3·5 per cent Rb_2O); only about fifty kilogrammes prepared annually; ^{87}Rb ($\sim$28 per cent of natural rubidium) is radioactive (β^-) with a half-life of $6{\cdot}3 \times 10^{10}$ years.

3.3.5 *Caesium*

7 p.p.m. total abundance; occurs in pollucite $CsAl_2(SiO_3)_2.xH_2O$; main source, however, is lepidolite which is used for lithium extraction.

3.3.6 *Francium*

The longest-lived of the known isotopes is francium-223, which has a half-life of only twenty-one minutes, hence it does not occur in nature in meaningful quantities, although it is formed from actinium

$$^{227}_{89}Ac \longrightarrow {}^{223}_{87}Fr + \alpha,$$
$$^{223}_{87}Fr \longrightarrow {}^{223}_{88}Ra + \beta^-.$$

3.4 **Extraction**

All the metals are highly reactive towards water and cannot, therefore, be prepared by electrolytic reduction of salts in aqueous media. Industrially the electrolysis of the fused chloride or, sometimes, the hydroxide can be used. However, this method is normally used only for lithium and sodium; in the Downs process, for example, the electrolysis of the fused chloride is carried out. The operating temperature is kept at about 500 °C by adding calcium chloride (up to 67 per cent) to the sodium chloride so that the electrolysis is actually carried out on the eutectic, m.p. 505 °C (the melting point of pure sodium chloride is 803 °C). The discharge potential for the sodium ion is lower than for calcium under these conditions and thus sodium is preferentially formed; the 1–2 per cent calcium liberated is insoluble in the (molten) sodium and precipitates back into the eutectic as it is formed. The low operating temperature does not cause difficulties due to the volatility of the sodium, b.p. 886·6 °C. The older Castner process, using molten sodium hydroxide, has a much lower current efficiency and is not now widely used.

Electrolysis of fused potassium chloride presents problems due to the high temperatures involved which cause vaporization of the isolated potassium; attempts to reduce the operating temperature by adding metal salts to the potassium chloride have not been very successful due mainly to potassium ions having a higher discharge potential than the added metal. The modern method is to employ a large fractionating tower filled with stainless steel Raschig rings; molten potassium chloride is fed into the centre of the tower and meets a counter-current of sodium vapour introduced from the bottom:

$$KCl + Na \longrightarrow K\uparrow + NaCl.$$

Further purification of the metal may be accomplished by fractional distillation; caesium is prepared in a similar manner. Sources of rubidium almost always contain caesium as impurity and, because metallic rubidium and caesium are very difficult to separate, the rubidium salt is normally purified by repeated fractional crystallization prior to its reduction by sodium or calcium. If a mixture of rubidium and caesium halides should be reduced (as for potassium chloride), then the resultant metal must be subjected to careful fractional distillation and even then the rubidium contains about 0·1 per cent of caesium.

3.5 Compounds of the alkali metals

3.5.1 *Hydrides*

All the metals react directly with hydrogen on heating to give colourless, saline (i.e. ionic) hydrides MH, which have the sodium chloride structure. Only in the case of lithium hydride is it possible to accurately determine the position of the weakly scattering hydride ions by X-ray diffraction. Lithium hydride, unlike the others, which decompose to metal and hydrogen, is stable in the molten state. The presence of hydride ions in the melt can be demonstrated by electrolysis when it is found that hydrogen is evolved at the *anode* in amounts required by Faraday's law. The hydrides react with water to give the hydroxide and hydrogen, and with the halogens or acids to give the corresponding salts.

$M + \frac{1}{2}H_2 \xrightarrow{T\,^\circ C} MH; \quad T(Li) = 700\text{–}800\ ^\circ C,\ T(Na\text{–}Cs) = 360\text{–}400\ ^\circ C$
LiH: m.p. 688 °C, stable at atmospheric pressure to 800–900 °C;
NaH: decomposes at 350 °C.

Sodium tetrahydroborate $NaBH_4$ and lithium tetrahydroaluminate $LiAlH_4$ are two complex, essentially ionic, hydrides which have assumed importance in recent years as exceedingly versatile reducing agents. They may be formally assumed to arise from the addition of borane (BH_3) or alane (AlH_3) to alkali-metal hydride, the hydride ion acting as an electron donor to the group III acceptor

$$:H^- + [BH_3] \longrightarrow [BH_4]^-$$

The tetrahedral ions BH_4^- and AlH_4^- are iso-electronic to methane and silane SiH_4 respectively. Lithium tetrahydroaluminate is the more widely used of the two, its ether-solubility being particularly useful to organic and inorganic chemists alike. Under the correct conditions a halide of most elements in the periodic table will react to form the element hydride (in the transition-metal series the metal usually has to be 'protected' by suitable organic ligands)

$$PCl_3 + LiAlH_4 \longrightarrow PH_3, \quad (100 \text{ per cent})$$
$$SiCl_4 + LiAlH_4 \longrightarrow SiH_4, \quad (100 \text{ per cent})$$
$$BrMn(CO)_5 + LiAlH_4 \longrightarrow HMn(CO)_5.$$

Typical of the reductions possible in organic chemistry are

$$RC(=O)H \longrightarrow RCH_2OH$$

$$RC(=O)OH \longrightarrow RCH_2OH$$

$$RR'C{=}O \longrightarrow RR'CH.OH.$$

3.5.2 *Oxides*

The alkali metals form three main types of oxide: the monoxide M_2O, the peroxide M_2O_2 and the superoxide MO_2. Lithium having a small, highly polarizing ion does not form a superoxide, the anion being destabilized by the high electric field gradient produced by the metal ion; similarly, although lithium peroxide can be prepared by special means, it is unstable to heat and readily evolves oxygen to form the monoxide

$$LiOH + H_2O_2 \xrightarrow[\text{solution}]{\text{alcohol}} Li_2O_2.H_2O_2 \xrightarrow[P_2O_5]{\text{desiccator}} Li_2O_2 \xrightarrow{300\,^{\circ}C} Li_2O + \tfrac{1}{2}O_2.$$

Monoxides. The metals will react with oxygen to form the monoxide but in all cases, except that of lithium, a deficiency of oxygen must be used otherwise higher oxides are also formed; the excess of metal remaining can be removed by heating the oxide in a vacuum when the metal sublimes away. The anti-fluorite structure is adopted by most of the monoxides (and sulphides, selenides and tellurides) of this group; the metals occupy what would normally be the fluoride sites and the oxygen (or chalcogen) occupies the calcium positions in the fluorite crystal (see

Figure 50, p. 105). Except for lithium, the monoxides are better made by reducing the nitrate or nitrite with the corresponding metal

$$6M+2MNO_2 \longrightarrow 4M_2O+N_2.$$

Peroxides. Lithium monoxide will not take up oxygen even under twelve atmospheres pressure at 480 °C, but the other monoxides readily absorb oxygen to give peroxides.

Superoxides. These are formed by heating the peroxides with oxygen (under pressure for sodium and potassium) or by treating solutions of the respective metals in liquid ammonia with an excess of oxygen at −78 °C; the crystal structures of the superoxides are similar to that of calcium carbide and contain O_2^- ions; see Figure 51, p. 106. The superoxide ion (see p. 214) has one unpaired electron and consequently the alkali-metal superoxides are found to be paramagnetic and have a magnetic moment close to the value of 1·73 Bohr magnetons calculated for a single electron.

Several of the alkali-metal oxides are quite deeply coloured due to lattice defects in the crystals (see Table 12).

Table 12

	Lithium	*Sodium*	*Potassium*	*Rubidium*	*Caesium*
M_2O	white	white	white, cold; yellow, hot	pale yellow; golden yellow at 200 °C	orange-red; black at 250 °C
M_2O_2	white	pale yellow	orange	dark brown	yellow; black when fused
MO_2	—	yellow	orange-yellow	dark orange	reddish-yellow

The three types of oxide, all of which react readily with water, can be distinguished by a study of their hydrolysis products,

$$M_2O+H_2O \longrightarrow 2MOH \quad \text{(no hydrogen peroxide; no oxygen),}$$
$$M_2O_2+2H_2O \longrightarrow 2MOH+H_2O_2 \quad \text{(hydrogen peroxide; no oxygen),}$$
$$MO_2+2H_2O \longrightarrow MOH+H_2O_2+\tfrac{1}{2}O_2 \quad \text{(hydrogen peroxide \textit{and} oxygen).}$$

The action of ozone on solutions of alkali-metal hydroxides in liquid ammonia is thought to give the paramagnetic ozonides MO_3, but nothing is known about their structure.

3.5.3 *Sulphides*

The metals react with sulphur to form compounds of formulae M_2S_x, where $x = 1, 2, 3, 4, 5$ or 6; for the higher sulphides the reaction is best carried out in liquid ammonia. The polysulphide ions are present in the crystal lattice as zig-zag chains; in Cs_2S_6 the S—S—S bond angle is 108·8° and there is an alternation in bond length in the chain with the terminal and central bonds being about 2·0 Å, the remainder 2·1 Å. Clearly the bonding in these sulphides is complex. Furthermore, there is a rather short van der Waals distance between the ends of the S_6 chains in Cs_2S_6 suggesting a weak interaction between the chains.

3.5.4 *Nitrides*

Lithium is unique among the alkali metals in reacting directly with nitrogen gas to give the ruby-red nitride

$$6Li + N_2 \xrightarrow{\text{below } 450\ °C} 2Li_3N \xrightarrow{H_2O} LiOH + NH_3.$$

This reaction presumably occurs because the product, made up of small ions (one of which is highly charged) has an extremely high lattice energy. Born–Haber calculations suggest that sodium nitride Na_3N should exist, but none can be detected when sodium is heated with nitrogen at temperatures up to 800 °C. An impure form of sodium nitride may be formed when active nitrogen is allowed to react with a sodium film.

The azides (salts of hydrazoic acid HN_3) are known for all the metals and can be made by passing dinitrogen monoxide through the molten amide

$$NaNH_2 + N_2O \longrightarrow NaN_3 + H_2O.$$

The N_3^- ion is linear, the two equal N—N bond lengths (1·15 Å) being indicative of multiple bonding; simple theory suggests a bond order of about two (see p. 190).

3.5.5 *Carbonates*

These salts are soluble in water but, as might be expected, lithium carbonate (small cation and doubly charged anion gives a high lattice energy) is the least soluble and can be precipitated from moderately concentrated solutions of lithium salts. Although the solubility of lithium carbonate may be increased by saturating the solution with carbon dioxide, no bicarbonate can be isolated; the other alkali metals form solid bicarbonates which decompose (Na > K > Rb > Cs) on heating,

e.g. $$2NaHCO_3 \longrightarrow Na_2CO_3 + H_2O + CO_2.$$

On stronger heating the carbonates will also decompose, giving the metal monoxide and carbon dioxide. At 1000 °C the dissociation pressure of carbon dioxide above the carbonates are: lithium carbonate, 90 mmHg; sodium carbonate,

19 mmHg; potassium carbonate, 8·3 mmHg. If the carbon dioxide is removed as it is formed by passing a stream of hydrogen over lithium carbonate, complete decomposition to lithium monoxide can be achieved at temperatures as low as 800 °C.

The decompositions of the bicarbonates and carbonates take place because the products are more stable, due mainly to the large increases in lattice energy when forming M_2CO_3 and M_2O respectively. The relative ease of decomposition parallels the decrease in ionic radius of the cations from Cs^+ to Li^+, which in turn increases the lattice energy of the solid reaction products. A kinetic effect probably also operates in that the high polarizing power of the smaller metal ions (especially of Li^+) helps to deform the large anions making them less stable.

3.5.6 *Nitrates*

Lithium nitrate and, perhaps unexpectedly, sodium nitrate are deliquescent. On heating, all the nitrates, except lithium nitrate, form the nitrite

$$2MNO_3 \longrightarrow O_2 + 2MNO_2,$$

and it is only on very strong heating that the nitrite is further decomposed to the oxide. Lithium nitrate decomposes directly to the monoxide without the intermediate formation of nitrite.

3.5.7 *Halides*

Only the halides of lithium form hydrates ($LiX.3H_2O$ for X = Cl, Br and I) and lithium fluoride, having the highest lattice energy of the series, is the most insoluble. In other solvents, where the solvent interacts more strongly with the metal ions, solvates of the other alkali-metal halides can be isolated. The best-known example is perhaps $NaI.4NH_3$ formed by dissolving sodium iodide in anhydrous liquid ammonia and evaporating the solution; its melting point is 3 °C so that this solvate is actually a liquid at room temperature and has an ammonia dissociation pressure of 420 mmHg at 25 °C. Like the alkali-metal hydrates, the coordinated ammonia is held by ion–dipole forces only, although this gives rise to quite a rigid system, the tetrahedral $Na(NH_3)_4^+$ ion having an infrared spectrum similar to that of tetramethyllead $Pb(CH_3)_4$. The fluorides crystallize from liquid hydrogen fluoride as solvates $MF.nHF$, in which the hydrogen fluoride is held to the anion by relatively strong hydrogen bonding (e.g. 3.7 and 3.8).

$[F—H—F]^-$

(3.7) $F[HF]^-$

$$F^- \text{ bonded to two H, each H bonded to F: } F—H \cdots F^- \cdots H—F$$

(3.8) $F[HF]_2^-$

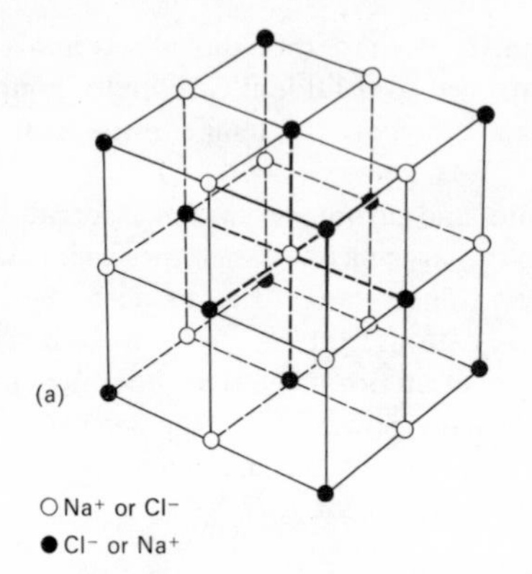

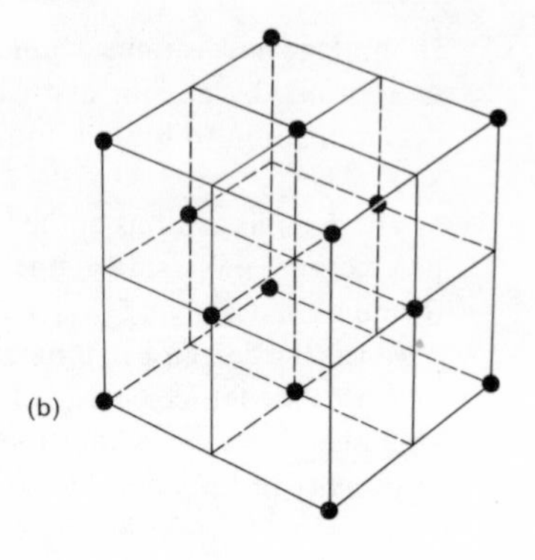

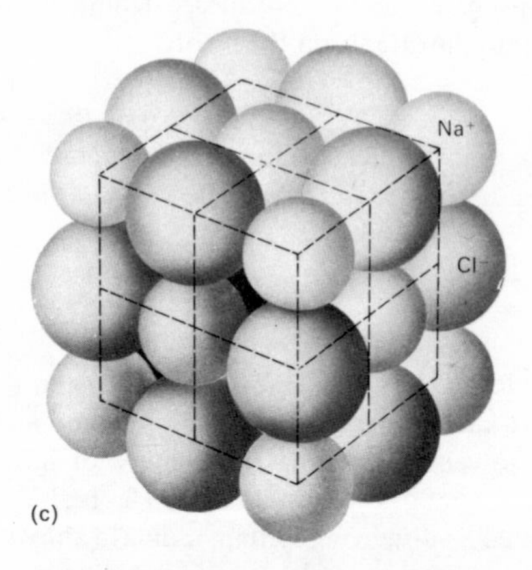

Figure 45 (a) The sodium chloride structure, showing the octahedral coordination of the ions. (b) Because identical ions (either sodium or chloride) are located at the corners and the face centres of a cube, sodium chloride is said to have a 'face-centred cubic' structure. (c) The packing of the sodium and chloride ions in crystalline sodium chloride

All the halides except caesium chloride, bromide and iodide normally adopt the face-centred cubic structure typical of sodium chloride (Figure 45), in which each metal and halide ion is surrounded octahedrally by six oppositely charged neighbours; that is, 6:6 coordination. This is somewhat unexpected considering the values of the ionic radius ratios (r_+/r_-) given in Table 13, since the sodium chloride structure is expected for ratios only between about 0·73 and 0·41; values outside these ratio limits represent a less efficient packing of the ions and a corresponding reduction in lattice energy. With the bigger metal ions the radius ratio is greater than 0·73 and the most stable arrangement of the ions becomes the 8:8 coordination of the body-centred cubic structure. This is the structure

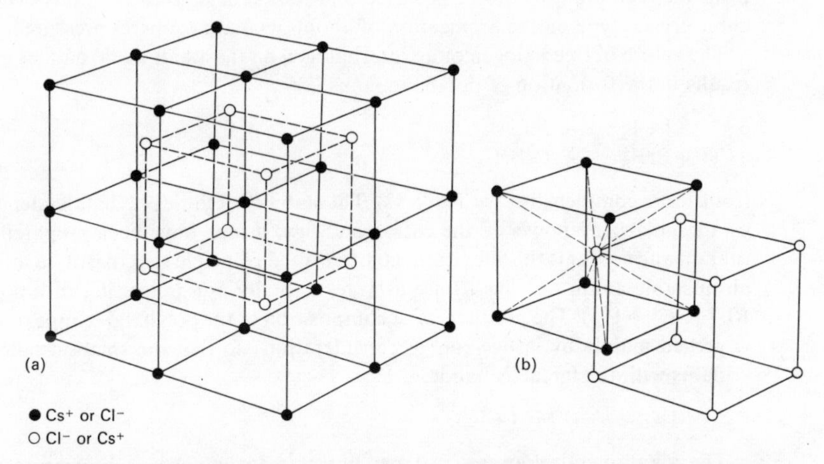

Figure 46 The caesium chloride structure. Identical ions are situated at the corners of a simple cube, at the centre of which is an ion of opposite charge (giving rise to the alternative, but slightly incorrect, name, 'body-centred cubic' structure). The structure as a whole may be regarded as arising from the interpenetration of two simple cubic lattices (b). Each ion, being at the centre of a cube, has a coordination number of eight

assumed by caesium chloride, bromide and iodide under normal conditions; see Figure 46. However, the lattice energies of the two types of cubic structure differ by only about 3 per cent for the rubidium and caesium halides and hence some of the halides are dimorphic, adopting one structure or the other depending

Table 13 Radius Ratios for the Alkali-Metal Halides

	F	Cl	Br	I
Li	0·44	0·33	0·31	0·28
Na	0·70	0·52	0·49	0·44
K	0·98	0·73	0·68	0·62
Rb	0·92	0·82	0·76	0·69
Cs	0·81	0·93	0·87	0·78

Ideally only those halides having radius ratios r_+/r_- between 0·73 and 0·41 should adopt the face-centred cubic (sodium chloride) structure.

on the conditions. At 445 °C caesium chloride changes to the sodium chloride structure and at low temperature rubidium chloride goes over to the body-centred cubic type. The change from face-centred to body-centred cubic results in a small reduction in volume and, as might be expected from le Chatelier's principle, rubidium chloride, bromide and iodide change to the body-centred cubic crystal type on the application of about 5000 atmospheres pressure.

The action of free halogen or an interhalogen on the alkali-metal halides often results in the formation of polyhalide ions,

e.g. $KI + I_2 \longrightarrow KI_3$,
$CsF + BrF_3 \longrightarrow CsBrF_4$

(for a more complete list, see Table 32). The stability of these polyhalides depends on the polarizing power of the cation and few, if any, have been prepared for lithium and sodium; this illustrates an often-used principle that to isolate salts of an unstable anion one has to use a large, virtually non-polarizing cation (e.g. Rb^+, Cs^+, NR_4^+). The products of decomposition of the polyhalides appear to be governed mainly by lattice-energy considerations so that the smallest halogen remains with the metal as halide,

e.g. $MICl_2 \longrightarrow MCl + ICl$.

The alkali-metal chlorates and perchlorates decompose on heating to give oxygen and the metal halide. The highly symmetrical perchlorate anion ClO_4^- is not strongly solvated in water, with the result that potassium, rubidium and caesium perchlorates are found to be insoluble.

3.5.8 *Compounds with carbon*

Only lithium reacts directly with carbon to give a carbide Li_2C_2. This compound reacts with water to give acetylene and probably contains $[C{\equiv}C]^{2-}$ ions since electrolysis in molten lithium hydride produces carbon at the anode.† The other alkali metals give similar carbides when heated in acetylene. The heavier metals (potassium, rubidium and caesium) form interstitial carbides when heated with graphite, the graphite lattice expanding to accommodate the metal atoms between the layers of carbon atoms (see p. 157). The compounds formed are highly coloured and non-stoichiometric, having compositions approximating to the following,

$$\underset{\text{dark grey}}{C_{60}K} \xrightarrow{K} C_{48}K \xrightarrow{K} \underset{\text{blue}}{C_{36}K} \xrightarrow{K} \underset{\text{bronze}}{C_{24}K + C_8K}.$$

The alkali metals are normally considered to undergo Würtz coupling reactions when treated with alkyl or aryl halides,

e.g. $2RX + 2Na \longrightarrow R{-}R + 2NaX$.

However, it is often possible to stop this reaction at an intermediate stage by using a chloride or bromide, when an organometallic derivative is formed,

$2Na + EtCl \longrightarrow NaEt + NaCl$.

†The C—C distance in the C_2^{2-} ion is 1·20 Å, identical to the C≡C bond in acetylene (1·204 Å).

For the alkali metals other than lithium, the alkyls and aryls are considered to be essentially ionic M^+R^- and in agreement with this they are exceedingly reactive, insoluble in most organic solvents and, when stable enough, have high melting points. In contrast to this, the lithium alkyls and aryls are typical covalent compounds, which can be sublimed or distilled in a vacuum and dissolve in many organic solvents. In the solid, gas and solution phases they have been shown to be polymeric; for example, ethyllithium can be tetrameric or hexameric depending on the conditions.

A more simple method of preparation for the alkyls of sodium, potassium, rubidium and caesium is to treat the metal with the corresponding mercury derivative, when a group-exchange reaction occurs,

$$2M + HgR_2 \longrightarrow Hg + 2MR.$$

A wide variety of lithium reagents can be made by starting with a simple alkyl derivative such as n-butyllithium (now commercially available) and carrying out lithium–halogen or lithium–hydrogen exchange reactions, the lithium being transferred to the alkyl or aryl group having the higher electronegativity,

e.g. $$LiBu + CH_2{=}CHBr \longrightarrow BuBr + CH_2{=}CHLi,$$
$$LiBu + C_6H_6 \longrightarrow BuH + C_6H_5Li.$$

The organolithium reagents are highly versatile reagents and are more reactive than Grignard reagents; a summary of some typical reactions of LiR compounds is given in Scheme 1.

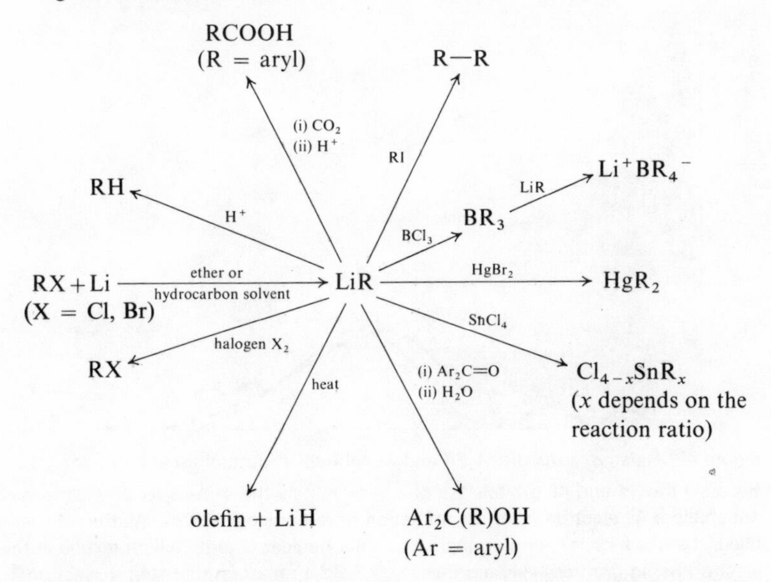

Scheme 1 Some typical reactions of organolithium compounds

3.6 A comparison of copper with the alkali metals

Table 14

		1st IE / kJ mol^{-1}	*2nd IE* / kJ mol^{-1}	*3rd IE* / kJ mol^{-1}	*Radius* M^+/Å	*Melting point*/°C
Na	$1s^2 2s^2 2p^6 3s^1$	496	4564	6911	0·95	98
K	$1s^2 2s^2 2p^6 3s^2 3p^6 4s^1$	418·8	2932	4602	1·33	63·5
Cu	$1s^2 2s^2 2p^6 3s^2 3p^6 3d^{10} 4s^1$	745·0	1957	3554	0·93	1083

Since both copper and the alkali metals have a single s-electron in their outer quantum shell, one might expect them to have rather similar properties. However, the presence of the filled 3d shell (which is similar in energy to the 4s shell at copper) modifies the chemistry of copper to a remarkable degree, and perhaps the only resemblance of copper to the alkali metals is that the copper(I) ion is diamagnetic and colourless. The 3d electrons participate in metallic bonding with the result that copper is much harder and has a higher melting point than any of the alkali metals; the stronger metallic bonding of copper also makes it much less reactive than the alkali metals.

The 4s electron in the copper atom is very poorly shielded from the nucleus by the filled 3d shell (see Figure 47) and is thus strongly attracted by the nuclear

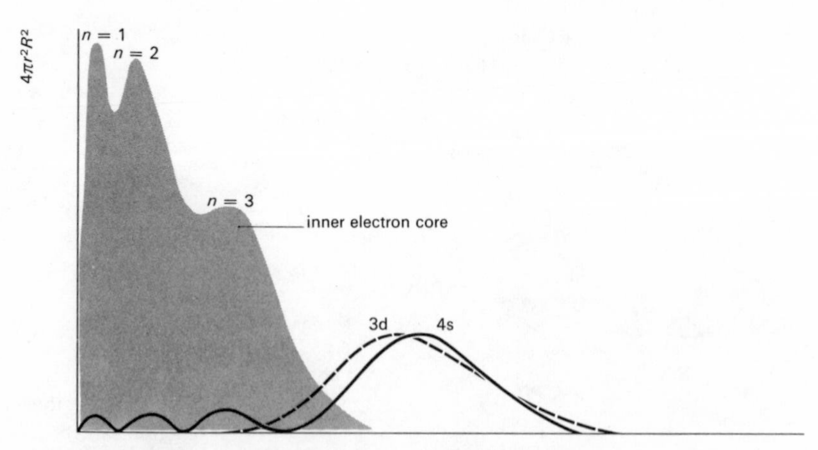

Figure 47 Relative sizes of the 3d and 4s orbitals at potassium and copper. Because the 3d and 4s orbitals are of approximately the same size, 3d electrons do not shield a 4s electron from the attraction of the nuclear charge. As the 3d shell is filled, there is a corresponding increase in the nuclear charge, which results in the 4s electron being progressively more strongly held, until at copper (ten d-electrons) the energy required to remove the 4s electron has increased to almost twice the ionization energy of potassium

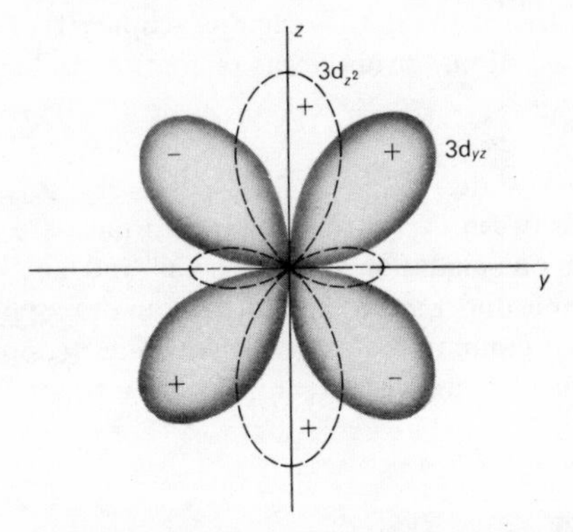

Figure 48 Poor mutual shielding of the 3d orbitals of the copper atom.
Due to their different spatial orientations, the 3d orbitals of the copper atom do not shield each other very effectively from the attraction of the nucleus. This is illustrated here for the $3d_{z^2}$ and $3d_{yz}$ orbitals. The lack of very extensive penetration of the d-orbitals into the inner core results in the core electrons being strongly attracted to the nucleus (the charge of which increases by +1 for each d-electron present). The combination of poor mutual shielding and lack of penetration of the 3d orbitals results in a contraction of all the electron shells at copper, thus making the copper(I) ion smaller than the potassium ion

charge. The effect of this is to make the first ionization energy of copper greater than that of an alkali metal; it is, in fact, about twice the first ionization energy of potassium. The lack of penetration and poor mutual shielding (see Figure 48) of the filled d-orbitals also cause an increased nuclear attraction in all the inner electron shells which results in the Cu^+ ion being much smaller than K^+ although ten elements occur between potassium and copper in the periodic table. (A contraction of atomic and ionic sizes due to penetration and shielding effects occurs across the first transition-metal series and is, of course, greatest at the last transition metal in the 3d group, copper.)

The second and third ionization processes at copper involve the loss of 3d electrons which, due to the lack of penetration close in to the nucleus, are more readily lost than the $(n-1)$p electrons of the alkali metals. Hence the second and third ionization energies of copper are considerably lower than those of the alkali metals. The combination of high first ionization energy and relatively low second ionization energy combine to make the copper(I) state highly unstable in aqueous media towards the disproportionation

$$2Cu^+(aq) \longrightarrow Cu + Cu^{2+}(aq); \qquad \frac{[Cu^{2+}(aq)]}{[Cu^+(aq)]^2} \simeq 10^6.$$

The high hydration energy (2121 kJ) of the doubly charged copper(II) ion is probably the driving force of this reaction, for in the gas phase the equilibrium

$$2Cu^{+}(g) \rightleftharpoons Cu^{2+}(g) + Cu(cryst)$$

is strongly in favour of the copper(I) state.

Therefore the main difference between copper and the alkali metals is that copper is able to exist in more than one oxidation state (Cu^{I}, Cu^{II} and Cu^{III}) and that the copper(II) state is more stable under many conditions than the copper(I) state. Furthermore, many copper(I) compounds, such as the halides (copper(I) fluoride is unknown), are covalent with the copper atom usually being tetrahedrally coordinated.

Chapter 4
Group IIa. The Alkaline-Earth Metals: Beryllium, Magnesium, Calcium, Strontium, Barium and Radium

4.1 Introduction

The M^{2+} ions of the alkaline-earth metals are much smaller than their M^+ counterparts in group I because of the increased nuclear charge which binds the remaining electrons much more tightly. This will be expected to lead to (a) more hydrate formation in group II due to stronger ion–dipole forces; (b) greatly increased lattice energies for corresponding salts resulting in lower solubilities, especially for salts containing multiply charged anions such as carbonate, sulphate and phosphate; (c) relatively lower thermal stability of salts containing polyatomic anions (e.g. carbonate, sulphate, nitrate) because of the increased polarizing power of the small, doubly charged cations.

The bare Be^{2+} ion appears to be too strongly polarizing to exist in salts, and all anhydrous beryllium compounds are found to be covalent. However, when hydration occurs to give the tetrahedral ion $[Be(H_2O)_4]^{2+}$, the effective size of the beryllium ion is increased and its polarizing power decreased. Under these conditions stable ionic salts can be formed; for example, $Be(H_2O)_4SO_4$, $Be(H_2O)_4Cl_2$ and $Be(H_2O)_4(NO_3)_2$.

From nuclear magnetic resonance studies using water enriched with the isotope oxygen-17, it has been demonstrated that four water molecules remain closely associated with the beryllium ion in solution. (An n.m.r. spectrum of the solution shows two ^{17}O resonance absorption peaks, one due to free solvent, the

Table 15 The Group II Elements

Element		*1st IE* / kJ mol^{-1}	*2nd IE* / kJ mol^{-1}	*3rd IE* / kJ mol^{-1}	*Radius of* M^{2+}/Å
Beryllium Be	$1s^2 2s^2$	899	1757	14 840	(0·31)
Magnesium Mg	[Ne]$3s^2$	737·5	1451	7731	0·65
Calcium Ca	[Ar]$4s^2$	589·4	1146	4942	0·99
Strontium Sr	[Kr]$5s^2$	549·4	1064	—	1·13
Barium Ba	[Xe]$6s^2$	502·9	965·2	—	1·35
Radium Ra	[Rn]$7s^2$	509·1	979·0	—	—

Table 16 Known Hydrates of the Group II Metal Sulphates, Nitrites, Nitrates and Chlorides

Salt	*Beryllium*	*Magnesium*	*Calcium*	*Strontium*	*Barium*	*Radium*
Sulphate	$BeSO_4$	$MgSO_4$	$CaSO_4$	$SrSO_4$	$BaSO_4$	$RaSO_4$
	$BeSO_4.2H_2O$					
	$BeSO_4.4H_2O$	$MgSO_4.H_2O$	$CaSO_4.0{\cdot}5H_2O$			
	$BeSO_4.5H_2O$	$MgSO_4.6H_2O$	$CaSO_4.2H_2O$			
		$MgSO_4.7H_2O$				
		$MgSO_4.12H_2O$				
Nitrite	(unknown)	$Mg(NO_2)_2.3H_2O$	$Ca(NO_2)_2$	$Sr(NO_2)_2$	$Ba(NO_2)_2$	
		$Mg(NO_2)_2.6H_2O$	$Ca(NO_2)_2.H_2O$	$Sr(NO_2)_2.H_2O$		
		$Mg(NO_2)_2.9H_2O$				
Nitrate	$Be(NO_3)_2$	$Mg(NO_3)_2$	$Ca(NO_3)_2$	$Sr(NO_3)_2$	$Ba(NO_3)_2$	$Ra(NO_3)_2$
	$Be(NO_3)_2.H_2O$	$Mg(NO_3)_2.2H_2O$	$Ca(NO_3)_2.2H_2O$	$Sr(NO_3)_2.4H_2O$		
	$Be(NO_3)_2.2H_2O$	$Mg(NO_3)_2.6H_2O$	$Ca(NO_3)_2.3H_2O$			
	$Be(NO_3)_2.3H_2O$	$Mg(NO_3)_2.9H_2O$	$Ca(NO_3)_2.4H_2O$			
	$Be(NO_3)_2.4H_2O$					
Chloride	$BeCl_2$	$MgCl_2$	$CaCl_2$	$SrCl_2$	$BaCl_2$	$RaCl_2$
	$BeCl_2.4H_2O$	$MgCl_2.4H_2O$	$CaCl_2.H_2O$	$SrCl_2.H_2O$	$BaCl_2.2H_2O$	$RaCl_2.2H_2O$
		$MgCl_2.6H_2O$	$CaCl_2.2H_2O$	$SrCl_2.2H_2O$		
		$MgCl_2.8H_2O$	$CaCl_2.4H_2O$	$SrCl_2.6H_2O$		
		$MgCl_2.12H_2O$	$CaCl_2.6H_2O$			

other due to $Be(H_2O)_4^{2+}$, the area under the peaks being directly proportional to the amount of water producing each resonance.) However, this inner hydration shell is surrounded by many more water molecules held by hydrogen bonds, the bonds being increased in strength compared to those in the body of the solvent due to the polarizing effect of the central Be^{2+} ion. Viscosity and ionic-mobility measurements on beryllium salt solutions show that Be^{2+} is the most heavily hydrated divalent cation known. Because of this strong ordering of the solvating water molecules, it is found that the value of the entropy of solution for the beryllium ion is both high and negative. The tendency for hydration falls off rapidly as the group II metal ion increases in size, but Mg^{2+} is found to have an inner shell of six coordinating water molecules in solution which it retains in some salts on crystallization; for example, the hydrate $MgCl_2 . 6H_2O$ consists of an array of discrete $Mg(H_2O)_6^{2+}$ ions and chloride ions. The relatively high affinity of the Mg^{2+} ion for water is presumably the main reason why magnesium sulphate is soluble in water and the sulphates of the larger calcium, strontium and barium are not.

In the hydrated beryllium ion the four water molecules are so strongly bound to the metal ion that the oxygen–hydrogen bonds are considerably weakened, thus making hydrolysis via transfer of a proton to a solvent water molecule a favourable process:

$$(H_2O)_3Be^{2+}\text{—}O\begin{matrix}\diagup H \\ \diagdown H\end{matrix} \xrightarrow{H_2O} (H_2O)_3\overset{+}{Be}OH + H_3O^+$$

$$\downarrow$$

$$Be(OH)_2 + H_3O^+$$

This reaction is accentuated in alkaline solution due to removal of the protons by hydroxyl ions; in pure water beryllium salts have a distinctly acid reaction. The other group II metal ions, which interact less strongly with the solvent, do not hydrolyse to any appreciable extent in solution.

Although the total energy required to form the M^{2+} ions is very high, being of the order of 2100 kJ mol^{-1} for magnesium, Born–Haber calculations show that both the lattice energy and hydration energy of the alkaline-earth salts are more than sufficient to overcome this large energy requirement. However, the high value of the second ionization energy does raise the question as to whether M^+ salts of these metals could possibly be made. A suitable cycle to investigate this might be that involving the reaction of magnesium with chlorine gas:

$$\begin{array}{ccc} Mg(s) + \tfrac{1}{2}Cl_2(g) & \xrightarrow[\frac{1}{2}\Delta H_{diss}]{\Delta H_{sub}} & Mg(g) + Cl(g) \\ \downarrow Q & & \downarrow IE_{Mg} \quad \downarrow -E_{Cl} \\ MgCl(s) & \xleftarrow{-L_{MgCl}} & Mg^+(g) + Cl^-(g) \end{array}$$

Assuming the lattice energy of MgCl (L_{MgCl}) to be equal to that of sodium chloride, the theoretical heat of reaction Q can be calculated:

$$Q = \Delta H_{sub} + \tfrac{1}{2}\Delta H_{diss} + IE_{Mg} - E_{Cl} - L_{NaCl}$$
$$= -125{\cdot}5 \text{ kJ mol}^{-1}.$$

Therefore the reaction

$$Mg + \tfrac{1}{2}Cl_2 \longrightarrow MgCl$$

is thermodynamically quite feasible. Why then have attempts to isolate MgCl, and other stable M^+ salts of the alkaline-earth metals, failed? The reason is that the above reaction is not the only method by which MgCl might be synthesized: it should also be possible to reduce $MgCl_2$ with magnesium

$$MgCl_2 + Mg \xrightarrow{Q'} 2MgCl.$$

When Born–Haber calculations are carried out on this reduction it is found that Q' is equal to about $+389$ kJ. In other words the reaction as written above is highly endothermic and should any MgCl happen to form it will be very unstable with respect to disproportionation

$$2MgCl \longrightarrow MgCl_2 + Mg; \quad -389{\cdot}3 \text{ kJ}.$$

Thus when we say a compound is 'stable' we must clarify the statement by asking 'Stable with respect to what?' MgCl is stable towards decomposition into its elements, but highly unstable towards disproportionation. There have been claims in the past to have made $M^{I}X$ derivatives of group II metals by heating the metal with its dihalide to 900 °C in an atmosphere of hydrogen. Later work shows that the compounds formed, which have strong reducing properties, are the hydride–halides $M^{II}HX$; they can be made more simply by heating the stoichiometric quantities of dihalide and dihydride together

$$MX_2 + MH_2 \longrightarrow 2MHX,$$
$$M = Ca, Sr, Ba;$$
$$X = Cl, Br, I.$$

The monohalides MX can be detected as short-lived species in the gas phase using spectroscopic techniques. A calculation based on Coulombic attraction between ion pairs shows that these species could be held together electrostatically. For example, the distance between Ca and Cl in CaCl(g) is 2·44 Å, which leads to an ionic heat of formation of $-255{\cdot}4$ kJ mol^{-1}, more than enough energy to hold the system together. It has proved possible to introduce, as impurities in crystals of the alkali-metal halides, small amounts of group II metals which enter the lattices as M^+. These monopositive ions are paramagnetic and have characteristic visible spectra by virtue of the lone electron in the outer quantum shell.

The third ionization energy for these elements is much too high to allow the formation of M^{3+} salts. The large increase from the second to the third ionization energy is due to the fact that the third electron has to be removed from an inner quantum shell; electrons in such a shell are nearer to the nucleus and thus very strongly attracted to it.

The vapour above the heated alkaline-earth metals consists entirely of monatomic species because the outer electron configuration of the metals, ns^2, does not allow the formation of dimers or polymers by any reasonable covalent bonding scheme. The situation is rather similar to that of helium, where the formation of a dimer He_2 would require the filling of an equal number of bonding and antibonding σ-orbitals.

It has been mentioned already that the simple compounds of beryllium are covalent. For example, anhydrous beryllium dichloride is polymeric in the solid state due to Be—Cl—Be bridging (4.1) and in the gas phase forms a mixture of monomer and dimer, both of which are covalently bonded (4.2, 4.3).

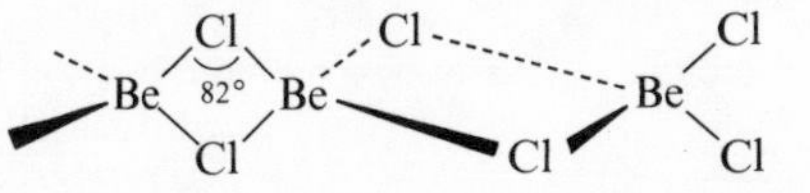

(4.1) (Be approximately sp^3 hybridized)

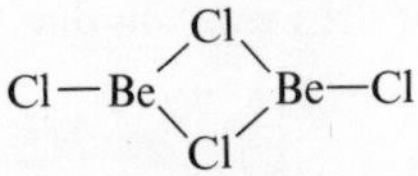

(4.2) planar (Be approximately sp^2 hybridized)

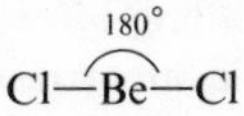

(4.3) (Be sp hybridized)

The other alkaline-earth halides sublime as monomers but, rather surprisingly, only those monomers having a relatively small metal and large halogen (e.g. $CaCl_2$, $CaBr_2$, $SrBr_2$, SrI_2) are linear; the others (e.g. CaF_2, $SrCl_2$, BaF_2, $BaBr_2$) are polar molecules and therefore non-linear. This suggests that these gaseous halides are covalently bonded because ion clusters would be expected to take up the symmetrical linear shape to minimize anion–anion repulsion. Coulson considers that in large metal–small halogen cases sd hybridization of the metal orbitals has occurred because s→d promotion is energetically easier to achieve in the heavier elements than the more normally expected s→p promotion. Probably the influence of the small halogen is to polarize the metal–halogen bond sufficiently to cause the d-orbitals to contract to a size which is suitable for effective hybridization with the outer s-orbital as in Figure 49. The linear molecules presumably represent cases of sp hybridization as found in $BeCl_2$. An ionic model of the gaseous oxides MO suggests them to be unstable, their heats of formation being large and positive. The *observed* heats of formation are high and negative, strongly suggesting the ionic model is not the correct one. In the cases of hydrates and complex ions such as MgF_3^-, MgF_4^{2-} and $BaCl_3^-$, it is perhaps safer to assume that the ligands are held only by ion–dipole forces and not by covalent bonds.

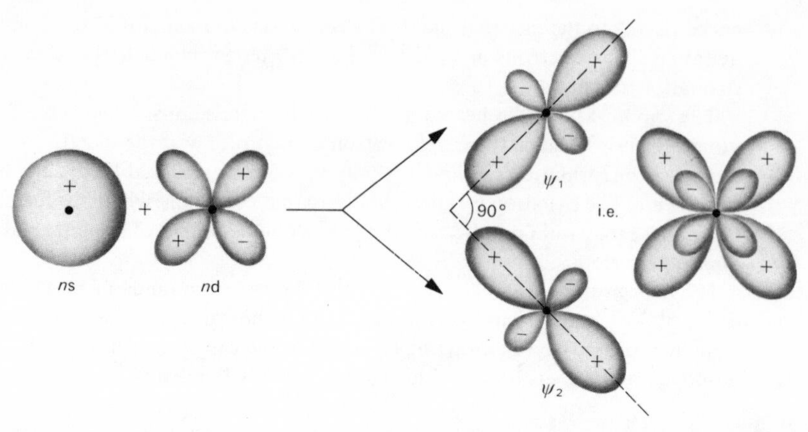

Figure 49 Hybridization of *n*s and *n*d orbitals to give two sd hybrid orbitals at 90° to each other

Complexes of the alkaline-earth elements are quite numerous, the tendency for complex formation falling off with increasing size of the metal ion due to the decrease in ion–dipole interactions

Be > Mg > Ca > Sr > Ba.

This is well illustrated in Table 16 where the known hydrates of several simple group II salts are listed, although the position of beryllium would appear to be anomalous due to the very high stability of the $Be(H_2O)_4^{2+}$ ion which forms at the expense of almost all other hydrates. The remaining metal ions have a tendency to be octahedrally coordinated to six water molecules as in $[Ca(H_2O)_6]Cl_2$ and hence, when a range of hydrates is known for a particular salt, the hexahydrate is often found to be the most stable. The oxygen-containing donor molecules which form complexes with the group II metals include ethers, alcohols, ketones, aldehydes, carboxylic acids, esters and phenols. Nitrogen, typified by ammonia, is also capable of complexing to give salts such as $[Be(NH_3)_4]Cl_2$ and $[Mg(NH_3)_6]Cl_2$. Those beryllium halide complexes having the formulae BeX_2D_2 (D = ethers, amines, nitriles and ketones) are anomalous in that they are tetrahedral and covalent, the metal atom being approximately sp^3 hybridized; for example, $BeCl_2[O(C_2H_5)_2]_2$ (4.4) has a dipole moment of 6·74 debyes over a range of concentrations in benzene showing that no dissociation occurs on dissolution.

$(C_2H_5)_2O$ ... Cl
Be
$(C_2H_5)_2O$ ... Cl

(4.4)

Remarkable among the many covalent compounds of beryllium are the basic carboxylates produced when $Be(OH)_2$ dissolves in a carboxylic acid. These compounds have the formula $Be_4O(OOCR)_6$, are volatile, are soluble in organic solvents, possess sharp melting points and can be sublimed or distilled without decomposition; thus the basic acetate ($R = CH_3$) melts at 285–6 °C, boils at 330 °C and sublimes at its melting point under reduced pressure. Their common structure has been determined by X-ray diffraction techniques and shown to contain a central oxygen bonded to four beryllium atoms arranged tetrahedrally around it; the six carboxylate groups act as bidentate ligands and bridge the metal atoms (4.5). This gives rise to very compact, symmetrical and non-polar molecules in agreement with their volatility and solubility in organic solvents.

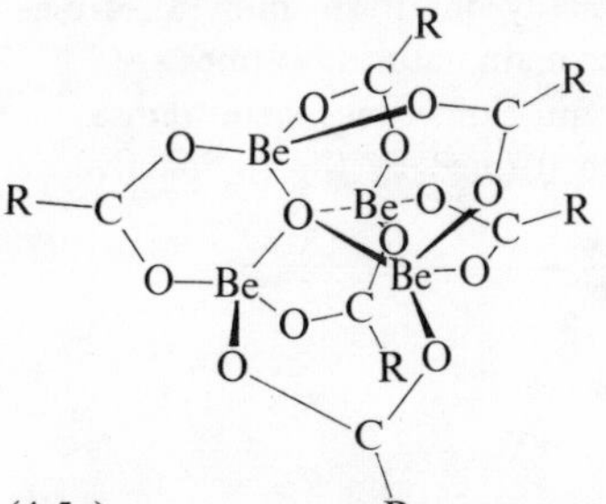

(4.5a) 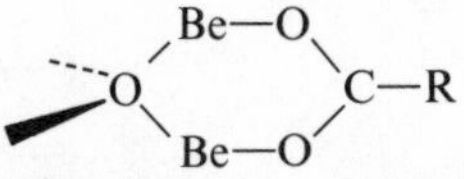 (4.5b)

Thus, as far as ionic properties are concerned, we should expect a gradual change in properties within the group of three metals calcium, strontium and barium, but a rather abrupt change between magnesium and calcium. Magnesium salts will be more soluble in the main due to the high solvation energy of Mg^{2+}; hydrate formation will be greater with magnesium salts than with those of calcium, strontium and barium; magnesium carbonate, nitrate, sulphate, hydroxide and peroxide will decompose much more readily than those of calcium whilst there will be a gradual increase in stability from calcium through strontium to barium for these salts. Beryllium never forms anhydrous ionic salts and hence there is a very abrupt change in properties (covalent to ionic) between beryllium and magnesium.

4.2 Occurrence of the alkaline-earth metals

4.2.1 *Beryllium*

2–6 p.p.m. of the Earth's crust; main minerals for the production of beryllium are beryl $Be_3Al_2Si_6O_{18}$ and phenacite Be_2SiO_4.

4.2.2 *Magnesium*

2·1 per cent total abundance, of which about 0·13 per cent occurs in sea-water (approximately 90 per cent of all magnesium produced in the USA is now obtained from the sea).

4.2.3 *Calcium*

3·63 per cent abundance; widely distributed in nature, often occurring as limestone mountain ranges ($CaCO_3$).

4.2.4 *Strontium*

42 p.p.m. of the Earth's crust; commercially important minerals are celestine $SrSO_4$ and strontianite $SrCrO_4$. The three main natural isotopes are ^{86}Sr 9·7 per cent, ^{87}Sr 7·0 per cent, ^{88}Sr 82·6 per cent. Sometimes strontium-87 is found virtually pure in a mineral, having been formed by the radioactive decay of rubidium-87

$$^{87}_{37}Rb \longrightarrow {}^{87}_{38}Sr + \beta^-.$$

4.2.5 *Barium*

39 p.p.m.; main source is barytes $BaSO_4$.

4.2.6 *Radium*

All isotopes of this element are radioactive, so it does not occur naturally in significant amounts; however, being a decay product of uranium, it occurs in pitchblende to the extent of about one gramme in seven tonnes of pitchblende.

4.3 **Extraction**

4.3.1 *Beryllium*

Beryl is heated with sodium fluorosilicate Na_2SiF_6 at 700–750 °C and the product extracted with water when the beryllium dissolves as either the fluoride or sodium fluoroberyllates. (The main impurity, aluminium, forms water-insoluble fluoride complexes and is removed at this stage by filtration.) Alkali is added to the solution to precipitate beryllium hydroxide, which is then dissolved in ammonium hydrogen fluoride solution

$$Be(OH)_2 + 2NH_4F.HF \longrightarrow (NH_4)_2BeF_4 + H_2O$$

$$\downarrow \text{evaporate and heat to 900 °C}$$

$$BeF_2$$

The metal is obtained by reducing beryllium fluoride with magnesium; the reaction proceeds rapidly at 900 °C, but it is necessary to increase the temperature to 1300 °C (i.e. above the melting points of magnesium fluoride and beryllium) to aid separation of the reaction products.

4.3.2 *Magnesium*

Sea-water is agitated with calcined (heated) dolomite (which also contains magnesium) and the precipitated magnesium hydroxide filtered off; every tonne of magnesia requires the processing of about three hundred tonnes of sea-water. Treatment of the hydroxide with hydrochloric acid produces magnesium chloride, which, after dehydration, is fused and electrolysed at temperatures above the melting point of magnesium (651 °C).

4.3.3 *Calcium*

The carbonate is calcined to give calcium oxide, which, after treatment with water and ammonium chloride, is converted to calcium chloride. The chloride is then fused and electrolysed.

4.3.4 *Strontium and barium*

Oxides are reduced with aluminium metal in a vacuum at such a temperature that the strontium and barium distil.

4.4 The elements

Two outer electrons are available for bonding in the solid group II metals, unlike the alkali metals which have only one; thus the alkaline-earth metals are much the harder, the hardness decreasing down the group (calcium, for example, is slightly tougher than lead). The melting points vary in an apparently irregular manner: beryllium, 1283 °C; magnesium, 651 °C; calcium, 851 °C; strontium, 800 °C; barium, 850 °C. However, this variation is mainly due to differences in the crystal structures adopted by the elements. Beryllium and magnesium have a close-packed hexagonal structure, calcium and strontium a face-centred cubic structure and barium a body-centred cubic structure; when the melting points are compared for elements having the same structure, the expected decrease in melting point with increasing size does occur.

The alkaline-earth metals dissolve in liquid ammonia to give blue solutions, but their solubilities are very much lower than those of the alkali metals. An interesting observation is that calcium, strontium and barium can be isolated from their ammonia solutions as solvates $M(NH_3)_x$, the value of x being somewhat variable but approximating to six. It is not clear what forces hold the ammonia molecules to the metal in these solvates, but their isolation does lend support to the idea that metals dissolve in liquid ammonia to give solvated species (see p. 78). If the solvates are heated to drive off the coordinated ammonia, some amide formation occurs

$$Ca(NH_3)_x \longrightarrow Ca(NH_2)_2 + H_2 + (x-2)NH_3.$$

4.5 Compounds of the alkaline-earth metals

4.5.1 *Hydrides*

Unlike the other alkaline-earth metals, beryllium does not form a hydride on being heated in hydrogen. However, BeH_2 can be prepared by the pyrolysis of di-*tert*-butylberyllium

$$(CH_3)_3CMgBr + BeCl_2 \xrightarrow{\text{ether}} [(CH_3)_3C]_2Be \xrightarrow{\text{heat}} 2(CH_3)_2C{=}CH_2 + BeH_2$$

It is a thermally stable, polymeric solid, in which the beryllium atoms are linked by hydrogen bridges (4.6). The other hydrides of this group are salt-like and contain the hydride ion H^-. All react with water liberating hydrogen.

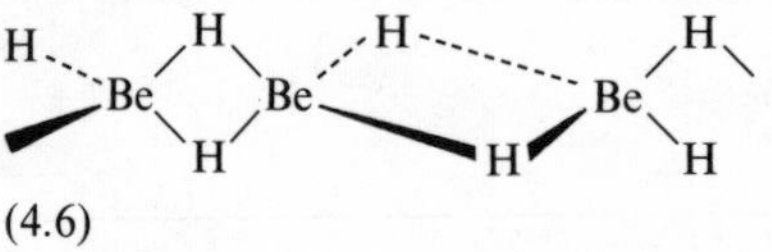

(4.6)

Beryllium hydride, like aluminium hydride, reacts with diborane to form a volatile, covalent tetrahydroborate $Be(BH_4)_2$ (4.7). The Be—H—Be bridges in

$$BeH_2 + B_2H_6 \longrightarrow$$

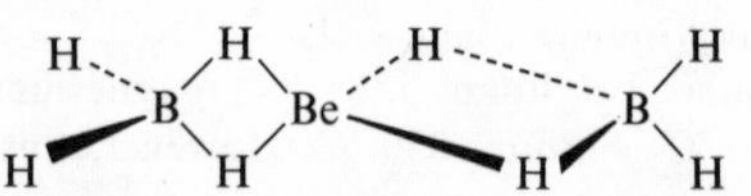

(4.7)

BeH_2 and the Be—H—B bridges in $Be(BH_4)_2$ are held together by two-electron, three-centre bonds very similar to those found in diborane (p. 55).

4.5.2 *Oxides*

The divalent alkaline-earth metal ions are sufficiently polarizing to make the peroxide, and especially the superoxide, ions unstable to heat, so that MO_2 and $M(O_2)_2$ do not form even when the metals are heated in pure oxygen; the monoxide MO is obtained under these conditions. Similarly this is the only oxide formed by the thermal decomposition of the carbonates, nitrates and hydroxides of these metals. However, strontium monoxide and barium monoxide absorb oxygen under pressure to give the peroxides. Peroxides of magnesium, calcium, strontium and barium can be also obtained when hydrogen peroxide is added to an aqueous solution of either their hydroxide or one of their salts,

$$Ca(OH)_2 + H_2O_2 \longrightarrow CaO_2.8H_2O \xrightarrow{130\,^\circ C} CaO_2 \xrightarrow{\text{heat}} CaO + \tfrac{1}{2}O_2.$$

Beryllium monoxide assumes the covalent wurtzite (4:4) structure unlike the other ionic monoxides which adopt the sodium chloride (6:6) lattice. The very high lattice energy of a sodium chloride structure containing *doubly* charged ions is reflected in the melting points of the monoxides: MgO, 2800 °C; CaO, 1728 °C; SrO, 1635 °C; BaO, 1475 °C; the fall in melting point along this series is due mainly to the increasing size of the cation from magnesium to beryllium which slightly reduces the lattice energy.

4.5.3 *Hydroxides*

Unlike the other alkaline-earth metals, beryllium has an amphoteric hydroxide, and as a consequence of this beryllium metal (cf. aluminium) dissolves in aqueous alkali evolving hydrogen and forming soluble beryllates. All the hydroxides can be obtained either by adding water to the monoxides or by adding alkali to an aqueous solution of a suitable salt. The hydroxides of magnesium, calcium, strontium and barium follow the familiar pattern of thermal decomposition, the heavier metals forming the more stable hydroxides (decrease in polarizing power of M^{2+} and lower lattice energy of the MO product). The temperatures at which the pressure of water above the hydroxides is 10 mmHg are found to be approximately: $Mg(OH)_2$, 300 °C; $Ca(OH)_2$, 390 °C; $Sr(OH)_2$, 466 °C; $Ba(OH)_2$, 700 °C;

$$M(OH)_2 \longrightarrow MO + H_2O.$$

4.5.4 *Nitrides*

All the metals react on heating in nitrogen to give the nitrides M_3N_2, which evolve ammonia on treatment with water. Only lithium of the alkali metals reacts in this way, presumably because the formation of N^{3-} ions from the very stable N_2 molecule requires a large amount of energy, which can only be counterbalanced by the formation of nitrides having very high lattice energies (i.e. those of small monovalent ions like Li^+ or doubly charged ions like those of the alkaline earths).

4.5.5 *Nitrates*

The hydrated nitrates can be obtained in the normal way by treating the oxides, hydroxides or carbonates with nitric acid and crystallizing the solutions.

Anhydrous beryllium nitrate results as a solvate, $Be(NO_3)_2.2N_2O_4$, when beryllium chloride is dissolved in dinitrogen tetroxide

$$BeCl_2 + N_2O_4 \xrightarrow{\text{solvolysis}} Be(NO_3)_2.2N_2O_4 \xrightarrow[\text{in vacuo}]{50\ ^\circ C} Be(NO_3)_2.$$

At about 125 °C in a vacuum $Be(NO_3)_2$ loses further quantities of nitrogen dioxide to give a basic nitrate, analysing as $Be_4O(NO_3)_6$, and which is thought to have a structure very similar to the basic beryllium carboxylates (p. 99); however, it is now the nitrate entity which acts as the bidentate ligand (4.8).

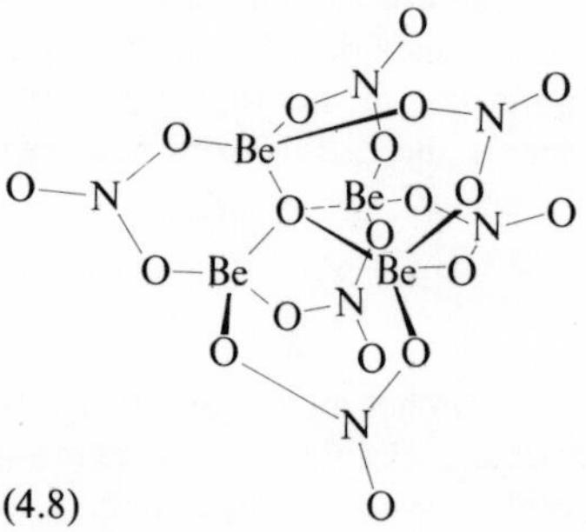

(4.8)

4.5.6 *Carbonates*

These increase in stability from beryllium to barium, beryllium carbonate being so unstable that it can only be precipitated from solution under an atmosphere of carbon dioxide. The approximate temperatures at which the pressure of carbon dioxide is one atmosphere are: $BeCO_3.4H_2O$, 25 °C; $MgCO_3$, 420 °C; $CaCO_3$, 810 °C; $BaCO_3$, 1280 °C. Metal carbonates are always ionic, so that the only way beryllium carbonate can exist at all is to incorporate the hydrated ion $Be(H_2O)_4^{2+}$ into the crystal structure.

4.5.7 *Sulphates*

On heating, the anhydrous sulphates lose sulphur trioxide; by observing the initial change in slope of the weight–temperature curve for samples held on a thermo-gravimetric balance, the relative decomposition temperatures of the various sulphates can be obtained: $BeSO_4$, 580 °C; $MgSO_4$, 895 °C; $CaSO_4$, 1149 °C; $SrSO_4$, 1374 °C. The solubilities in water of the group II sulphates are in the order Be > Mg ≫ Ca > Sr > Ba, presumably reflecting the high solvation energies of the small and strongly polarizing beryllium and magnesium ions.

4.5.8 *Halides*

As already mentioned on p. 97 the anhydrous beryllium halides are polymeric, covalently bonded substances which do not conduct electricity in the fused state

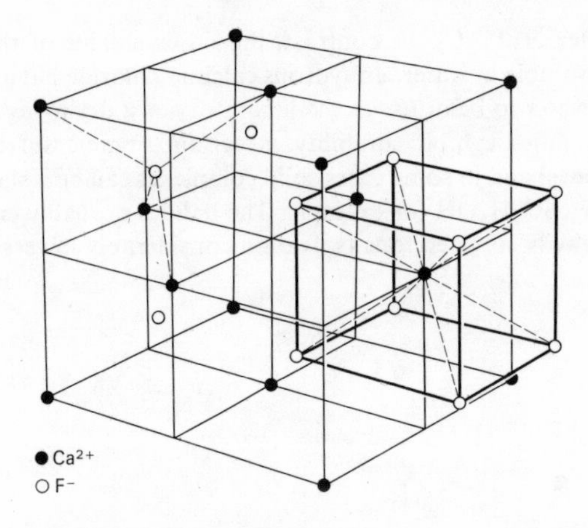

Figure 50 The fluorite structure.

The fluorite structure can be considered as derived from the caesium chloride lattice (Figure 46) by removal of four cations from opposite corners of the faces in each cation cube; this gives rise to the 1:2 cation–anion ratio of the crystal as a whole. The anions are coordinated tetrahedrally to four calcium ions and the cations are surrounded by eight fluoride ions at the corners of a cube. The crystal is therefore said to show 4:8 coordination.

Several compounds having the M_2X stoichiometry, for example, the alkali-metal monoxides M_2O, adopt the 'anti-fluorite' structure, in which the doubly charged anions replace the calcium cation and the singly charged cations replace the fluoride anion in the normal fluorite lattice.

In the titanium–hydrogen, zirconium–hydrogen and hafnium–hydrogen systems, hydrogen occupies some of the fluoride ion sites in the fluorite structure, whilst the other fluoride ion sites remain vacant. The fluorite structure is not stable up to the composition MH_2 (see p. 60)

and which sublime quite readily on heating. The addition of even small quantities of an alkali halide to the melts considerably increases their electrical conductivity because of the formation of complex ions such as $BeCl_4^{2-}$. With the exception of beryllium fluoride, the beryllium halides are soluble in polar organic media due to a strong tendency towards complex formation with the solvent. For example, beryllium chloride is monomeric in pyridine because of the presence of the tetrahedral complex $Cl_2Be(NC_5H_5)_2$. The anhydrous halides readily absorb water to give the ionic tetrahydrates, $Be(H_2O)_4X_2$ (except the iodide, which hydrolyses to hydrogen iodide) and, as expected of ionic compounds, these hydrated salts are insoluble in organic solvents.

Probably due to their high lattice energies, the fluorides of magnesium, calcium, strontium and barium are only slightly soluble in water; increasing the size of the cation decreases the lattice energy and produces a slight increase in solubility along the series, so that barium fluoride is the most soluble (0·209 g dissolves in 100 g of water at 25 °C). In contrast, the other halides of these elements are exceedingly soluble in water, anhydrous calcium chloride having a great enough affinity for water to be of use in the laboratory as a drying agent. Anhydrous magnesium halides exhibit solubility in certain organic solvents (alcohols, ethers and ketones) and in some cases solid complexes can be isolated from the solutions, $MgCl_2 . 6C_2H_5OH$ for example. The halides probably enter the organic solvents as heavily solvated ions (solvation considerably lowers the

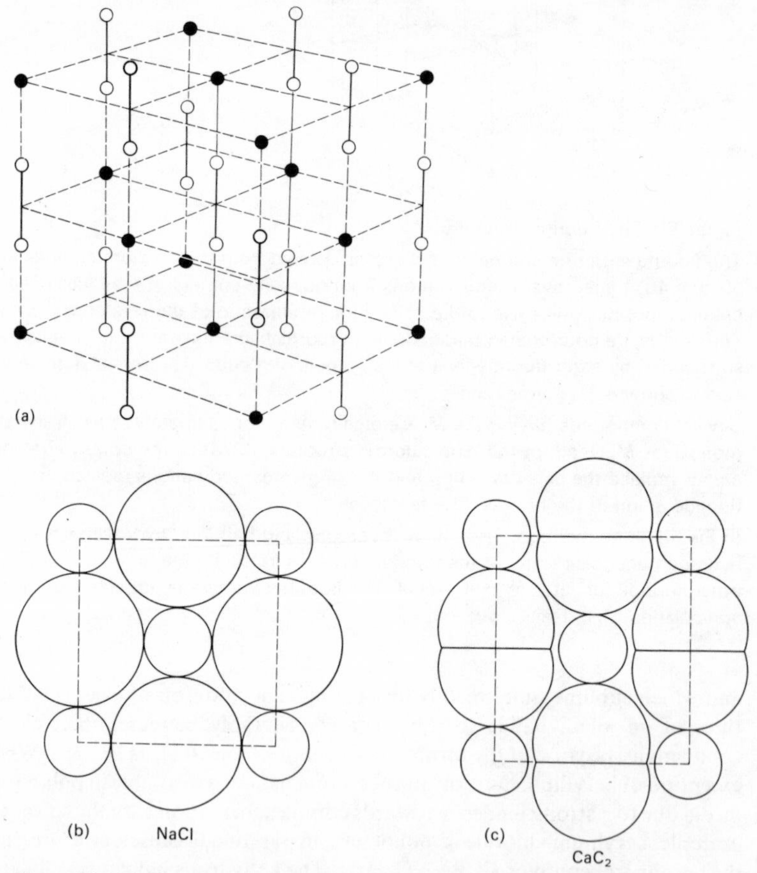

Figure 51 (a) The crystal structure of calcium carbide CaC_2. (b) Packing of the ions in sodium chloride compared with (c) packing of the ions in calcium carbide (reproduced from A. F. Wells, *Structural Inorganic Chemistry*, Clarendon Press, 1971, p. 761, with permission).

The structure of calcium carbide is very similar to that of sodium chloride except that one axis of the crystal has lengthened to accommodate the non-spherical, but parallel-aligned, carbide C_2^{2-} ions. Crystalline potassium superoxide KO_2 has an almost identical structure lattice energy of the halide making the process of solution energetically more feasible); this differs from the case of the beryllium halides which enter the organic solvents as molecular, as opposed to ionic, complexes.

4.5.9 *Carbides*

Several carbides are formed by the group II metals, viz. M_2C, MC_2 and M_2C_3.

Only beryllium is small enough to be accommodated within the limited space available between the carbons of an M_2C carbide having the antifluorite structure. Beryllium carbide Be_2C, prepared by direct reaction at about 1900 °C, varies in colour from amber to dark brown and slowly liberates methane on hydrolysis.

The MC_2 carbides are formed by all the group II elements though the preparative conditions differ slightly: beryllium and magnesium require heating in acetylene, whereas calcium, strontium and barium react directly with carbon at 2000 °C. Calcium carbide CaC_2, which can also be made by heating calcium carbonate with carbon, is a salt containing the $[C{\equiv}C]^{2-}$ ion; typically of a carbide containing the C_2^{2-} ion, it liberates acetylene on treatment with water and is therefore an important commercial source of this hydrocarbon. Calcium carbide absorbs nitrogen from the air at red heat giving calcium cyanamide $Ca(N{=}C{=}N)$, widely used as a nitrogenous fertilizer; the CN_2^{2-} ion is isoelectronic to CO_2.

Magnesium metal gives MgC_2 when heated in acetylene at around 400 °C, but at 600 °C, a new carbide, Mg_2C_3, is obtained; this liberates methylacetylene $CH_3C{\equiv}CH$ on hydrolysis and is considered to contain the C_3^{4-} ion.

4.6 **Organometallic compounds**

The most important of these derivatives are the Grignard reagents made by treating magnesium with either an alkyl or aryl halide (never the fluoride) in an anhydrous organic solvent, usually diethyl ether,

$$RX + Mg \xrightarrow{\text{ether}} RMgX.$$

They are normally given the simple formula RMgX, but this is a gross over-simplification. Solvation is important, as is shown by the crystallization from Grignard solutions of solid etherates, of which (4.9–11) are typical. However, at

Et_2O, Et_2O, Mg, Br, C_2H_5

(4.9) tetrahedrally coordinated magnesium

Et_2O, Et_2O, Mg, Br, C_6H_5

(4.10) tetrahedrally coordinated magnesium

CH_3, $H_8C_4O{-}Mg$, OC_4H_8, OC_4H_8, Br

(4.11) trigonal-bipyramidally coordinated magnesium (OC_4H_8 is tetrahydrofuran)

concentrations above about 0·3 mol l^{-1}, the diethyl ether solutions are found to contain polymeric species in which the magnesium atoms are probably linked by halogen bridges (4.12). The extent of polymerization depends not only on the

Et_2O X X X Et_2O Mg Mg Mg R O R O R Et_2 Et_2

(4.12)

concentration but also on the type of solvent used. In strongly coordinating solvents such as triethylamine and tetrahydrofuran, Grignard reagents are monomeric (i.e. RMgX (solvated)) over a wide range of concentrations, presumably because the Mg—N and Mg—O bonds in the solvated monomers are very much stronger than the magnesium–halogen bridges.

A further complicating factor is the possibility of a disproportionation

$$2RMgX \rightleftharpoons MgR_2 + MgX_2,$$

but in simple ethers there is little or no evidence for the formation of significant amounts of either MgR_2 or MgX_2. On the other hand a solvent which complexes much more strongly with one member of the above equilibrium than the other two can cause the equilibrium, which is normally well over to the left, to shift to

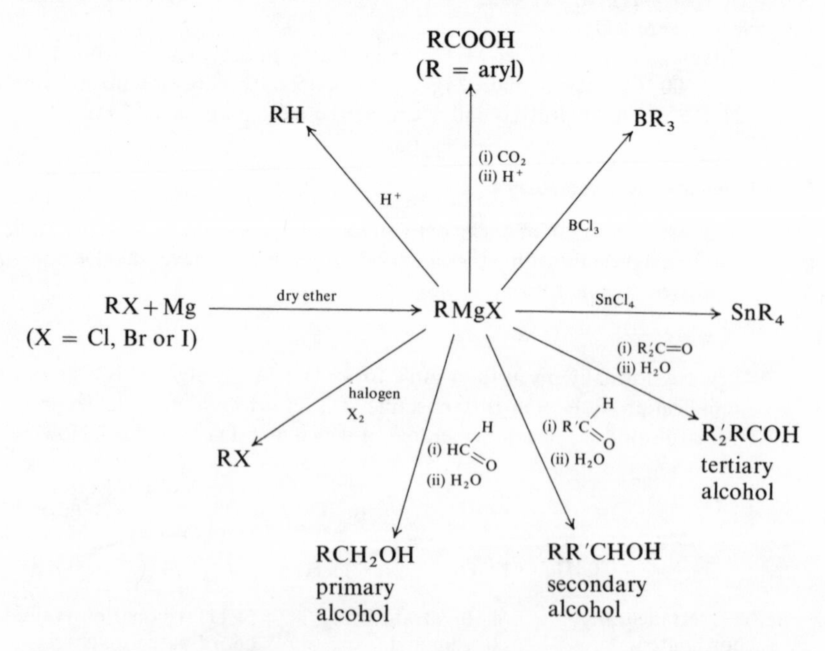

Scheme 2 Some general reactions of Grignard reagents

the right-hand side. Such a displacement occurs when dioxan is added to an ether solution of a Grignard reagent, a precipitate of $MgX_2.2C_4H_8O_2$ being formed

$$2RMgX + 2C_4H_8O_2 \longrightarrow R_2Mg + MgX_2.2C_4H_8O_2\downarrow.$$

Grignard reagents find wide application in organic chemistry, a few general reactions being illustrated in Scheme 2. In inorganic chemistry they are invaluable in the synthesis of a wide variety of organometallic compounds, normally accomplished by adding to the Grignard in ether solution an anhydrous covalent metal halide,

e.g. $BeCl_2 + 2RMgX \longrightarrow BeR_2 + 2MgXCl.$

Dimethylberyllium and dimethylmagnesium are polymeric white solids linked by M—C—M bridges (4.13). There are clearly not enough electrons available

(4.13)

to allow the formation of normal two-centre molecular orbitals between the bonded atoms. It is thought that the bridges involve three-centre molecular orbitals formed by combination of sp^3 hybrid atomic orbitals from the two metals and the carbon atom in the bridge. The three-centre system with two electrons in the lowest energy orbital constitutes a strongly bonding configuration as shown in Figure 52.

Other alkyls of beryllium normally appear to be dimeric, unless steric inter-

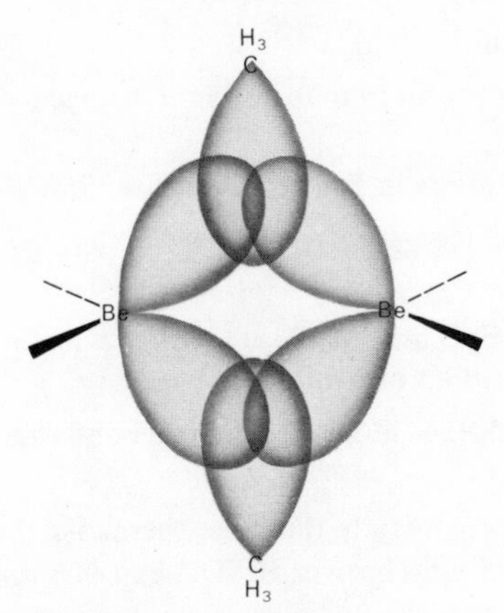

Figure 52 The interaction of atomic sp^3 orbitals in the methyl bridges of dimethylberyllium polymer

actions make the bridge system unstable and force the alkyl to be monomeric. An example of the latter effect is to be found in di-*tert*-butylberyllium (4.14),

```
H3C         CH3
   \       /
H3C—C—Be—C—CH3
   /       \
H3C         CH3
```

(4.14)

which is a monomer under normal conditions, being the only known compound of beryllium to adopt the linear (sp hybridized) configuration at room temperature.

The organo-derivatives of calcium, strontium and barium are very poorly characterized and difficult to make, but a few highly reactive alkyl metal halides can be obtained by treating the metals with an alkyl halide in a suitable solvent. Little is known about their structures or chemistry.

4.7 Similarities between beryllium and aluminium

The ratio Z/r, where Z is the charge on an ion of radius r, is similar for Be^{2+} (6·5) and Al^{3+} (6·0) suggesting that the two elements might have rather similar properties. Certainly it is found that beryllium resembles aluminium in many respects, more so than it does magnesium. The main points of similarity are summarized below.

(a) The metals are soluble in alkalis evolving hydrogen.

(b) Both have amphoteric hydroxides which dissolve in alkalis giving beryllates and aluminates.

(c) The metals are rendered passive by nitric acid.

(d) The anhydrous halides have a strong tendency to form metal–halogen–metal bridges.

(e) The anhydrous halides are useful catalysts in Friedel–Craft reactions.

(f) Both elements form a carbide having what approximate to C^{4-} ions present in the lattice.

(g) Although the heats of formation of BeO and Al_2O_3 are high, the metals are protected from aerial oxidation by a thin film of oxide.

(h) Their alkyls contain M—C—M bridges held by three-centre bonds and the molecules are therefore 'electron deficient'.

(i) Their anhydrous sulphates have approximately the same thermal stability; vapour pressure of SO_3 at 750 °C is 365 mmHg over $BeSO_4$, and 900 mmHg above $Al_2(SO_4)_3$.

Chapter 5
Group IIb. Zinc, Cadmium and Mercury

5.1 Introduction

Although zinc, cadmium and mercury have an ns^2 outer electron configuration, the presence of a poorly shielding $(n-1)d^{10}$ shell makes their first and second ionization energies considerably larger than those of the corresponding alkaline-earth metals calcium, strontium and barium. The filled 4f shell increases the binding energy of mercury's electrons still further and in fact the first ionization energy of mercury is greater by about 130 kJ than that of *any* other metal and is even higher than the *second* ionization energy of barium (see Figure 53). Mercury is unique among the metals in being a liquid at room temperature, a property which is presumably due to the tightly bound nature of the 6s electrons

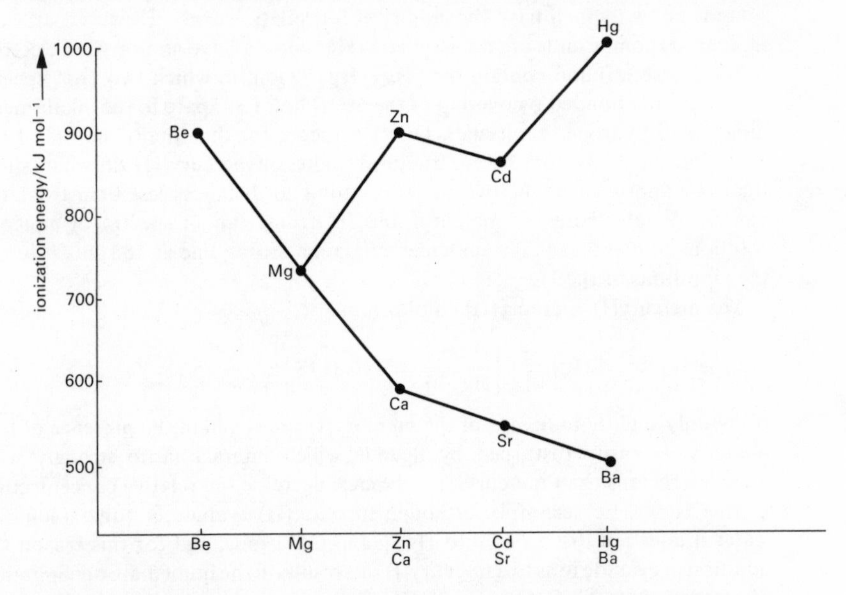

Figure 53 The effect of poor d-electron and f-electron shielding on the first ionization energies of zinc, cadmium and mercury. The second ionization energies follow a similar trend

making them relatively inaccessible for metallic bonding. All three elements have very high third ionization energies, which preclude the formation of the M^{3+} state. The combination of the high ionization energies and the relatively low $ns^2 \rightarrow ns^1np^1$ promotion energy of mercury results in the formation of many covalent compounds in which mercury is sp hybridized and coupled to only two ligands: X—Hg—X.

Table 17 Zinc, Cadmium and Mercury

Element		*1st IE* kJ mol^{-1}	*2nd IE* kJ mol^{-1}	*3rd IE* kJ mol^{-1}	*Radius of* M^{2+}/Å	$ns^2 \rightarrow ns^1np^1$ *Promotion energy*/kJ mol^{-1}
Zinc Zn	[Ar]$3d^{10}4s^2$	906·4	1733	3831	0·69	431·7
Cadmium Cd	[Kr]$4d^{10}5s^2$	867·7	1631	3616	0·92	407·6
Mercury Hg	[Xe]$4f^{14}5d^{10}6s^2$	1006	1810	3302	0·93	524·2
cf.						
Calcium Ca	[Ar]$4s^2$	589·4	1146	4942	0·99	
Strontium Sr	[Kr]$5s^2$	549·4	1064	—	1·13	
Barium Ba	[Xe]$6s^2$	502·9	965·2	—	1·35	

The high second ionization energies of zinc, cadmium and mercury suggest the possibility of M^+ derivatives, but only in the case of mercury are stable compounds known which have the empirical formulation HgX. However, all such mercury(I) compounds are diamagnetic (Hg^+ would have an unpaired electron in the 6s orbital) and contain the $[Hg—Hg]^{2+}$ ion, in which two Hg^+ species are covalently bonded by overlap of the 6s orbitals. (compare to the alkali-metal dimers and hydrogen molecule). Direct evidence for the dimeric nature of the Hg_2^{2+} ion comes (a) from X-ray structural studies on mercury(I) salts which show the close approach of the two mercury atoms to distances less than twice the van der Waals radius of mercury, and (b) from Raman spectra of aqueous solutions of mercurous nitrate which contain a strong line at 182 cm^{-1} due to the vibrations of the Hg—Hg bond.

The mercury(I)–mercury(II) equilibrium

$$Hg^{2+} + Hg \underset{\text{soln}}{\overset{\text{aq}}{\rightleftharpoons}} Hg_2^{2+}; \quad \frac{[Hg_2^{2+}]}{[Hg^{2+}]} \simeq 116 \text{ at } 18\ °C$$

being only slightly in favour of the mercury(I) state even in the presence of free mercury, is easily disturbed by ligands which interact more strongly with mercury(II) ions than mercury(I) and hence decrease the relative concentration of free Hg^{2+}. For example, although mercury(II) cyanide is quite soluble in water it does not dissociate into Hg^{2+} and CN^- ions, and for this reason the addition of cyanide ions to a mercury(I) salt results in the immediate precipitation of mercury and the formation of $Hg(CN)_2$. Similarly, ammonia and amines, which form very stable complexes with the mercury(II) ion, cause the disproportionation of mercury(I) salts.

The vapour above heated mercury(I) chloride, although having a density corresponding to HgCl, is diamagnetic. This is because mercury(I) chloride disproportionates completely in the gas phase

$$Hg_2Cl_2(g) \longrightarrow HgCl_2(g) + Hg(g).$$

There is no evidence for Zn_2^{2+} or Cd_2^{2+} ions in aqueous solution although the cadmium(I) ion has been detected in Cd–$CdCl_2$–$AlCl_3$ melts using Raman spectroscopy and isolated from such systems as the tetrachloroaluminate complex $Cd_2(AlCl_4)_2$. When this green cadmium(I) derivative is added to water, an immediate disproportionation occurs with the production of cadmium metal

$$Cd_2^{2+} \longrightarrow Cd\downarrow + Cd^{2+}.$$

Spectroscopic studies on the Cd_2^{2+} ion suggest that the Cd—Cd bond is considerably weaker than the Hg—Hg bond in the mercury(I) ion. This is, of course, against the normal trend within any group of the periodic table: in almost all other cases the bonds formed by the heaviest elements in a group are weaker than those involving the earlier members of the same group, and this is especially true when homonuclear (i.e. M—M) bonds are being considered. It has been argued that the strengths of the M—M bonds in these two M_2^{2+} ions will be proportional to the electron affinities of the M^+ ions (the electron affinity of M^+ being the negative of the first ionization energy of M). The first ionization energy of mercury (and thus the electron affinity of Hg^+) is 138 kJ larger than that of cadmium due to the poor shielding effects of the filled 4f shell in mercury; hence the Hg—Hg bond in the mercury(I) ion is expected, on these grounds, to be stronger than the metal–metal bond in Cd_2^{2+}.

Recent work has shown that the addition of mercury to a solution of mercury(I) chloride in an $AlCl_3$–NaCl melt apparently results in the formation of the Hg_3^{2+} ion. When the melts are cooled in attempts to isolate solid derivatives containing this ion, rapid disproportionation to mercury and the mercury(I) ion occurs. By using other solvents it has proved possible to prepare a few salts, one of which is the hexafluorophosphate,

$$PF_5 + Hg \xrightarrow{SO_2(l)} Hg_3(PF_6)_2.$$

The fluorides of divalent zinc, cadmium and mercury are ionic, adopting either the rutile structure (ZnF_2; the small Zn^{2+} ion being octahedrally surrounded by six fluoride ions, see Figure 54) or the calcium fluoride structure (CdF_2, HgF_2; the larger Cd^{2+} and Hg^{2+} ions having eight coordination in this structure). The other anhydrous halides of zinc and cadmium crystallize with a layer structure in which the cations are surrounded by six halide ions (see Figure 55); mercury(II) chloride and bromide are not ionic and their crystal lattices contain linear X—Hg—X molecules. Although molten zinc and cadmium halides conduct electricity, thus demonstrating the presence of ions, the vapours above such melts contain linear X—M—X molecules. Zinc forms several hydrated oxy-salts in which the Zn^{2+} ion is octahedrally coordinated to six water molecules, the crystals often being iso-structural with the corresponding magnesium

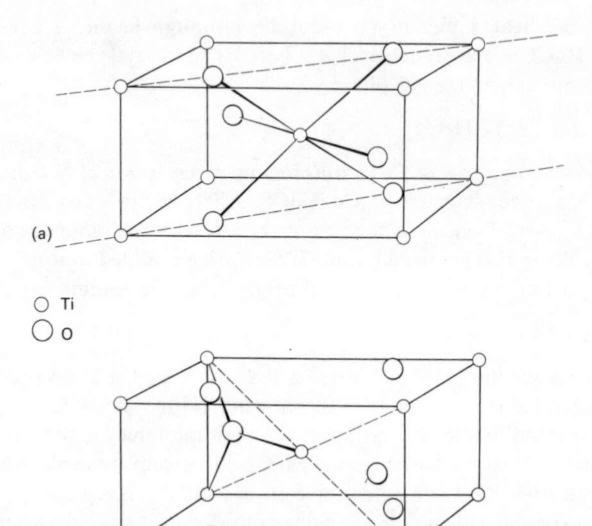

Figure 54 The rutile structure, showing (a) the octahedral coordination of the cations and (b) the planar triangular coordination of the anions.
The rutile structure is normally adopted when the radius ratio (cation–anion) has values between 0·732 and 0·414. Magnesium fluoride MgF_2 and zinc fluoride ZnF_2 crystallize with this structure

salts: for example, $MgSO_4.7H_2O$ and $ZnSO_4.7H_2O$; $Mg(BrO_3)_2.6H_2O$ and $Zn(BrO_3)_2.6H_2O$; $Mg(ClO_4)_2.6H_2O$ and $Zn(ClO_4)_2.6H_2O$.

Raman spectra of aqueous zinc chloride solutions show that the process of dissolution is complex and involves the formation of at least four species:

$Zn(H_2O)_6^{2+}$, $Zn(H_2O)_2Cl_4^{2-}$, $ZnCl_2(aq)$ and $ZnCl^+(aq)$

the anion $Zn(H_2O)_2Cl_4^{2-}$ predominating. Solutions of zinc and cadmium salts have a decidedly acid reaction due to hydrolysis which gives rise to ions such as MOH^+ and $M_2(OH)^{3+}$.

All three elements form a wide variety of complexes, the stabilities of which are normally in the order Hg > Cd > Zn. The $MX_2.2L$ complexes of zinc and cadmium are covalent and contain tetrahedrally coordinated metals (5.1),

X L
M
X L

(5.1) e.g. X = Cl, L = NH_3, pyridine and thioethers

whereas mercury tends to form linear $(L—Hg—L)^{2+}$ ions when treated with two molecules of a donor such as ammonia. The tetrammines of zinc and cadmium, on the other hand, contain the tetrahedral ion $M(NH_3)_4^{2+}$.

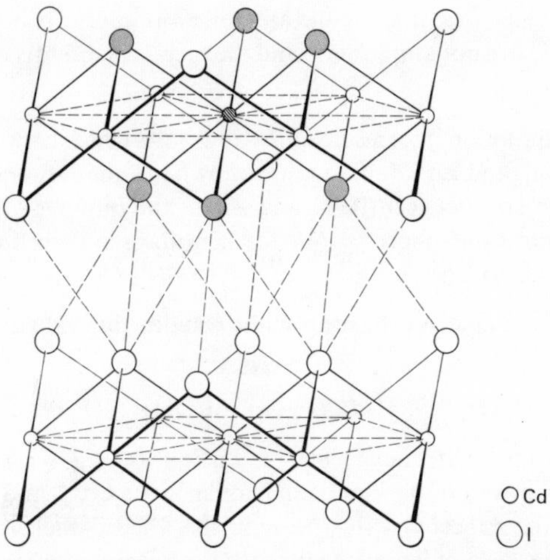

Figure 55 The cadmium iodide structure, showing parts of two layers.
The six iodide ions octahedrally surrounding the hatched cadmium ion are shown as shaded circles. Each iodide ion is coordinated on one side only by its three cadmium neighbours. Since adjacent ions in the neighbouring layers are identical (i.e. iodide), the forces between the layers are weak and are essentially of the van der Waals type. Not unexpectedly, it is found that the crystals show a preference to cleavage parallel to the layers.
The crystal can be regarded as having the iodide ions in a hexagonal close-packed arrangement with half of the octahedral holes occupied by cadmium ions. In cadmium chloride the anions are in a cubic close-packed array with half the octahedral holes again occupied by cadmium ions

5.2 A comparison of zinc with beryllium

The first and second ionization energies of beryllium and zinc are very similar and might be expected to lead to similarities in chemical properties, although it should be borne in mind that the zinc atom is bigger than that of beryllium, which may lead to differences in the coordination numbers of the two elements. As we have seen above, the high first and second ionization energies of zinc arise from the poor shielding properties of the filled 3d shell; those of beryllium are high because the electrons, coming from a small 2s orbital, experience a strong Coulombic attraction to the nucleus.

Some points of similarity between zinc and beryllium are:

(a) The free metals are iso-structural and adopt a slightly distorted hexagonal close-packed arrangement (cadmium and magnesium have the same structure).

(b) Both metals dissolve in strong bases, evolving hydrogen owing to the formation of soluble beryllates or zincates; their oxides are thus amphoteric. (Cadmium oxide and magnesium oxide are not amphoteric and the metals do not dissolve in strong bases.)

(c) Although beryllium (unlike zinc) forms no anhydrous salts which contain the bare M^{2+} ion, both $Be^{2+}(aq)$ and $Zn^{2+}(aq)$ ions are known in aqueous solutions. Because of the disparity in size between Be^{2+} and Zn^{2+}, the number of water molecules in the inner hydration sphere is four for beryllium and six for zinc (Cd^{2+} and Mg^{2+} are similar to Zn^{2+}).

(d) Solutions of beryllium and zinc salts have an acidic reaction due to hydrolyses of the type

$$M^{2+}(H_2O)_n \longrightarrow HOM^+(H_2O)_{n-1} + H^+(aq) \longrightarrow M(OH)_2 + H^+(aq).$$

This type of reaction is somewhat unexpected for an ion as large as zinc. It appears that the polarizing power of the Zn^{2+} ion (similarly the Cd^{2+} and Hg^{2+} ions) is increased above the expected value because the filled d-shell is quite readily deformed by the presence of ligands and hence the ligands 'see' an effective ionic charge which is somewhat greater than +2.

(e) The $MX_2.2D$ complexes of both beryllium and zinc are tetrahedral and covalent; those of stoichiometry $MX_2.4D$ are ionic and contain the tetrahedral cation MD_4^{2+}. By virtue of its greater size, zinc is also able to form complexes (e.g. hydrates and ammines) which contain the octahedral cation ZnD_6^{2+} (cadmium resembles zinc).

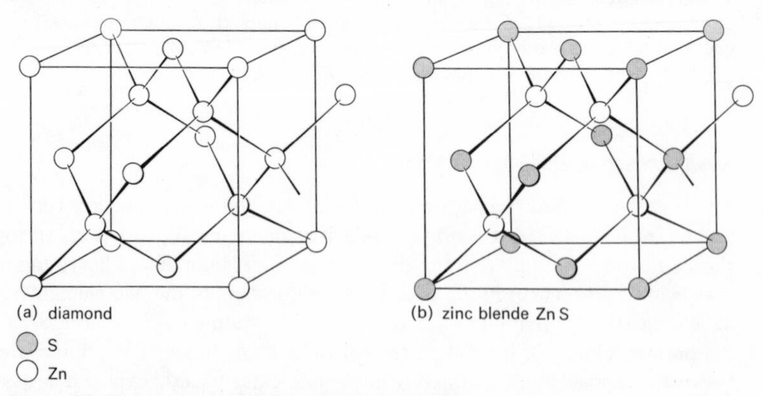

Figure 56 The structures of (a) diamond and (b) zinc blende.

Zinc blende can be visualized as being derived from the diamond structure by replacing alternate carbon atoms with zinc and sulphur atoms. Each zinc atom and each sulphur atom is then tetrahedrally coordinated to its four neighbours.

The zinc blende structure may also be considered to be a cubic close-packed arrangement of sulphur atoms with zinc atoms occupying half of the tetrahedral holes

(f) Both beryllium and zinc form a basic acetate having the formulation $M_4O(O_2CCH_3)_6$; for the structure of these unusual basic acetates see p. 99. (Cadmium and magnesium do not form this type of acetate.)

(g) Beryllium oxide and zinc oxide adopt the covalent wurtzite structure in which the metal atoms have four coordination (unlike magnesium oxide and cadmium oxide, which have the ionic sodium chloride structure and octahedral metal coordination). Four coordination of beryllium and zinc also occurs in the orthosilicates Be_2SiO_4 and Zn_2SiO_4, whereas magnesium is six coordinate in Mg_2SiO_4.

5.3 Occurrence of zinc, cadmium and mercury

5.3.1 *Zinc*

130 p.p.m. of the Earth's crust; major source is zinc blende ZnS, which is often associated with sulphides of other metals, such as lead, copper, cadmium and iron.

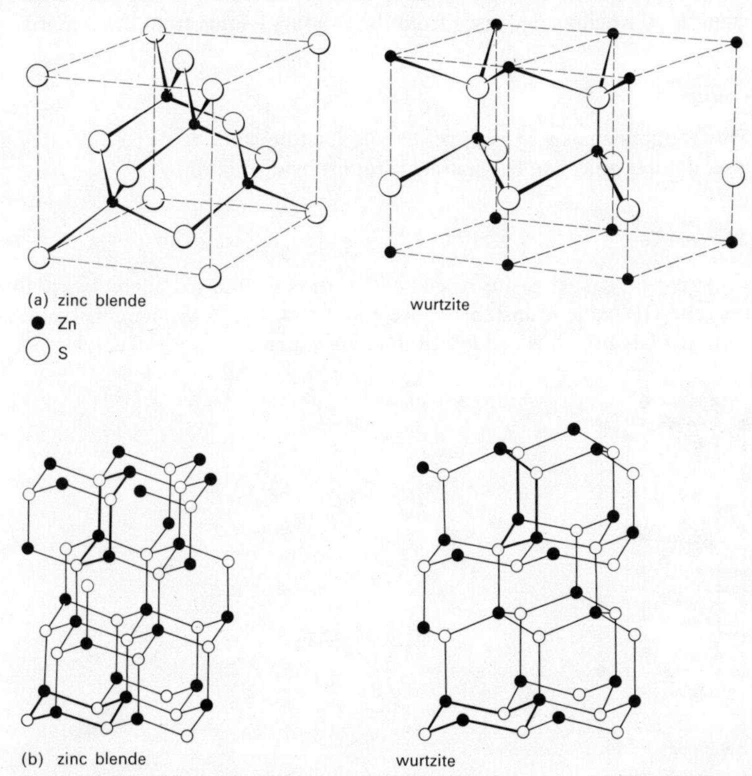

Figure 57 Two different representations of the zinc blende and wurtzite structures of zinc sulphide. In both structures the zinc and sulphur atoms are tetrahedrally coordinated

5.3.2 *Cadmium*

0·15 p.p.m.; cadmium minerals are rarely found alone, more often they are associated with zinc ores (e.g. zinc sulphide often contains 0·1–0·2 per cent of cadmium).

5.3.3 *Mercury*

0·5 p.p.m.; occurs native to a slight extent, but the main ore is cinnabar HgS.

5.4 Extraction

5.4.1 *Zinc*

Zinc blende is roasted to give zinc oxide and sulphur dioxide, the latter being oxidized to sulphur trioxide and converted into the valuable by-product, sulphuric acid. The zinc oxide is then reduced in a blast furnace using coke, the metallic zinc being condensed from the vapours issuing from the furnace.

5.4.2 *Cadmium*

This is obtained as a by-product of zinc smelting and, being considerably more volatile than zinc, can be separated from it by distillation.

5.4.3 *Mercury*

Cinnabar is roasted in air, when sulphur dioxide and free mercury are formed (mercury(II) oxide is unstable above about 500 °C); as the mercury distils away from the hot zone it is condensed in large, water-cooled metal condensers.

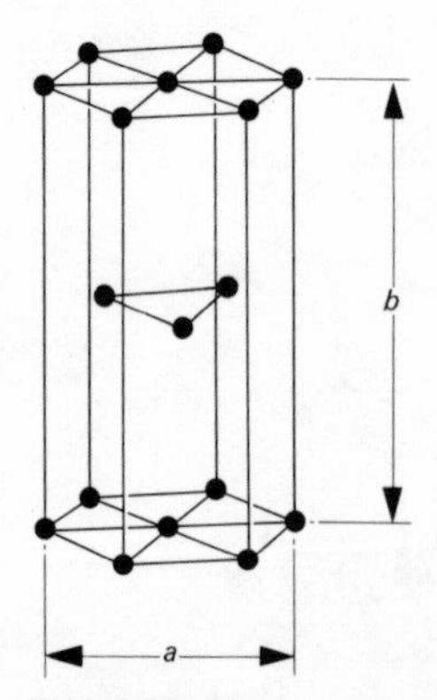

Figure 58 The hexagonal close-packed structure adopted by metallic zinc and cadmium. The structure is slightly distorted, in that the axial ratio b/a is about 1·87 instead of the ideal value for this arrangement of 1·63

5.5 The elements

Both zinc and cadmium have a distorted hexagonal close-packed structure, the distortion being an elongation of the distance between the close-packed layers (see Figure 58). Mercury is unique among the metals in being a liquid at room temperature; it freezes at −39 °C to a rhombohedral structure in which all the metal atoms have a coordination number of six. The tightly bound nature of the two *ns* electrons makes them relatively unavailable for bonding in the metals and as a consequence zinc, cadmium and mercury are among the most volatile of the heavy metals; the normal boiling points are: zinc, 906 °C; cadmium, 765 °C; mercury, 357 °C. The vapours are essentially monatomic, as one would expect from the electron configuration of the metal atoms.

Table 18 Standard Reduction Potentials at 25 °C

$Zn^{2+} + 2e \rightleftharpoons Zn$	−0·76 V
$Cd^{2+} + 2e \rightleftharpoons Cd$	−0·40 V
$2H^{+} + 2e \rightleftharpoons H_2$	±0·00V
$Hg_2^{2+} + 2e \rightleftharpoons 2Hg$	+0·79 V
$Hg^{2+} + 2e \rightleftharpoons Hg$	+0·85 V
$2Hg^{2+} + 2e \rightleftharpoons Hg_2^{2+}$	+0·91 V
cf.	
$Sr^{2+} + 2e \rightleftharpoons Sr$	−2·89 V
$Ca^{2+} + 2e \rightleftharpoons Ca$	−2·87 V
$Mg^{2+} + 2e \rightleftharpoons Mg$	−2·37 V

Notwithstanding their high ionization energies, the electrode potentials of zinc and cadmium have quite high negative values; these two metals, therefore, dissolve in non-oxidizing acids to give the corresponding salts and hydrogen. In apparent contradiction to this it is found that zinc can be electrically deposited from an aqueous solution of its salts; this is because the hydrogen overvoltage at a zinc surface amounts to one volt (at a current of about 0·1 A). Mercury will dissolve only in oxidizing acids.

5.6 Compounds of zinc, cadmium and mercury

5.6.1 *Hydrides*

The transient diatomic species MH and MH^+ have been detected spectroscopically for all three elements. The reduction of zinc and cadmium alkyls with lithium tetrahydroaluminate in ether at low temperatures gives the hydrides MH_2 as involatile, white solids,

e.g. $Zn(CH_3)_2 + 2LiAlH_4 \longrightarrow ZnH_2 + 2LiAlH_3(CH_3)$.

They are not very stable: zinc hydride decomposes slowly at room temperature, whereas cadmium hydride cannot even be warmed to room temperature without complete decomposition into cadmium and hydrogen.

5.6.2 *Oxides and hydroxides*

The oxides MO result either when the metals are heated in air or when the nitrates are decomposed thermally. Mercury(II) oxide begins to dissociate to mercury and oxygen at temperatures above 400 °C. Pure zinc oxide is colourless when cold but becomes yellow at about 250 °C due to the formation of lattice defects; it has the covalent wurtzite structure, unlike cadmium oxide, which adopts the ionic sodium chloride lattice.

Zinc hydroxide $Zn(OH)_2$ and cadmium hydroxide $Cd(OH)_2$ are precipitated from solutions of the respective salts by the addition of alkali; zinc hydroxide, unlike cadmium hydroxide, is soluble in an excess of base due to formation of complex hydroxy-ions such as $Zn(OH)_4^{2-}$. Both zinc hydroxide and cadmium hydroxide dissolve in aqueous ammonia to give ammines.

The addition of base to an aqueous solution of a mercury(II) compound results in the precipitation of yellow mercury(II) oxide; this yellow oxide differs from the red form (prepared as above) only in particle size and is not a different crystal modification.

5.6.3 *Nitrates*

Zinc, cadmium and mercury dissolve in nitric acid to give the hydrated nitrates; in the case of mercury an excess of acid is required to ensure the formation of mercury(II) nitrate, because if any metallic mercury were to be left in excess it is capable of reducing the Hg^{2+} ion to Hg_2^{2+} in aqueous solution. Mercury(II) nitrate is extensively hydrolysed and in dilute solution it breaks up completely into mercury(II) oxide and nitric acid.

5.6.4 *Carbonates*

Mercury forms only a basic carbonate. Zinc carbonate and cadmium carbonate resemble magnesium carbonate in being rather unstable to heat. (The dissociation pressure of carbon dioxide is one atmosphere at about 350 °C for both salts.) This is in line with the unexpectedly high polarizing power of the Zn^{2+} and Cd^{2+} ions; the pressure of carbon dioxide above calcium carbonate and barium carbonate only reaches one atmosphere at 900 °C and 1330 °C, respectively.

5.6.5 *Halides*

Zinc fluoride ZnF_2 (rutile structure) and cadmium fluoride CdF_2 (calcium fluoride structure) are not very soluble in water and can be precipitated from solutions of zinc and cadmium salts using a soluble fluoride (cf. the low solubility of magnesium fluoride). Mercury(II) fluoride (calcium fluoride structure) is completely hydrolysed in aqueous solution.

In sharp contrast, the other halides of zinc and cadmium are not only highly soluble in water but are also reasonably soluble in alcohols, ethers, amines, ketones, esters and nitriles (compare beryllium chloride, bromide and iodide). Aqueous solutions of zinc chloride have been shown by Raman spectroscopy to contain mainly $Zn(H_2O)_2Cl_4^{2-}$ ions together with $Zn(H_2O)_6^{2+}$, $ZnCl_2$(aq) and $ZnCl^+$(aq), but no evidence could be obtained for the presence of either $ZnCl_3^-$ or $ZnCl_4^{2-}$. Complexes between Cd^{2+} and halide ions are more stable than those of Zn^{2+} and as a result cadmium halides are found to dissolve in water giving Cd^{2+}(aq), CdX^+(aq), CdX_2, CdX_3^- and CdX_4^{2-}. The tetrahedral ZnX_4^{2-} and octahedral CdX_6^{4-} ions can be stabilized in the solid state (e.g. as in Cs_2ZnBr_4 and K_4CdCl_6).

Mercury(II) chloride, bromide and iodide are only slightly soluble in water (Cl > Br > I); little or no dissociation to Hg^{2+}(aq) occurs in such solutions. Halide complexes of the mercury(II) ion are relatively stable and in aqueous solution the common ones are HgX^+, HgX_2 and HgX_4^{2-}. The stability of the iodide complexes causes some interesting effects when a soluble iodide is added to a mercury(II) salt: although mercury(II) iodide is very insoluble (solubility $\sim 10^{-4}$ mol l^{-1}) it does not precipitate from solution until about 0·2 mol of iodide ions have been added to the system because of the formation of the stable HgI^+ ion. In the presence of a large excess of iodide ion the red mercury(II) iodide dissolves to give the colourless HgI_4^{2-} ion.

In the vapour state the anhydrous chlorides, bromides and iodides of all three metals exist as the linear, covalent molecules X—M—X, in which the metal atoms are presumably sp hybridized. An interesting fact is that the zinc halides are appreciably more volatile in the presence of metallic zinc; it has been suggested that this may be due to the formation of gaseous ZnX or $(ZnX)_2$ molecules, but attempts to condense out any lower halides by quenching the vapour leads only to a mixture of zinc and ZnX_2.

5.7 Complexes of zinc, cadmium and mercury

The presence of the filled $(n-1)$d, and to a lesser extent 4f, shells in the electronic configuration of these elements causes two effects which contribute to the higher stability of complexes formed by zinc, cadmium and mercury relative to their immediate counterparts, the alkaline-earth metals calcium, strontium and barium. The first is that the poor shielding characteristics of the d- and f-orbitals reduce the size of Zn^{2+}, Cd^{2+} and Hg^{2+} ions relative to Ca^{2+}, Sr^{2+} and Ba^{2+}, because the outer electrons in zinc, cadmium and mercury experience the higher effective nuclear charge (the radius of Hg^{2+} is smaller even than Ca^{2+}); the smaller the ion forming a complex, the higher will be the ion–dipole contribution to the bonding. The second effect is that filled $(n-1)$d orbitals are more easily polarized ('pushed aside') by ligand electrons than are filled $(n-1)$p orbitals; in this way a ligand on zinc, cadmium or mercury experiences a nuclear charge which is somewhat greater than the simple $+2$ charge on the ion.

Because the d-orbitals of these elements are completely filled, there is no ligand-field stabilization effect to contribute to the overall stability of their complexes.

The coordination numbers adopted by zinc, cadmium and mercury in their complexes can be two, four, five or six. Although the stability sequence is often Hg $\gg$ Cd $>$ Zn for tetrahedral complexes (e.g. the anionic halide derivatives MX_4^{2-}), mercury has a peculiar reluctance to form octahedral complexes.

5.7.1 *Two-coordinate complexes*

Cations of stoichiometry ML_2^{2+}, where L is a monodentate ligand, are rare; the ammine $Hg(NH_3)_2Cl_2$ contains $Hg(NH_3)_2^{2+}$ ions which have a linear N—Hg—N system (compare this to the diammines of zinc and cadmium halides, which are covalent and tetrahedral; see below). The halide complexes MX^+ mentioned on p. 121 are formed in solution and are undoubtedly solvated, so that the coordination number of M is certainly greater than one; in the specific case of the HgX^+ complexes, it is thought that the species in aqueous solution is the linear monohydrate $[X—Hg—OH_2]^+$.

A number of covalent compounds of these metals ($HgCl_2$, $HgBr_2$, HgI_2, the gaseous halides of zinc and cadmium, alkyls and aryls of zinc, cadmium and mercury) have linear X—M—X structure, the metal being sp hybridized. These are not considered here as 'complexes'.

5.7.2 *Four-coordinate complexes*

The metal atoms in these complexes have the four ligand groups arranged tetrahedrally around them. This gives rise to three main types of derivative (5.2–4).

L_2MX_2	$[ML_4]^{2+}$	$[MX_4]^{2-}$
(5.2) neutral	(5.3) cationic	(5.4) anionic
e.g. $Zn(py)_2Cl_2$	e.g. $[Zn(NH_3)_4][Cl]_2$	e.g. $[Cs]_2[ZnBr_4]$
$Cd(py)_2Cl_2$	$[Cd(NH_3)_4][Cl]_2$	$[Li]_2[Zn(CH_3)_4]$
(py = pyridine)	$[Hg(NH_3)_4][NO_3]_2$	$[K]_2[HgI_4]$

5.7.3 *Five-coordinate complexes*

There are very few complexes of this type; three zinc compounds shown by X-ray crystallography to have a distorted trigonal bipyramidal shape are $ZnCl_2$.terpyridyl, bis(acetylacetonato)zinc monohydrate and $NaZn(OH)_3$. The terpyridyl complex has the tridentate ligand occupying one equatorial and the two axial positions of the trigonal bipyramid (5.5).

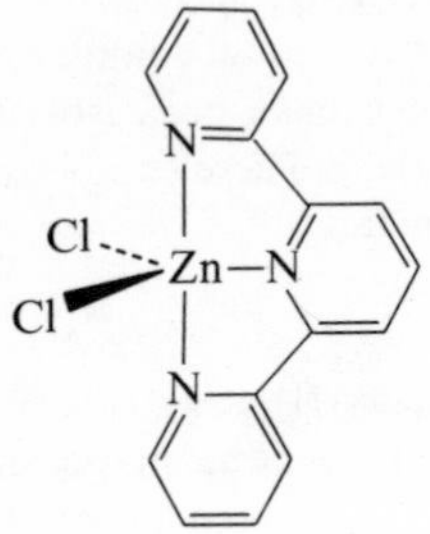

(5.5)

5.7.4 *Six-coordinate complexes*

Many zinc and cadmium salts of oxy-acids contain the octahedral $Zn(H_2O)_6^{2+}$ and $Cd(H_2O)_6^{2+}$ ions: e.g. some common hydrates containing $Zn(H_2O)_6^{2+}$ ions are $ZnSO_4.7H_2O$, $Zn(BrO_3)_2.6H_2O$ and $Zn(ClO_4)_2.6H_2O$. Hexammines such as $M(NH_3)_6X_2$ and $M(en)_3X_2$, where en is ethylenediamine and X is Cl, Br or I, are also known for zinc and cadmium, but they are not particularly stable; the ammonia complexes exhibit a considerable dissociation pressure of ammonia at room temperature and must be stored in sealed tubes to prevent loss of the ligand.

Mercury forms very few octahedral complexes. Two recently reported examples are [Hg(pyridine-*N*-oxide)$_6$][ClO$_4$]$_2$ and [Hg(dimethylsulphoxide)$_6$][ClO$_4$]$_2$, which are also unusual in being oxygen complexes, because mercury is usually considered to have a low affinity for oxygen or oxygen-containing donor molecules.

5.8 **Organometallic compounds**

The monomeric dialkyls and diaryls of zinc, cadmium and mercury are readily obtained by the action of Grignard reagents or lithium reagents on ethereal solutions of the anhydrous metal halides,

e.g. $$CdCl_2 + 2RMgX \longrightarrow CdR_2 + 2MgXCl,$$
$$HgCl_2 + 2LiR \longrightarrow HgR_2 + 2LiCl.$$

Treatment of MR_2 with one mole of a dihalide MX_2 in an organic solvent results in the formation of the organometal halide RMX. Many alkyl zinc halides, like the Grignard reagents which they closely resemble, can be made directly from the corresponding alkyl halides and zinc dust.

Organozinc and organocadmium compounds are oxygen and water sensitive, the lower dialkyls of zinc being pyrophoric in air. The high stability of organomercurials towards air and water appears to be due to the fact that mercury has a very low affinity for oxygen and not to any enhanced strength of mercury–carbon σ-bonds. Although the zinc alkyls do not conduct electricity, they dissolve alkyls of the alkali metals to give conducting solutions which contain complex anions such as ZnR_3^- and ZnR_4^{2-}; typical derivatives which have been isolated are $Na[Zn(C_2H_5)_3]$, $Li_2[Zn(CH_3)_4]$ and $K_2[Zn(C{\equiv}CH)_4]$. These complexes also result when an alkylzinc reacts with the free alkali metal

$$2Rb+3Zn(C_2H_5)_2 \longrightarrow 2Rb[Zn(C_2H_5)_3]+Zn.$$

When aromatic compounds are treated with mercury(II) acetate, it is found that mercury will substitute one of the hydrogen atoms in the aromatic ring; the reactive species is thought to be $HgOAc^+$

$$ArH+HgOAc^+ \longrightarrow [ArHHgOAc]^+ \longrightarrow ArHgOAc+H^+.$$

Many RHgX derivatives have been widely used as pesticides and such a simple route to a potentially wide variety of products is of obvious industrial importance. However, the high toxicity of these mercurials is causing some concern at the present time: the high stability of the mercury–carbon bond towards air and water means that the rate of degradation of pesticides is so slow that 'pesticide residues' are beginning to find their way (still as RHgX) into the human food cycle.

5.9 Mercury(I) derivatives

The chemistry of the mercury(I) state is not very extensive, mainly because the equilibrium

$$2Hg_2^{2+} \rightleftharpoons Hg+Hg^{2+}$$

is forced to the right under the conditions of many chemical reactions. Furthermore, the standard potentials at 25 °C for the two reactions

$$Hg_2^{2+}+2e \rightleftharpoons 2Hg \qquad +0{\cdot}79\ V,$$
$$Hg^{2+}+2e \rightleftharpoons Hg \qquad +0{\cdot}85\ V$$

are so close that when liquid mercury is treated with an excess of oxidizing agent the mercury(II) state always results because there are no known oxidizing agents having a potential lying between 0·79 and 0·85 V. In such reactions it is only by having metallic mercury in excess that the mercury(I) state can be obtained, by virtue of the reaction

$$Hg+Hg^{2+} \longrightarrow Hg_2^{2+}.$$

A typical example of this behaviour is the reaction of mercury with nitric acid.

Mercury(I) oxide and hydroxide are unknown; when alkali is added to a mercury(I) salt the black solid obtained has been shown to be an intimate mixture of mercury(II) oxide and mercury. All the halides exist, but the fluoride is completely hydrolysed by water to mercury and mercury(II) oxide. When mercury(I) chloride and bromide sublime, their vapours have densities corresponding to 'HgX', due to complete disproportionation

$$Hg_2Cl_2 \longrightarrow HgCl_2 + Hg.$$

The iodide is so unstable that it forms mercury and mercury(II) iodide simply on warming in water.

Mercury(I) nitrate is one of the most soluble salts, making it a convenient starting point for the preparation, by precipitation, of insoluble mercury(I) derivatives (e.g. Cl, Br, I, BrO_3, IO_3). It crystallizes from solution as the dihydrate $Hg_2(NO_3)_2.2H_2O$, which contains the linear ion $[H_2OHgHgOH_2]^{2+}$.

The number of known mercury(I) complexes is small because of the disproportionation of the mercury(I) ion in the presence of common ligands including ammonia, many amines, CN^-, OH^-, SCN^-, thioethers and acetylacetone. The magnitude of those stability constants which have been measured for mercury(I) complexes suggests that the mercury(I) ion does not form weak complexes; the scarcity of mercury(I) complexes must be due to disproportionation because of the very strong complexing ability of mercury(II). It appears probable that the usual coordination numbers of Hg_2^{2+} are two and four. The two-coordinate complexes contain linear $[L—Hg—Hg—L]^{2+}$ ions as in the dihydrate mentioned above. Four-coordinate complex ions $Hg_2.4L^{2+}$ are formed when triphenylphosphine oxide or pyridine-*N*-oxide are added to mercury(I) perchlorate solutions, but their stereochemistry is not known.

Chapter 6
Group III. Boron, Aluminium, Gallium, Indium and Thallium

6.1 Introduction

Among the group I and II elements, the ionization energies were found to decrease with increasing atomic number; this is not the case in group III, where the ionization energies vary down the group in an apparently erratic way. On closer inspection, however, it is seen that the inner electronic configurations of the group III elements are not identical: instead of a rare-gas electron structure between them and the nucleus, the ns^2np^1 electrons of gallium, indium and thallium are outside a filled set of *n*d orbitals and have what might be called a 'd^{10} core'. The inner electron structure of thallium also contains fourteen electrons in the 4f orbitals filled during the building up of the preceding lanthanide series of elements.

Table 19 The Group III Elements

Element		*1st IE* / kJ mol^{-1}	*2nd IE* / kJ mol^{-1}	*3rd IE* / kJ mol^{-1}	*4th IE* / kJ mol^{-1}	*Radius of* M^+/Å	*Radius of* M^{3+}/Å
Boron B	$1s^22s^22p^1$	800·3	2427	3658	25 030		(0·20)
Aluminium Al	[Ne]$3s^23p^1$	564·2	1816	2744	11 580		0·50
Gallium Ga	[Ar]$3d^{10}4s^24p^1$	564·2	1979	2962	6193	1·13	0·62
Indium In	[Kr]$4d^{10}5s^25p^1$	558·3	1820	2705	5230	1·32	0·81
Thallium Tl	[Xe]$4f^{14}5d^{10}6s^26p^1$	589·0	1970	2975	4896	1·45	0·95

In groups I and II, the atomic size increases down the groups so that the outer orbitals are progressively further from the nucleus and are also relatively well shielded from the nuclear charge by the inner s- and p-electron shells: thus there is a lowering of the energy required to remove the outer *n*s electrons as the atomic number increases. On the other hand, filled d-orbitals are much less efficient at shielding the outer electrons from the Coulombic attraction of the nucleus, and hence electrons outside a d^{10} core are relatively strongly held, resulting in ionization energies for gallium, indium and thallium which are rather higher than expected by direct extrapolation from their group II counterparts. This shielding effect also makes the elements with d^{10} cores smaller than their immediate group I and group II neighbours by causing the outer electron shells to be pulled in closer towards the nucleus. The increase in ionization energies from indium to thallium occurs because the 6s and 6p electrons of thallium, besides being outside a $5d^{10}$ core, also have the fourteen poorly shielding 4f electrons between them and the nucleus. (The shielding effects of the various orbitals within a given inner quantum shell are in the order $s > p > d > f$).

The M^{3+} state for gallium, indium and thallium will be energetically less favourable than Al^{3+}, because the high ionization energies of these three elements cannot always be balanced by the lattice energies of possible reaction products (the lattice energies decrease down the aluminium, gallium, indium and thallium group owing to the increasing size of the M^{3+} cations) and, for the group as a whole, the M^{3+} state is the exception and not the rule. More commonly these elements achieve an oxidation state of three by forming covalent compounds via $ns \rightarrow np$ promotion.

Increasingly down the group there is a tendency toward the production of M^+ (with thallium this is the more stable oxidation state) probably because the energy of the M^{III}—X covalent bonds, which become weaker down the group, is less able to compensate for the energies of $s \rightarrow p$ promotion, hybridization and atomization entailed in covalent bond formation; that is, the free energy of formation of M^+X^- salts becomes higher than for $M^{III}X_3$ or similar covalent derivatives as the atomic number of M increases.

The ionization energies suggest that the formation of salts containing M^{2+} ions is feasible and, indeed, compounds analysing as GaX_2 can be isolated by heating gallium metal with a gallium trihalide. The compounds are, however, diamagnetic (Ga^{2+} would have one unpaired electron and be paramagnetic) and are actually mixed-valency salts $Ga^+[Ga^{III}X_4]^-$. The nearest approach to M^{II} derivatives occurs in the compounds GaS, GaSe and GaTe made by direct reaction at high temperature. Structural analysis using X-ray diffraction techniques shows that these chalcogenides contain $[Ga—Ga]^{4+}$ units arranged in a layer lattice; the coupling of the gallium atoms in this manner explains why the compounds are diamagnetic.

The vast amount of energy required to remove three electrons from boron and the high polarizing power of the small B^{3+} ion combine to make the formation of salts containing bare B^{3+} cations impossible; even water of hydration would be too highly deformed by this cation and hence $B^{3+}(aq)$ is unknown in aqueous solution. Much less energy is required for $s \rightarrow p$ promotion and boron compounds are consequently covalent, the boron atomic orbitals being hybridized to either the sp^2 or sp^3 configuration. In a very few cases it has proved possible to increase the effective size of the B^{3+} ion, and at the same time to recoup some of the expended ionization energy in the form of strong ion–dipole forces, by complexing round it a number of suitable donor molecules; this has the effect of reducing the cation's polarizing power and so makes its salts more stable. Such a salt is formed when pyridine reacts with the addition compound, trimethylamine–boron tribromide

$$(CH_3)_3N{-}BBr_3 + 4C_5H_5N \longrightarrow \left[\begin{matrix} H_5C_5N & & NC_5H_5 \\ & B & \\ H_5C_5N & & NC_5H_5 \end{matrix}\right]^{3+} 3Br^-$$

(tetrahedral ion)

Solutions of the reaction product conduct electricity and have a conductivity expected of a salt giving rise to four ions in the solution.

Although simple M^{3+} cations are relatively uncommon in anhydrous compounds of group III elements, the hydrated ions of aluminium, gallium, indium and thallium are quite well known in aqueous solution. Nuclear magnetic resonance studies show that there are six water molecules held strongly by these ions and often their salts can be crystallized from solution as hexahydrates. The high charge on the central ion induces hydrolysis via ionization of protons on the coordinated water resulting in the formation of basic salts

$$M(H_2O)_n^{3+} \rightleftharpoons HOM(H_2O)_{n-1}^{2+} + H^+(aq) \rightleftharpoons (HO)_2M(H_2O)_{n-2}^{+} + H^+(aq)$$
$$\rightleftharpoons M(OH)_3 + H^+(aq).$$

The addition of acid to the solution will depress such hydrolytic processes, but at high acid concentrations complex anions are sometimes formed, especially with the aqueous hydrogen halides,

e.g. $Ga^{3+}(aq) + HX(conc) \longrightarrow GaX_4^-$ $(X = Cl, Br, I)$.

In the gas phase at high temperatures all the group III elements form diatomic halides MX either by dissociation of the trihalide or, more often, by reduction of the trihalide with the free element

$$B + BF_3 \xrightarrow{2000\ ^\circ C} 3BF,$$

$$Al + AlCl_3 \xrightarrow{800\text{–}1000\ ^\circ C} 3AlCl.$$

Although many of these species (especially for boron, aluminium and gallium) are unstable in the solid state, they are stable in the gas phase at high temperatures because of the increase in the entropy term $T\,\Delta S$ for the reaction. All the gaseous monohalides are covalently bonded with the exception of thallium(I) fluoride, which, from its very large dipole moment of 7·6 debyes (compare Table 8), must be considered as an ion pair Tl^+F^-.

The electrical conductivity of aluminium trichloride in the solid state increases rapidly as the melting point is approached, at which temperature the conductivity suddenly falls to zero. This phenomenon occurs because solid aluminium trichloride forms an ionic lattice but, on melting, the forces in the liquid are insufficient to compensate for the ionization energy required to sustain Al^{3+}, and the compound changes to the covalent state, the liquid then consisting entirely of dimeric Al_2Cl_6 molecules. In crystalline aluminium trichloride the aluminium is six coordinated but in the dimer the coordination number has fallen to four, resulting in the density of the liquid being much lower (by about 45 per cent) than that of the solid at the melting point. The rather delicate balance between ionic and covalent bonding even for aluminium(III) (the most favourable case for ionic bonding in the group) can be appreciated when it is realized that, whilst crystalline aluminium trifluoride is ionic like the chloride, aluminium tribromide forms a molecular crystal made up of Al_2Br_6 dimers. The loss of lattice energy suffered by substituting either the larger bromine for chlorine or gallium for aluminium in aluminium trichloride is sufficient to make the Al^{3+} and Ga^{3+} ions thermodynamically unstable. The fluorides of gallium, indium and thallium are all ionic by virtue of both the low dissociation energy of F_2 and the small size of the F^- ion.

The anhydrous trihalides are monomeric in the vapour phase at sufficiently high temperatures and have the trigonal, planar structure (6.1). In such a

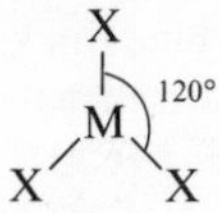

X
120°
M
X X

(6.1)

molecule, M has only a sextet of electrons in its outer quantum shell and readily accepts electron density from many molecules having an available lone pair of electrons (i.e. from Lewis-type bases such as ethers, amines and phosphines),

e.g. $GaBr_3 + NMe_3 \longrightarrow$ Br, Br, Ga, Br, $(CH_3)_3N$

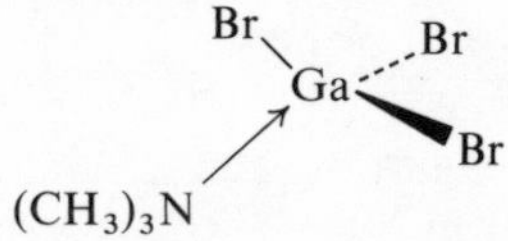

(central atom is sp^3 hybridized)

The product of this reaction, trimethylamine–gallium tribromide, is normally called an 'addition compound' or, more simply, an 'adduct'. In the absence of a base, the halides of aluminium, gallium, indium and thallium tend to dimerize by utilizing lone pairs on the halogen atoms (6.2, for example). One possible reason

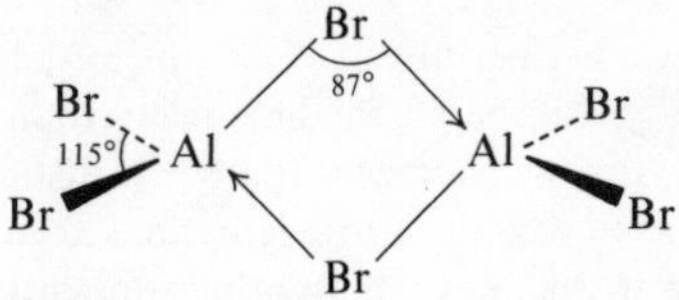

Br 87° Br 115° Al Al Br Br Br Br

bridging effect: Cl > Br > I

(6.2)

why the trihalides of boron are monomeric under all conditions is that boron is too small to allow the formation of the necessary four-membered ring system in the dimers without the occurrence of severe steric interactions (covalent radii are: boron, 0·88 Å; aluminium, 1·26 Å; beryllium, 1·06 Å). It has also been suggested that π-bonding between filled p-orbitals on the halogens and boron 2p orbitals may be partially responsible for the stabilization of the monomer (6.3).

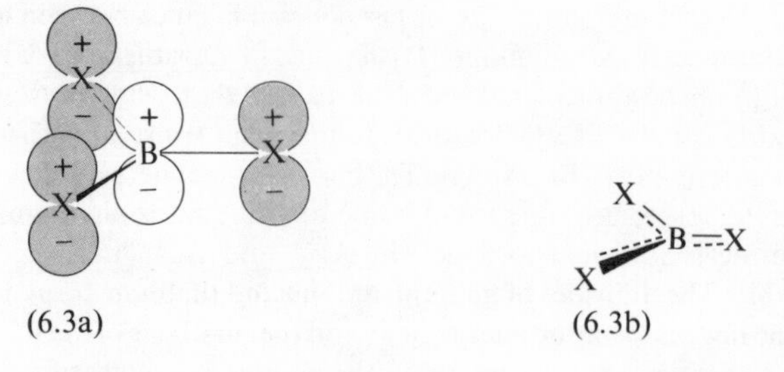

(6.3a) (6.3b)

Account has also to be taken of the fact that aluminium, gallium, indium and thallium have *n*d orbitals available for bonding purposes and it is conceivable that these elements use at least a small amount of d-character in the orbitals making up the dimers. It is possible that these elements also use their d-orbitals when forming adducts of the type $MX_3.base_2$ (6.4) and the complex ions MF_6^{3-} (6.5).

(6.4) trigonal bipyramid
(sp^3d hybridization at M)

(6.5) regular octahedron
(sp^3d^2 hybridization at M)

The π-bonding suggested in the boron trihalides should result in a shortening of the B—X bond lengths compared to the B—X distance in, say, BX_4^-, where π-bonding is impossible as no vacant orbitals are left on the boron atom. Such a shortening does occur (e.g. the B—F distance in BF_3 is 1·30 Å and in BF_4^- is 1·43 Å) but at least two factors other than π-bonding can also lead to the observed difference in the B—X bond lengths: (a) the steric crowding in BX_4^- will slightly increase this B—X distance, (b) in BX_3 there are only six electrons in the outer quantum shell and these will experience a greater pull from the nucleus than when two extra electrons are added as in BX_4^-; this will decrease the B—X distance in BX_3 relative to BX_4^-. Thus the questions as to why the boron trihalides are always monomeric and why the B—X distance in them is apparently 'too short' are exceedingly difficult ones to answer, both occur because of a balance between a variety of subtle effects.

6.2 Occurrence of the group III elements

6.2.1 *Boron*

About 3 p.p.m. of the Earth's crust; main sources are the sodium borates $Na_2B_4O_7.4H_2O$ and $Na_2B_4O_7.10H_2O$.

6.2.2 *Aluminium*

7·5 per cent; the most common of all metals; the most commercially important ore is the hydrated oxide, bauxite $Al_2O_3.H_2O$.

6.2.3 *Gallium*

15 p.p.m.; widely distributed in nature, but only in minute concentrations; zinc blende ZnS often contains 0·1–0·5 per cent of gallium.

6.2.4 *Indium*

0·1 p.p.m.; no concentrated deposits have been found and, like gallium, it is obtained from the residues of zinc and lead smelting processes.

6.2.5 *Thallium*

0·3 p.p.m.; obtained from the flue dust of sulphuric acid works.

6.3 Extraction

6.3.1 *Boron*

Amorphous boron of about 95 per cent purity is obtained by reducing boric oxide with either sodium or magnesium

$$Na_2B_4O_7.10H_2O(\text{borax}) \xrightarrow{\text{acid}} B(OH)_3 \xrightarrow[\text{by heat}]{\text{dehydrate}} B_2O_3 \xrightarrow[\text{or Mg}]{\text{Na}} B.$$

Pure, crystalline boron can be obtained in gramme quantities by passing a hydrogen–boron trihalide mixture over a tantalum wire held at 1100–1300 °C.

6.3.2 *Aluminium*

Bauxite is purified by dissolving it in sodium hydroxide solution (to give sodium aluminate) when the oxides of iron present precipitate out. If the solution is held at 25–35 °C and a little $Al_2O_3.3H_2O$ added to it, most of the bauxite present crystallizes out as the trihydrate; this is heated to 1200 °C, then dissolved in molten cryolite Na_3AlF_6 and electrolysed between carbon electrodes.

6.3.3. *Gallium, indium and thallium*

As mentioned above these metals are obtained as by-products in the production of zinc, lead and sulphuric acid. Gallium may be purified by electrolysis of an alkaline solution of one of its salts (gallium hydroxide is amphoteric and dissolves in alkali to give gallates).

6.4 The elements

The melting points within this group of elements vary widely: boron, ~2300 °C; aluminium, 639 °C; gallium, 29·6 °C; indium, 156 °C; thallium, 302·5 °C. Gallium, with a boiling point of about 2030 °C, has the longest liquid range of all the elements and has been used in high-temperature thermometers.

Boron is notable for the complexity of its crystalline forms, all of which are based on icosahedra of boron atoms; for example, the α-rhombohedral modification consists of an approximately cubic close-packed arrangement of icosahedral B_{12} groups bound to each other by covalent bonds (Figure 59). Aluminium has a face-centred cubic structure, but there is some evidence to suggest that all three outer electrons on the aluminium atoms are not used in bonding, which would account for the comparatively low melting point (only 8 °C above that of magnesium). Gallium has a complex structure in which the atoms have the following set of neighbours: one at 2·44 Å, two at 2·70 Å, two at 2·73 Å and two at 2·79 Å. It is thought that gallium may also contain some pairs of atoms in the liquid state because the X-ray diffraction pattern is different from that of a simple liquid like mercury. Indium has a slightly distorted face-centred cubic structure. Thallium is dimorphic, having hexagonal (α) and cubic (β) forms.

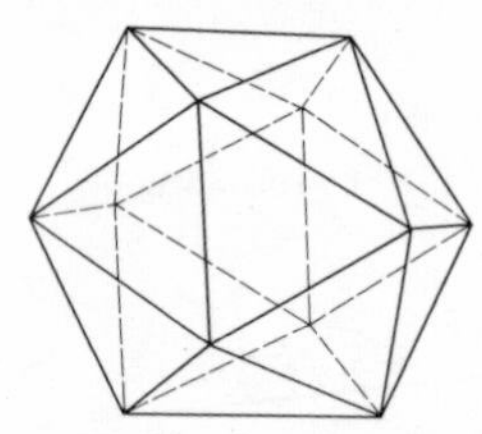

Figure 59 The icosahedron, a solid figure having twenty equilateral triangular faces and a total of twelve vertices. The icosahedral B_{12} unit is the basic 'building block' in the three known forms of crystalline boron

Elemental boron is highly inert and is not affected by acids, even boiling hydrofluoric;.fused sodium hydroxide slowly attacks it, but only above 500 °C. By contrast, the other elements are quite reactive and dissolve readily in acids giving salts, and aluminium, gallium and indium, being amphoteric, dissolve in alkalis evolving hydrogen and forming aluminates, gallates and indates, $M^IM^{III}O_2$.

6.5 Compounds of the group III elements

6.5.1 *Hydrides*

(See section 2.6 for a discussion of the hydrides of boron.) The expected monomeric hydrides of this group, MH_3, are unstable and can only be detected at low pressures where the polymerization rate is slow. For example, when aluminium is slowly vaporized off a tungsten filament in a current of hydrogen, the formation of

AlH_3 is revealed by mass-spectral analysis; if the amount of AlH_3 in the vapour phase is increased (by evaporating the aluminium metal more quickly) then the dimer Al_2H_6 can also be detected. Under similar conditions the much less thermally stable GaH_3 and InH_3 can be produced, but there is no indication of either Ga_2H_6 or In_2H_6. In the solid state alane is highly polymeric, the aluminium atoms being linked together by hydrogen bridges in such a way that each metal atom is octahedrally surrounded by six hydrogen atoms. The bridges in Al_2H_6 and $(AlH_3)_n$ are thought to involve three-centre, two-electron bonds (Figure 60) similar to those found in diborane (p. 55).

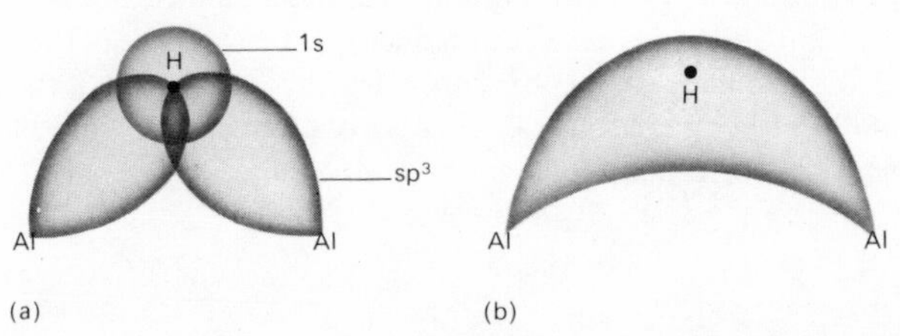

(a) (b)

Figure 60 (a) The atomic orbitals used in forming the aluminium–hydrogen–aluminium bridge in aluminium hydride Al_2H_6. (b) The bonding three-centre molecular orbital formed from the atomic orbitals shown in (a) contains two electrons

Although the free monomeric hydrides of group III do not exist under normal conditions their complexes with suitable donor molecules, such as amines, are well known,

e.g. $B_2H_6 + 2D \longrightarrow 2H_3B \leftarrow D$,
$(AlH_3)_n + nD \longrightarrow n(H_3Al \leftarrow D)$.

H, H, H, D around M (tetrahedral)

(6.6) tetrahedral molecule
(M = B, Al, Ga; sp^3 hybridized)

The donor species D can also be the hydride ion H^-, in which case the reaction product is the tetrahedral anion MH_4^-

$$AlCl_3 + LiH \xrightarrow{\text{ether}} (AlH_3)_n \xrightarrow{\text{LiH}} LiAlH_4.$$

These ionic hydrides of group III are used extensively in both inorganic chemistry and organic chemistry as reducing agents. Lithium aluminium hydride (lithium tetrahydroaluminate) reacts with many covalent metal and metalloidal halides to form the corresponding hydrides, often in very high yield,

e.g. $SiCl_4 + LiAlH_4 \longrightarrow SiH_4$ (100 per cent yield).

Sodium tetrahydroborate $NaBH_4$ is soluble in water and in a variety of organic solvents. It is widely used in organic chemistry to reduce carbonyl groups to alcohols; for example, aldehydes are very rapidly reduced to the corresponding primary alcohols

$$4\left[\begin{array}{c}H\\|\\-C=O\end{array}\right]+BH_4^- \xrightarrow[\sim 2\ \text{min}]{20\ ^\circ C} 4\left[\begin{array}{c}H\\|\\-C-OH\\|\\H\end{array}\right]$$

The specificity of sodium tetrahydroborate for the aldehydic carbonyl group is shown by the fact that aldehydes have been reduced in the presence of the following functional groups: olefin, ester, epoxy, nitro and carboxylic acid. Ketones are reduced less rapidly, although the reaction time is usually less than ninety minutes, the product being the corresponding secondary alcohol,

$$\gt C=O \longrightarrow \gt C\lt^{OH}_{H}$$

6.5.2 *Oxides*

The normal oxide M_2O_3 is formed by heating the elements in oxygen; on strong heating, $Tl^{III}_2O_3$ loses oxygen to form Tl^{I}_2O, the dissociation pressure of oxygen being about 115 mmHg at 700 °C. Gallium also gives a lower oxide when Ga_2O_3 is heated with the free metal at 700 °C,

$$Ga_2O_3+4Ga \underset{800\ ^\circ C}{\overset{700\ ^\circ C}{\rightleftharpoons}} 3Ga_2O.$$

6.5.3 *Hydroxides*

Down any group of the periodic table the metallic character (and hence basicity) is found to increase. Thus in group III, $B(OH)_3$ is acidic; $Al(OH)_3$, $Ga(OH)_3$ and $In(OH)_3$ are amphoteric; $Tl(OH)_3$ apparently does not exist, but the oxide Tl_2O_3 is basic and dissolves in acids to give thallic salts.

Boric oxide is hygroscopic and readily takes up water to form boric acid $B(OH)_3$. This is a very weak acid with a first dissociation constant of only 6×10^{-10} at 25 °C; complex formation with polyhydroxyl compounds such as glycerol increases the ionization of one proton, making the acid stronger, and it can then be titrated against sodium hydroxide using phenolphthalein as the indicator

$$B(OH)_3 + \begin{array}{c}-\overset{|}{C}-OH\\ -\underset{|}{C}-OH\end{array} \rightleftharpoons \begin{array}{c}-\overset{|}{C}-O\\ -\underset{|}{C}-O\end{array}\!\!>B-OH + H_2O$$

$$\Updownarrow$$

$$\begin{array}{c}-\overset{|}{C}-O\\ -\underset{|}{C}-O\end{array}\!\!>B^{-}\!\!<\begin{array}{c}O-\overset{|}{C}-\\ O-\underset{|}{C}-\end{array} + H^+(aq) + H_2O$$

Aluminium hydroxide dissolves in concentrated solutions of alkalis to give the linear ion $[O—Al—O]^-$, and not the tetrahedral $Al(OH)_4^-$ ion as might have been expected.

6.5.4 *Sulphides*

All the elements react on heating with an excess of sulphur to give the trisulphides M_2S_3. One of the lower sulphides of gallium, GaS, has a layer structure containing $[Ga—Ga]^{4+}$ units and is therefore not a derivative of gallium(II); the sulphide TlS is a completely different type of compound in that half the thallium is present as thallium(I) and half as thallium(III): $Tl^{I}[Tl^{III}S_2]$. Gallium exists as gallium(I) in the sulphide Ga_2S.

6.5.5 *Nitrides*

The boron–nitrogen bond is iso-electronic to a carbon–carbon bond and hence it might be anticipated that boron nitride $(BN)_x$ could exist in two forms: one having a layer lattice like graphite and consisting of sheets of hexagonal (3B, 3N) rings (Figure 61), the other having a giant, three-dimensional lattice like diamond (see Figure 56). Both types of boron nitride have been realized. The graphitic form is the ultimate product formed by heating a wide variety of boron–nitrogen compounds. On the application of high temperatures and high pressures, the layer form of boron nitride is converted to the diamond-like variety called borazon. Borazon is exceedingly hard and is the only known compound capable of cutting diamond.

The nitrides of aluminium, gallium and indium are known; aluminium nitride AlN produces ammonia on treatment with water or acid and for a time was considered as a means of 'fixing' atmospheric nitrogen as ammonia for use in agriculture

$$\text{Liquid air} \xrightarrow[\text{distillation}]{\text{fractional}} N_2 \xrightarrow[\text{heat}]{\text{Al}} AlN \xrightarrow{H_2O} NH_3.$$

Borazine. Although the final product of heating diborane with ammonia is hexagonal boron nitride, an intermediate in the decomposition is a colourless,

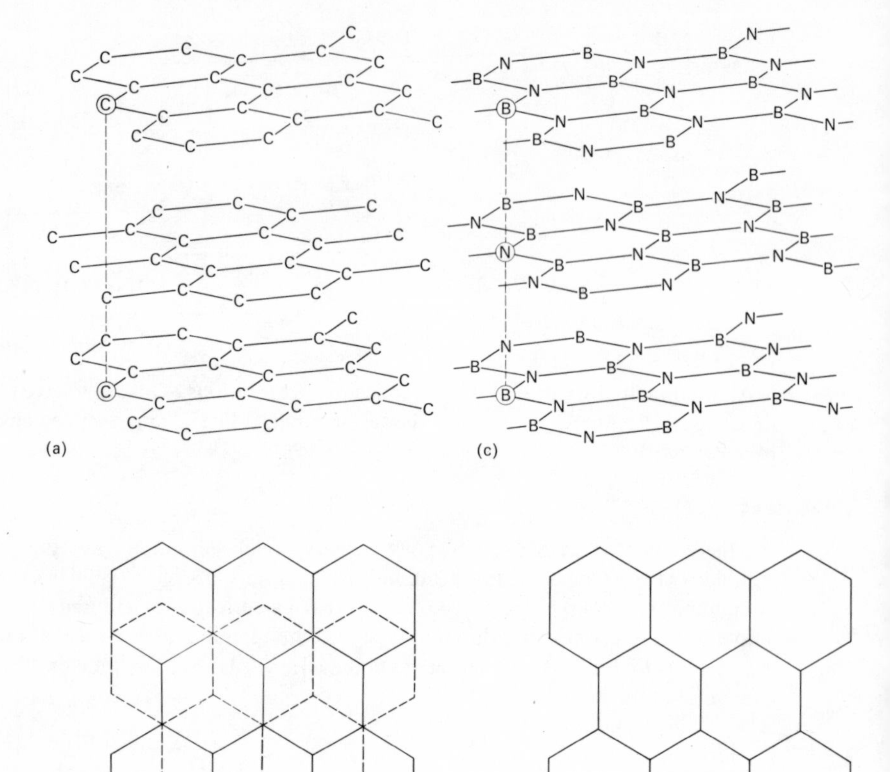

(b) graphite
(successive layers out of phase)

(d) hexagonal boron nitride
(successive layers in phase)

Figure 61 (a) and (b) The structure of graphite (successive layers out of phase). (c) and (d) The structure of hexagonal boron nitride (successive layers in phase). Both structures consist of layers of hexagonal (either C_6 or B_3N_3) rings. In graphite the rings are in phase only at every other layer and hence each carbon atom has only carbon atoms from the same plane as its neighbours. All the layers of boron nitride have their rings in phase, each boron atom having two nitrogen atoms (from neighbouring layers) above and below it. Because of this weak boron–nitrogen interaction, the layers in boron nitride do not slide over each other as readily as those in graphite

volatile liquid called borazine $B_3N_3H_6$. This compound is iso-electronic to benzene and, like benzene, it has a cyclic structure (6.7a). All the boron–nitrogen

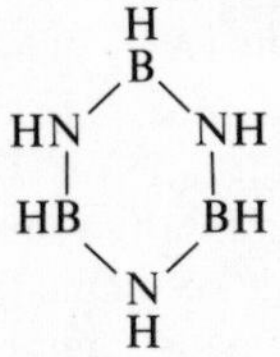

(6.7a)

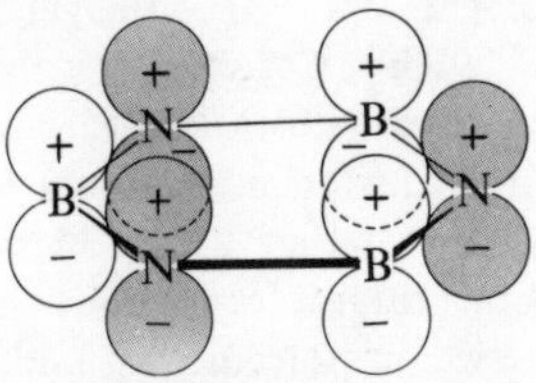

(6.7b)

bonds are equal in length at 1·44 Å, which is to be compared to the boron–nitrogen distance of 1·60 Å in the ammonia–boron trifluoride complex $F_3B \leftarrow NH_3$, which contains a single boron–nitrogen bond. From this evidence it is concluded that π-bonding occurs between filled nitrogen $2p_z$ orbitals and empty $2p_z$ orbitals on the boron atoms.

Although benzene and borazine have some very similar physical properties their chemical properties are very dissimilar. For example, borazine readily undergoes addition reactions whereas such reactions are relatively uncommon in the chemistry of benzene. Typically, three molecules of hydrogen chloride (or methyl iodide, methanol, water) are added to the borazine ring at room temperature, the ring losing its aromatic character and reverting to a saturated, cyclohexane-like cyclic system (6.8). In such addition reactions the most electronegative half of the attacking molecule becomes attached to the boron atoms,

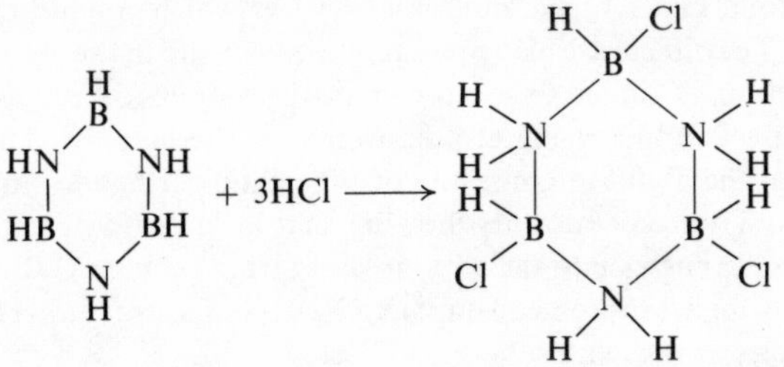

+ 3HCl ⟶

(6.8)

probably because the boron atoms in borazine are positively charged relative to the nitrogens.

6.5.6 *Salts of oxy-acids*

The metals aluminium, gallium and indium dissolve in the stronger oxy-acids to give, for example, the M^{III} nitrates, sulphates, selenates, halates and perhalates, which may be crystallized from solution as hydrates. Thallium dissolves to give thallous salts.

Aluminium sulphate forms a series of essentially isomorphous double sulphates (called 'alums') with a wide variety of sulphates containing monovalent cations. The alums have the general formula $M^{I}Al(SO_4)_2 . 12H_2O$, where M^{I} is Na^+, K^+, Rb^+, Cs^+, NH_4^+, substituted ammonium ions and Tl^+; gallium and indium can be substituted for aluminium, but the Tl^{3+} ion appears to be too large for inclusion in the alum lattice. The water of crystallization is divided equally between the M^+ and M^{3+} ions, which probably explains why a small ion such as Li^+ cannot form alums because it cannot satisfactorily accommodate six separate water molecules around itself; the maximum coordination number for the lithium ion appears to be only four.

6.5.7 *Trihalides*

All the trihalides of this group, with the exception of TlI_3, may be prepared by treating the respective element with the free halogen (or HF may be used in place of fluorine). The product formed on treating thallium with an excess of iodine has the formula TlI_3 but, as it is isomorphous with RbI_3, it must be thallous tri-iodide $Tl^+[I_3]^-$ and contain the linear $[I{-}I{-}I]^-$ anion. It is interesting to note that thallium can be oxidized to thallium(III) by iodine in the presence of iodide ions, when the tetrahedral complex ion $[Tl^{III}I_4]^-$ is formed.

The anhydrous trifluorides and trichlorides of aluminium, gallium, indium and thallium (with the exception of $GaCl_3$) adopt crystal structures in which the metals are octahedrally coordinated to six halogens, suggesting that these halides are essentially ionic. On the other hand, the crystalline bromides and iodides of these four elements (and $GaCl_3$) have molecular lattices which contain the covalent dimers M_2X_6. The dimers are also present in the melt and in the gas phase, although the proportion of dimer to monomer in the vapour decreases with increasing temperature and increasing atomic weight of the halogen. This behaviour is reflected in the Trouton constants of the gallium trihalides: molten Ga_2Cl_6 has a normal Trouton constant, showing that little or no dissociation occurs when the chloride passes into the vapour state; whereas both $GaBr_3$ and GaI_3 have appreciably high Trouton constants, indicative of a structural change in the molecular species as they enter the gas phase.

6.5.8 *Lower halides*

All the group III elements form diatomic halides in the gas phase at elevated temperatures which, with the exception of Tl^+F^-, are covalently bonded

$$2M + MX_3 \xrightarrow{\text{heat}} 3MX.$$

Unlike the other members of this group, boron has several volatile lower halides which contain covalently linked boron atoms (6.9, 6.10).

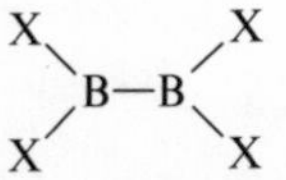

(6.9) diboron tetrahalides
(decompose slowly at
room temperature)

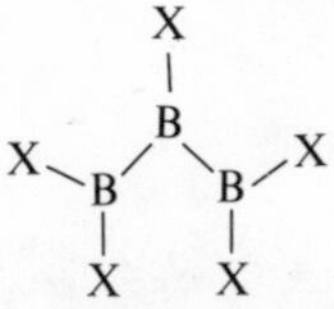

(6.10) triboron pentahalides
(only the fluoride has been
isolated and this decomposes at −30 °C)

The most extensively studied boron subhalide is B_2Cl_4, which can be made by passing boron trichloride through an electrical discharge between mercury electrodes

$$2BCl_3 + 2Hg \xrightarrow[\text{pressure}]{\text{low}} Hg_2Cl_2 + B_2Cl_4 .$$

Rotation about the boron–boron bond occurs easily: in the gaseous and liquid states at room temperature the B_2Cl_4 molecules adopt the 'staggered' structure (6.11a), whereas crystal forces in the solid make the molecules assume a planar shape (6.11b) for ease of packing.

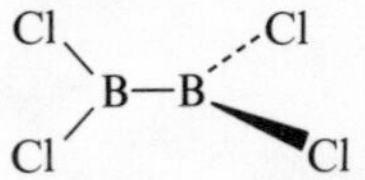

(6.11a)

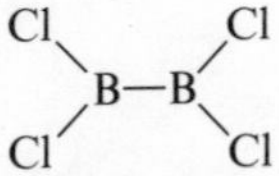

(6.11b)

(boron atoms are sp^2 hybridized in both structures)

Boron also forms three crystalline monochlorides, B_4Cl_4, B_8Cl_8 and B_9Cl_9, all of which possess closed cages of boron atoms (Figure 62). The bonding in these molecules is complex and probably involves multicentred σ-bonds extending over all the boron atoms in the cage. Low yields of B_4Cl_4 are obtained by passing B_2Cl_4 through an electric discharge between mercury electrodes, whereas B_8Cl_8 and B_9Cl_9 result when B_2Cl_4 slowly decomposes at room temperature

$$B_4Cl_4 \xleftarrow[\text{discharge}]{\text{mercury}} B_2Cl_4 \xrightarrow[\text{decomposition}]{\text{thermal}} B_8Cl_8 + B_9Cl_9 + B_{10}Cl_{10} + B_{11}Cl_{11} + B_{12}Cl_{12}.$$

The structures of the red compounds $B_{10}Cl_{10}$, $B_{11}Cl_{11}$ and $B_{12}Cl_{12}$ are not yet known.

The halides of empirical formula MX_2 ($X \neq F$), which exist for gallium, indium and thallium, are not volatile dimers like those of boron, but are ionic solids having the composition $M^I[M^{III}X_4]$. The two main methods used for their preparation are either the reduction of the trihalide with the free element

$$MCl_3 + M \longrightarrow M^+[M^{III}Cl_4]^-,$$

or the stoichiometric halogenation of the element using mercury(II) halides

$$2M + 2HgX_2 \longrightarrow M^+[M^{III}X_4]^- + 2Hg.$$

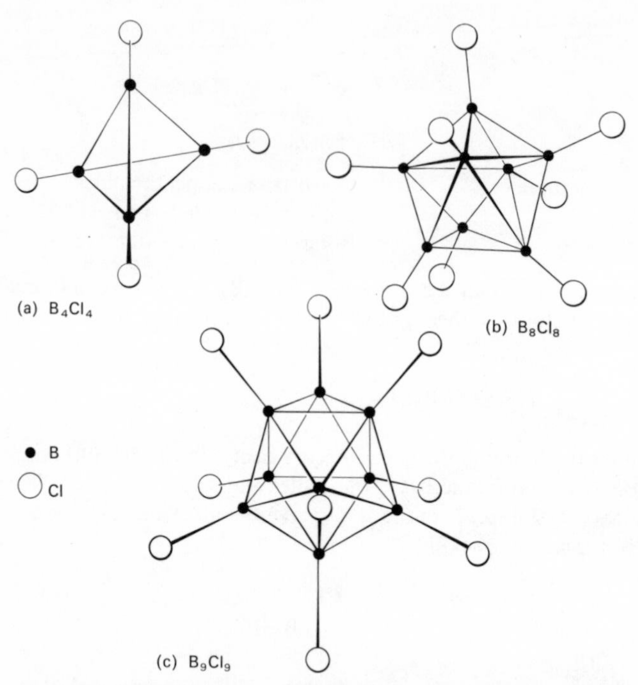

Figure 62 Polyhedral boron cages in the boron monochlorides B_nCl_n. (a) B_4Cl_4: tetrahedron of boron atoms. (b) B_8Cl_8: triangular-faced dodecahedron of boron atoms. (c) B_9Cl_9: tri-capped trigonal prism of boron atoms

The latter reaction can also be carried out in the presence of aluminium (for M = Ga or In) when a Ga^+ or In^+ salt of the tetrahaloaluminate ion is produced

e.g. $Ga + Al + 2HgCl_2 \longrightarrow Ga^+AlCl_4^- + 2Hg$.

Although compounds containing the Ga^+ or In^+ ions are rather reactive and, for example, are decomposed vigorously by water, they can be stabilized by complexing the M^+ ion with a variety of ligands

$$4L + GaGaCl_4 \longrightarrow [Ga^IL_4]^+[Ga^{III}Cl_4]^-.$$

The structures of the ML_4^+ ions are not known with certainty, but are thought to be based on the trigonal-bipyramidal shape, the *ns* lone-pair electrons of the free M^+ ion now occupying a sterically important orbital in the equatorial plane (6.12).

L, L, M^+—L, L

(6.12)

The reduction of the trihalides of gallium, indium and thallium with an excess of the free element results in the formation of solid monochlorides MX. (See section 6.8 for a brief summary of thallium(I) chemistry, including that of the halides.)

6.6 Organometallic compounds

Grignard and lithium reagents readily react with the group III trihalides to give the trialkyls or triaryls,

$$\text{e.g.}\quad 3CH_3MgI + BF_3 \xrightarrow{\text{ether}} (CH_3)_3B,$$
$$3C_6H_5Li + GaCl_3 \longrightarrow (C_6H_5)_3Ga.$$

Aluminium appears to occupy a unique place within the group in that its triphenyl and lower trialkyls are dimeric in the solid, liquid and gaseous states (dissociation to the monomers occurs on heating the dimers in the vapour) (6.13). The bonding

H_3C, H_3C — Al — (CH_3 bridge, 75°; CH_3 bridge) — Al (105°) — CH_3, CH_3

(6.13)

in the Al—C—Al bridges is thought to involve three-centre molecular orbitals formed from essentially sp^3 orbitals on the carbon and aluminium atoms (Figure 63). This type of three-centre interaction in the bridge necessarily results in an acute Al—C—Al angle (75°); the fact that the boron alkyls and aryls are monomeric in all phases is a direct result of the small size of the boron atom, which does not allow the formation of a stable four-membered ring in the bridge system. (For example, the two carbon atoms would have to approach to within about 2·5 Å of each other which is much less than twice the van der Waals radius for

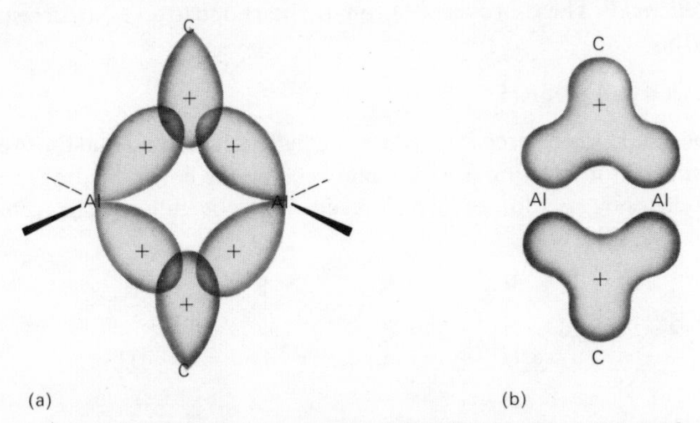

Figure 63 The trimethylaluminium dimer $Al_2(CH_3)_6$: (a) the sp^3 atomic orbitals used in bonding, (b) the approximate shape of the resulting two bonding molecular orbitals, each of which holds two electrons

carbon, $\sim$3·8 Å). The organo-derivatives of gallium, indium and thallium, although apparently weakly associated in the solid state, do not form dimers like the aluminium compounds; this may be due to adverse metal–metal steric interaction which would occur in the M—C—M bridge system.

6.7 Hydroboration

Many olefins and acetylenes react with diborane under very mild conditions to give the corresponding alkylborane or vinylborane in high yield. The diborane may be bubbled through an ether solution of the olefin or may be generated *in situ* by adding boron trifluoride–etherate $BF_3.O(C_2H_5)_2$ to an ether solution of sodium tetrahydroborate and the olefin. In both cases the reaction must be carried out under an atmosphere of dry nitrogen because of the reactive nature of both the diborane and the alkylboranes which are formed

$$RCH{=}CH_2 + B_2H_6 \longrightarrow (RCH_2CH_2)_3B,$$

$$RC{\equiv}CR + B_2H_6 \longrightarrow \left[\begin{matrix} R & & R \\ & C{=}C & \\ H & & \end{matrix} \right]_3 B$$

This simple hydride-addition reaction would not have assumed much importance were it not for the fact that the organic groups can be removed from the boron atom by a variety of reagents; by a suitable choice of reagent the boron–carbon bond can be cleaved to give a paraffin, olefin, alcohol, ketone or carboxylic acid. As the alkylborane need not be isolated before treatment with the cleaving reagent, this 'hydroboration' process provides an extremely simple route for the conversion of olefins and acetylenes into the products mentioned. Typical reagents which are normally employed are:

(a) *Carboxylic acids*. These cleave the boron–carbon bond to give the corresponding hydrocarbon

$$R_3B + 3R'COOH \longrightarrow 3RH.$$

If the alkylborane has been prepared from an acetylene, it is found that the product from this protonation reaction is a *cis* olefin because the boron–hydrogen bond adds to acetylenes by *cis* addition, which results in the overall *cis* hydrogenation of the original acetylene

$$RC{\equiv}CR + H{-}B\langle \longrightarrow \begin{matrix} R & & R \\ & C{=}C & \\ H & & B\langle \end{matrix} \xrightarrow{R'COOH} \begin{matrix} R & & R \\ & C{=}C & \\ H & & H \end{matrix}$$

(b) *Alkaline hydrogen peroxide*. This reagent cleaves the boron–carbon bond to give an alcohol and boric acid

$$(RCH_2)_3B+3H_2O_2 \xrightarrow{OH^-} 3RCH_2OH+B(OH)_3.$$

Normally boron–hydrogen bonds add to an olefin in such a manner that the boron atom becomes attached to the least substituted of the two carbon atoms making up the double bond

$$RCH{=}CH_2 \xrightarrow{>B-H} RCH_2CH_2B< \xrightarrow[OH^-]{H_2O_2} RCH_2CH_2OH,$$

and hence the overall reaction is the *cis* anti-Markovnikov addition of the elements of water to the olefin.

(c) *Chromic acid.* Oxidation with this reagent gives either ketones from secondary alkyl groups attached to boron (i.e. $RR'CH-B<$) or carboxylic acids from primary alkyl groups attached to boron.

(d) *Alkaline silver nitrate.* This causes the direct coupling of the alkyl groups on boron

$$2R_3B \longrightarrow 3R-R.$$

6.8 The chemistry of thallium(I)

As was mentioned at the beginning of the chapter, there is an increasing tendency down the group for the elements to be found in the monovalent state with the ns^2 electrons taking no part in compound formation. For thallium, the ionic M^I state is the more important state and is formed when thallium metal dissolves in strong acids.

The thallous ion (radius 1·45 Å) is unique in having an $(n-1)d^{10}ns^2$ outer electronic structure and hence there are no other singly charged ions to which its chemistry might be directly compared. The nearest approaches to such an electronic structure in ions of comparable size are to be found in Ag^+ ($4d^{10}$ core, ionic radius 1·13 Å) and Rb^+ ($5s^25p^6$ core, ionic radius 1·48 Å). In much of its chemistry, the colourless thallous ion resembles the heavier alkali metals and, for example, thallous chloride and thallous bromide are ionic halides which normally have the caesium chloride structure but can be forced under suitable conditions to adopt the sodium chloride structure. On the other hand, thallous halides show a remarkable resemblance in colour and solubility to the silver halides: thallous fluoride (like silver fluoride) is very soluble in water, whereas thallous chloride, bromide and iodide are insoluble and darken on exposure to light.

Thallous hydroxide is a strong base, is soluble in water and absorbs carbon dioxide from the air to give thallous carbonate (compare potassium hydroxide and rubidium hydroxide). The closeness of the ionic radii of Tl^+ and Rb^+ makes many thallous salts (e.g. sulphate, chromate, nitrate, chloride, bromide and iodide) isomorphous to the corresponding rubidium salts. Also Tl^+ is able to replace Rb^+ in many double salts such as the alums.

Chapter 7
Group IV. Carbon, Silicon, Germanium, Tin and Lead

7.1 Introduction

The ionization energies follow the same trends as were noted for the group III elements; the irregular changes down the group are due to the presence of poorly shielding d-orbitals (germanium, tin, lead) and f-orbitals (lead). The poor shielding effect of the 3d electrons of germanium is also responsible for the relatively small increase in covalent radius between silicon and germanium (0·05 Å) because the outer electrons in germanium are more strongly attracted to the nucleus than might otherwise have been expected; the *n*s and *n*p electrons of germanium and tin are both outside filled $(n-1)$d shells and hence the change in covalent radius between the two elements (0·19 Å) is much more normal. Likewise, the difference in radius between tin and lead of only 0·13 Å is due to the presence of the filled 4f shell at lead.

The exceedingly high energies required to form the M^{4+} ions, coupled to the fact that such ions will be very small and strongly polarizing for the earlier members of the group, means that the M^{4+} state will only be found in tin and lead salts of small anions; for example, SnF_4, PbF_4, SnO_2 and PbO_2. In favourable cases, the high ionization energy can be offset by complexing ligands round the M^{4+} ion; as well as recouping energy by M–ligand bond formation, this increases the effective ionic size and so reduces the polarizing power of the highly charged cation, thus making its salts more stable. In this way even carbon, with its estimated ionic radius of only 0·15 Å, can be coaxed into forming ionic complexes†.

$$CX_4 + 2\,C_6H_4(As(CH_3)_2)_2 \rightarrow [C^{4+}(As\frown As)_2][X^-]_4$$

(X = Br, I)

†Alternatively, these compounds can be regarded as quaternary salts of the 'ligand' atoms, the arsenic derivative in the equation thus being a complex arsonium salt in which each arsenic atom carries one positive charge.

Table 20 The Group IV Elements

Element		1st IE / kJ mol^{-1}	2nd IE / kJ mol^{-1}	3rd IE / kJ mol^{-1}	4th IE / kJ mol^{-1}	Covalent radius/Å
Carbon C	$1s^2 2s^2 2p^2$	1086	2352	4619	6222	0·77
Silicon Si	$[Ne]3s^2 3p^2$	786·1	1576	3227	4355	1·17
Germanium Ge	$[Ar]3d^{10} 4s^2 4p^2$	761·5	1537	3302	4393	1·22
Tin Sn	$[Kr]4d^{10} 5s^2 5p^2$	708·5	1412	2943	3929	1·41
Lead Pb	$[Xe]4f^{14} 5d^{10} 6s^2 6p^2$	715·5	1450	3081	4083	1·54

From their outer electron configuration of ns^2np^2, one might expect the group IV elements to form covalent compounds in which they bond only to two other atoms or groups using their half-filled p-orbitals; this would lead to the production of MX_2 molecules in which the angle X—M—X would be approximately 90°. However, this is contrary to observation because, in the vast majority of their compounds, these elements (and more especially carbon) are tetrahedrally surrounded by four other groups. This occurs because it is relatively easy for an *ns* electron to be 'promoted' to the unfilled *np* orbital; for carbon the $ns \rightarrow np$ promotion energy is 405·8 kJ mol^{-1}, that of the other elements probably being somewhat smaller.

However, although the $ns \rightarrow np$ promotion certainly leaves the element with four unpaired electrons, the electrons have *all their spins parallel.* To obtain the element in its valence (i.e. reacting) state, work must be done to 'randomize' the electron spins – and for carbon this absorbs about a further 250 kJ mol^{-1}. (The hypothetical process of hybridization of the one s- and three p-orbitals to sp^3 tetrahedral orbitals is simply a mathematical step and requires no energy absorption on the part of the reacting element.) Therefore, to form CX_4 from X_2 molecules and graphite, the standard state of carbon, the following energy steps must be considered:

$$C_{(s)} \xrightarrow[715\ \mathrm{kJmol^{-1}}]{\Delta H_{sub}} C_{(g)} \xrightarrow{P} C_{(g)}\ (s^{\uparrow}p_x^{\uparrow}p_y^{\uparrow}p_z^{\uparrow}) \xrightarrow{R} C_{(g)}(\text{valence state})$$

$$2X_2 \xrightarrow{2\Delta H_{diss}} 4X \qquad\qquad \longrightarrow CX_4$$

where P is the promotion energy and R is the energy required to randomize the electron spins. The total energy input is thus $\Delta H_{sub} + P + R + 2\Delta H_{diss}$, which for the production of methane is about 2260 kJ mol^{-1}; this energy has to be regained by the formation of four strong C—X bonds. The production of CX_2 requires fewer energy-consuming steps

$$C_{(s)} \xrightarrow{\Delta H_{sub}} C_{(g)}$$

$$X_2 \xrightarrow{\Delta H_{diss}} 2X \qquad \longrightarrow CX_2$$

but only two C—X bonds are formed to compensate for this energy.

The question arises as to which of these two processes leads to the more thermodynamically stable molecule. At room temperature the former process resulting in the formation of MX_4 molecules is almost always the more satisfactory both for carbon and the other members of the group; for example, the enthalpies of formation at 25 °C for CH_2 and CH_4 are about +343 and −74·9 kJ mol^{-1} respectively. Thus, although the utilization of the ns^2 electrons requires a higher initial input of energy, this is more than offset by the formation of two extra M—X bonds. This does not mean that MX_2 molecules cannot be made at all; they often result as highly reactive intermediates from either the reduction of MX_4 molecules by the free element at high temperature

$$SiF_4 + Si \xrightarrow{2000\ °C} [SiF_2],$$

or by the photolysis of unstable M^{IV} compounds

$$CH_2N_2 \xrightarrow{h\nu} [CH_2] + N_2.$$

The M—X covalent bond strength decreases as the atomic number of M increases (Table 21) with the result that at lead not all Pb—X bonds are capable of supplying the energy required to stabilize the lead(IV) state with respect to lead(II). For example, lead tetrafluoride, tetrachloride and tetrabromide decompose readily on heating

$$PbCl_4 \longrightarrow PbCl_2 + Cl_2,$$

whilst lead tetraiodide is too unstable to exist at room temperature.

Gaseous tin and lead dihalides are non-linear molecules in which the X—M—X bond angles are close to the value of 90° expected if pure p-orbitals on M are used in bonding. However, CH_2 ('carbene') is found to be linear in the ground state and contains two unpaired electrons; the carbon atom is sp hybridized, the

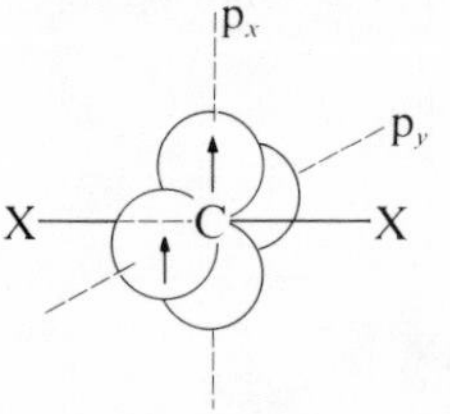

(7.1)

two unpaired electrons being housed in the remaining p-orbitals. This triplet ground state of carbene has only slightly less energy than a singlet state in which the CH_2 molecule is bent, the H—C—H angle being 102·4°, and has all its electrons spin paired. Substituents other than hydrogen on the carbon can invert the stability sequence of the triplet and singlet states, an example of this being CF_2, which is a bent diamagnetic molecule having an F—C—F angle of 105°; SiF_2 is similar and has an F—Si—F angle of 102°.

Table 21 Bond Energies of the Group IV Elements/(kJ mol^{-1})

	Carbon	*Silicon*	*Germanium*	*Tin*
M—M	347	226	188	151
M=M	611	—	—	—
M≡M	841	—	—	—
M—H	414	318	285	251
M—F	490	598	473	—
M—Cl	326	402	339	314
M—Br	272	331	280	268
M—I	—	—	213	—

An alternative to the covalent M^{II} state is of course the formation of M^{2+} by ionization of the *n*p electrons. However, there are no compounds known of carbon, silicon, germanium or tin in which M^{2+} ions have been reported with certainty and of the lead(II) derivatives only lead(II) fluoride, which has the rutile structure, can be said to be a typical ionic compound – although $PbCl_2$, $PbBr_2$ and PbClF are often considered to be ionic also. The most favourable compounds in which to look for metallic ions are the fluorides – due to the low dissociation energy of the F_2 molecule and the small radius of the fluoride ion, which would lead to the high lattice energies required to compensate for the considerable first and second ionization energies of these elements. However, the structure of GeF_2 (7.2) has been shown to be that of a fluorine-bridged chain polymer with weak fluorine bridging *between* the chains giving cross-linking: the germanium atoms have an approximately trigonal-bipyramidal arrangement of fluorine atoms around them, one of the equatorial positions being occupied by a sterically active lone pair of electrons. The germanium atom may use its d-orbitals in the bonding, because the lone electron pair is certainly not residing in a spherical s-orbital. The F—Ge—F angle of 163° presumably differs from 180° due to lone-pair–bond-pair repulsions as indicated (7.2).

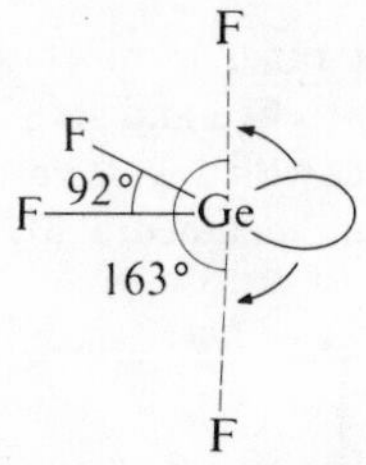

F
F
92°
F
Ge
163°
F

(7.2)

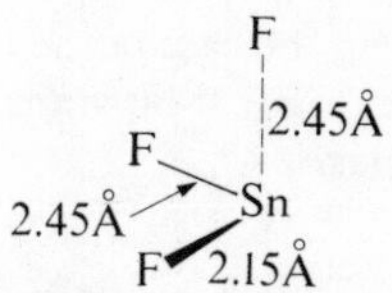

F
2.45Å
F
2.45Å
Sn
F
2.15Å

(7.3)

In the orthorhombic form of SnF_2, the tin atom has three fluorine atoms as nearest neighbours with the F—Sn—F bond angles approximately 90° as expected if the tin atom were using pure p-orbitals for bonding (7.3). Three other fluorines at a distance greater than 2·80 Å are the next nearest neighbours to tin. This type of trigonal-pyramidal coordination round tin appears to be common to many tin compounds, for example (7.4–6).

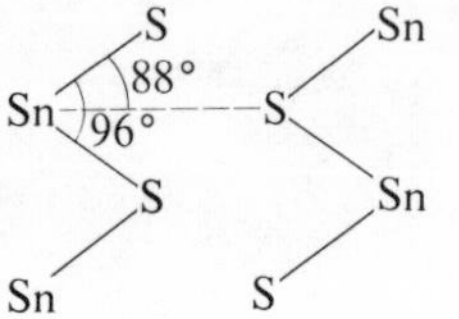

(7.4)
SnS; zig-zag chains cross-linked by Sn—S interactions; the selenide is isostructural

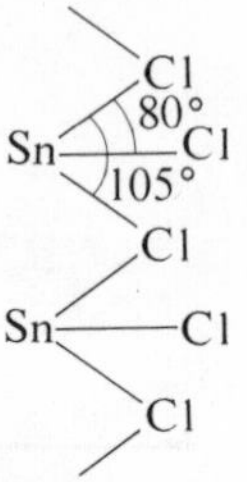

(7.5)
$SnCl_2$; chain polymer

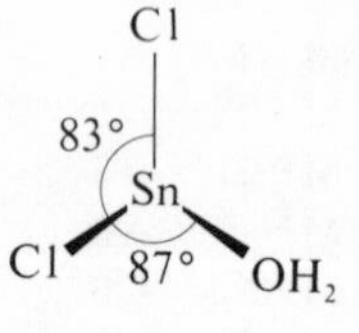

(7.6)
$SnCl_2 . 2H_2O$; $SnCl_2 . H_2O$; second H_2O in the dihydrate is hydrogen-bonded to the coordinated water molecule

In aqueous solution there is again little evidence for the Sn^{2+}(aq) ion, extensive hydrolysis of tin compounds occurring on their dissolution in water. The chloride, for example, appears to dissolve in small amounts of water as $SnCl_2$(aq) but on dilution the basic chloride SnClOH precipitates as a white solid; in the presence of an excess of halide ion, complex polyhalostannates(II) can be formed,

$$\text{e.g.}\quad SnF_2 + F^- \xrightarrow[\text{solution}]{\text{aqueous}} SnF_3^- .$$

There appears to be more evidence for Pb^{2+}(aq) in water, but such solutions apparently contain a significant proportion of $PbOH^+$ ions formed by hydrolysis.

The covalency maximum for carbon is four, but the other members of the group are able to increase their coordination numbers to five and six. This is especially true in the case of the tetrahalides, which react with a wide variety of ligands to form trigonal-bipyramidal $MX_4 . L$ (7.7, 7.8) and octahedral $MX_4 . 2L$ (7.9–11) complexes.

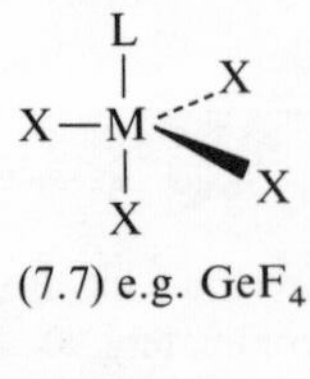

(7.7) e.g. $GeF_4 . N(CH_3)_3$

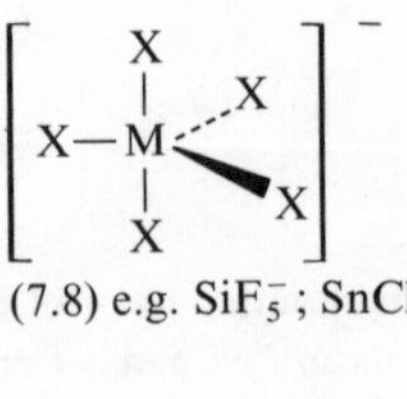

(7.8) e.g. SiF_5^-; $SnCl_5^-$

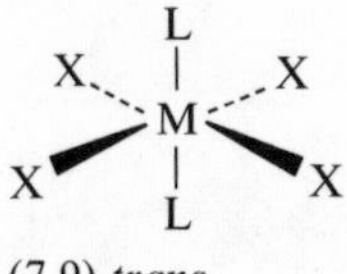

(7.9) *trans*
e.g. $SiCl_4 . 2P(CH_3)_3$

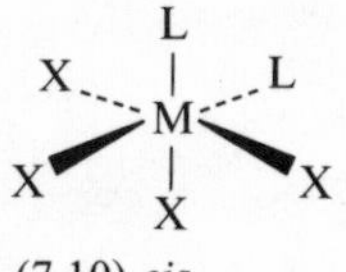

(7.10) *cis*
e.g. $SiF_4 . 2$(amine)

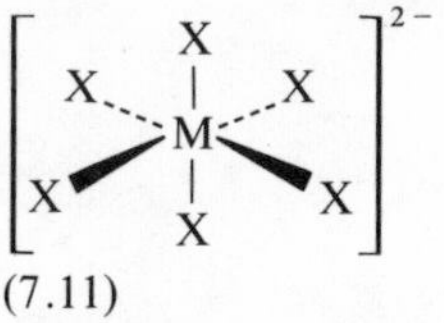

(7.11)
e.g. SiF_6^{2-}

To achieve this increase in coordination number, it is often assumed that the *n*d orbitals of silicon, germanium, tin and lead can participate in both sp^3d and sp^3d^2 hybridization, but direct evidence of this is lacking; although no low-lying d-orbitals are available in the $n = 2$ shell of carbon, it should be realized that a carbon atom is probably too small to accommodate even the one extra ligand required for five coordination (p. 32). The fact that silicon, germanium, tin and lead are able to increase their coordination number in this way probably accounts for the ready hydrolysis of their tetrahalides on treatment with water.

$$MX_4 + OH^- \xrightarrow[\text{process}]{S_N2} [X_4MOH]^- \longrightarrow X_3MOH + X^-$$

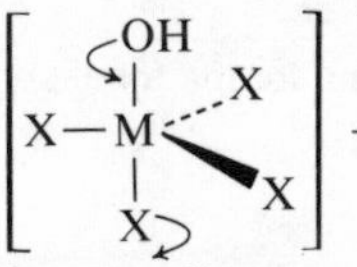

five-coordinate intermediate

Although the carbon tetrahalides are thermodynamically unstable with respect to their hydrolysis products,

e.g. $CF_4(g) + 2H_2O(g) \rightarrow CO_2(g) + 4HF(g), \quad \Delta G = -151 \text{ kJ mol}^{-1}$,

they are unaffected by water under normal conditions because of a kinetic block in the reaction mechanism due to carbon not being able to form a five-coordinate hydrolysis intermediate.

Carbon is unique among the group IV elements both in its ability to catenate (i.e. to make chains with itself as in the paraffin hydrocarbons) and in forming multiple p_π–p_π bonds with itself, oxygen and nitrogen. The strength of M—M bonds (Table 21) decreases rapidly down the group, so that, although many hundreds of carbon atoms may be linked together in 'polythene', thermal stability severely limits the maximum number of M—M bonds in straight-chain molecules of hydrides and halides of the other group IV elements to eleven or twelve for silicon, nine or ten for germanium and only two for tin and lead.

When carbon forms multiple p_π–p_π bonds, as in ethylene, it is considered that a normal σ-bond is initially formed by overlap of two carbon sp^2 hybrid orbitals and that this bond is then augmented by overlap of the two remaining 2p orbitals (Figure 64).

In acetylene C_2H_2, two bonding π molecular orbitals are formed at right angles to each other and each contains two electrons, leaving the two degenerate

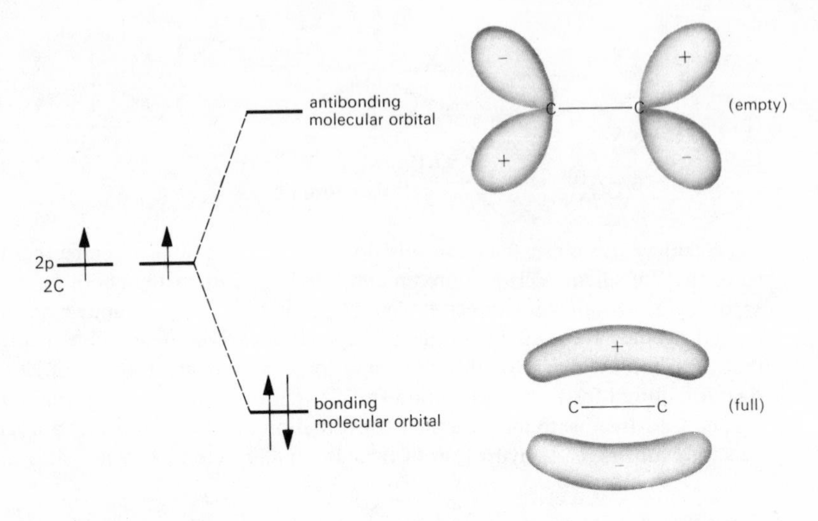

Figure 64 The energy-level diagram for the formation of a π-bond from two carbon 2p orbitals

(i.e. equal-energy) antibonding π-orbitals empty. Many attempts have been made to isolate compounds containing M=M and M≡M multiple bonds of the other elements, but so far no success has been achieved. Two main factors are thought to account for this: (a) *n*p orbitals for quantum numbers greater than $n = 2$ have nodes (i.e. zones having zero electron density) in some of the regions of space where electron density build-up is necessary for strong bond formation; (b) such p-orbitals are larger than carbon 2p orbitals and consequently, to achieve the same amount of overlap as in a C=C bond, the atoms would have to approach rather closely to each other, so introducing strong repulsions between the inner electron shells of the two atoms; see p. 35. The outcome is that an atom gains more energy by forming σ-bonds with two other atoms than by forming a $\sigma+\pi$ double bond with only one other atom. Poor orbital overlap would also occur if a multiple bond were to be formed between an *n*p orbital on silicon, germanium, tin or lead and the small 2p orbitals of elements like carbon, nitrogen or oxygen, with the result that no compounds are yet known which contain M=C, M≡C, M=O, M=N or M≡N bonds. The effect of this can be seen in the physical properties of many carbon compounds which differ in a dramatic way from their silicon, germanium, tin and lead counterparts, although both may have the same stoichiometry; two typical examples are carbon dioxide, which has a sublimation point of −78 °C, and silicon dioxide, which has a melting point of 1700 °C. In the carbon dioxide molecule the carbon atom is bound to the two oxygens by both σ and p_π–p_π bonds (see Figure 73, p. 164) and hence the solid contains molecules of CO_2

held in the lattice only by weak van der Waals forces. Silicon, however, gains more energy by forming four Si—O single bonds than by being multiply bonded to two oxygen atoms only, and this results in a polymeric structure for solid SiO_2 based on O—Si—O bridges; on melting, many strong Si—O bonds must be broken and this requires a very high temperature.

The formation of Si—O—Si single bonds in preference to Si=O bonds is the fundamental basis of the 'silicone' industry. Silicone polymers, which can be either cyclic or long chain molecules, have the R_2SiO group as the unit building block (7.12). They are formed by hydrolysis of the dialkylsilicon dichlorides, the

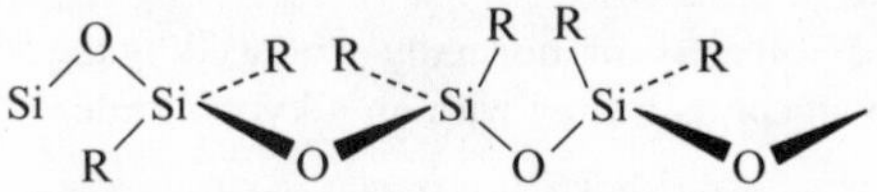

(7.12) the alkyl group R is usually CH_3

intermediate dihydroxyl derivative being highly unstable

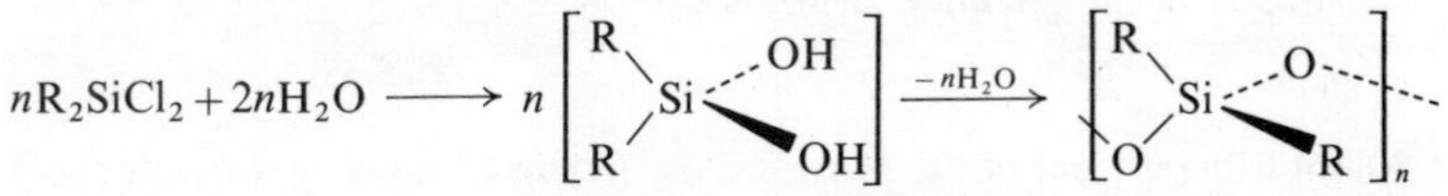

$$nR_2SiCl_2 + 2nH_2O \longrightarrow n\left[R_2Si(OH)_2\right] \xrightarrow{-nH_2O} \left[R_2SiO\right]_n$$

(The analogous carbon diol $R_2C(OH)_2$ is also unstable and loses a molecule of water, but in this case the product is a ketone, $R_2C=O$, in which the carbon and oxygen atoms are linked by a $\sigma + \pi$ double bond.)

To limit the length of the polymer to some desired value of n, the above hydrolysis is carried out in the presence of R_3SiCl, which stops polymer growth by 'sealing off' the end of the chain with an $—SiR_3$ group

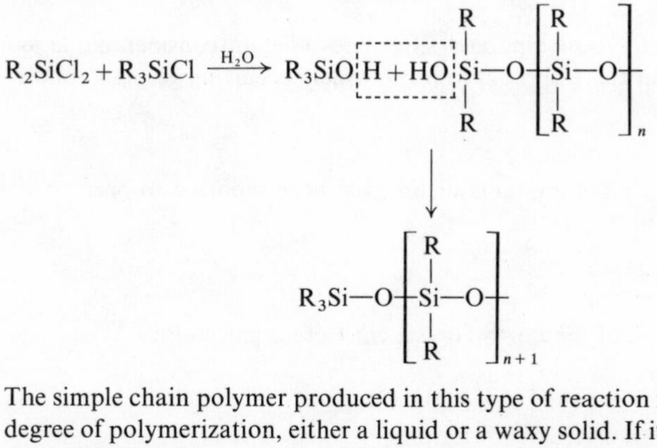

$$R_2SiCl_2 + R_3SiCl \xrightarrow{H_2O} R_3SiO\,H + HO\,SiR_2—O\left[SiR_2—O\right]_n$$

$$\downarrow$$

$$R_3Si—O\left[SiR_2—O\right]_{n+1}$$

The simple chain polymer produced in this type of reaction is, depending on the degree of polymerization, either a liquid or a waxy solid. If it is desired to have a rigid, solid polymer this can be achieved by 'cross-linking' the chains of silicon and oxygen atoms by adding a controlled quantity of $RSiCl_3$ to the hydrolysis reaction mixture

$$\mathrm{RSi}\begin{matrix}\diagup\mathrm{OH}\\ -\mathrm{OH}\\ \diagdown\mathrm{OH}\end{matrix} + \begin{matrix}\mathrm{HO}\!\left[\begin{matrix}\mathrm{R}\\ -\mathrm{Si}-\mathrm{O}-\\ \mathrm{R}\end{matrix}\right]_n\\ \mathrm{HO}\!\left[\begin{matrix}\mathrm{R}\\ -\mathrm{Si}-\mathrm{O}-\\ \mathrm{R}\end{matrix}\right]_m\\ \mathrm{HO}\!\left[\begin{matrix}\mathrm{R}\\ -\mathrm{Si}-\mathrm{O}-\\ \mathrm{R}\end{matrix}\right]_o\end{matrix} \longrightarrow \mathrm{RSi}\begin{matrix}\diagup\mathrm{O}\!\left[\begin{matrix}\mathrm{R}\\ -\mathrm{Si}-\mathrm{O}-\\ \mathrm{R}\end{matrix}\right]_n\\ -\mathrm{O}\!\left[\begin{matrix}\mathrm{R}\\ -\mathrm{Si}-\mathrm{O}-\\ \mathrm{R}\end{matrix}\right]_m\\ \diagdown\mathrm{O}\!\left[\begin{matrix}\mathrm{R}\\ -\mathrm{Si}-\mathrm{O}-\\ \mathrm{R}\end{matrix}\right]_o\end{matrix}$$

The starting materials for these polymers are normally produced by the 'direct-synthesis' technique, in which silicon is heated with an alkyl chloride,

e.g. $Si + CH_3Cl \xrightarrow{Cu} CH_3SiCl_3 + (CH_3)_2SiCl_2 + (CH_3)_3SiCl$,

and the mixture of alkylsilicon chlorides separated by fractional distillation.

7.2 Occurrence of the group IV elements

7.2.1 *Carbon*

About 0·08 per cent of the Earth's crust; crude oil and coal provide the highest natural concentrations of carbon.

7.2.2 *Silicon*

27·6 per cent; the second most common element after oxygen in the Earth's crust; occurs widely in silicates and as silica SiO_2 in quartz and sandstone.

7.2.3 *Germanium*

About 7 p.p.m.; some zinc and silver ores contain considerable amounts of germanium, but there appears to be no widely occurring germanium ore.

7.2.4 *Tin*

0·004 per cent; small amounts of tin have been reported to occur native; the oxide SnO_2 is mined as 'tinstone'.

7.2.5 *Lead*

About 16 p.p.m. of the Earth's crust; chief ore is galena PbS.

7.3 Extraction

7.3.1 *Carbon*

Artificial graphite is made by heating either powdered anthracite or coke with sand in an electric furnace for about twenty-four hours. This is the thermodynamically stable form of carbon at room temperature, but the more dense diamond can be made directly from graphite by the application of 125 000 atmospheres pressure at about 3000 K; in the presence of a transition metal as catalyst, the transformation can be accomplished at 70 000 atmospheres and 2000 K (see Figure 66).

7.3.2 *Silicon, germanium and tin*

Reduction of the dioxides with carbon at high temperatures gives the free elements; silicon and germanium may be made ultra-pure for use in semiconductors by zone refining (see Figure 76, section 7.12).

7.3.3 *Lead*

Galena PbS is roasted in air to give the oxide, which is then reduced by carbon monoxide in a blast furnace.

7.4 **The elements**

There is an increase in metallic character of the elements in going down the group: carbon is a typical non-metal; silicon and germanium are 'metalloidal', having properties between the true metals and non-metals; tin and lead are essentially metallic. The behaviour is reflected in the structures adopted by the elements. Carbon, silicon and germanium adopt diamond-like structures, the crystals of which are actually giant molecules with each atom tetrahedrally surrounded by four neighbours (Figure 65). Tin exists in three forms, the low-temperature α-form having the non-metallic diamond structure, whereas the

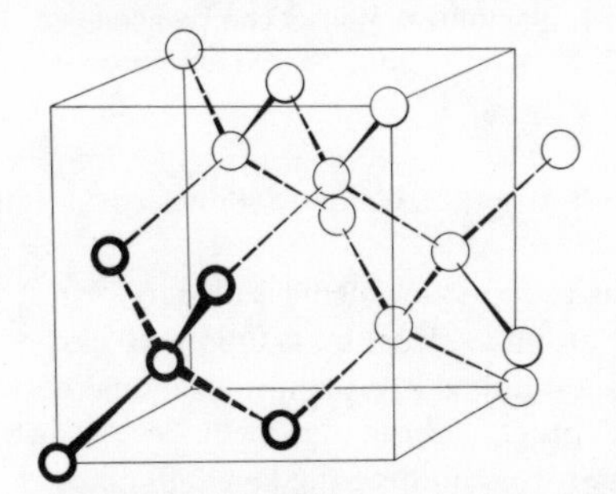

Figure 65 The structure of diamond. Each atom is tetrahedrally surrounded by four neighbours. The structure is adopted by carbon, silicon and germanium

two high-temperature modifications each have a typically metallic, close-packed structure. The transition temperatures for the tin allotropes are

$$\underset{\text{(grey tin)}}{\alpha\text{-Sn}} \underset{}{\overset{13{\cdot}2\ ^\circ\text{C}}{\rightleftharpoons}} \underset{\text{(white tin)}}{\beta\text{-Sn}} \overset{161\ ^\circ\text{C}}{\rightleftharpoons} \underset{\text{(brittle tin)}}{\gamma\text{-Sn}}.$$

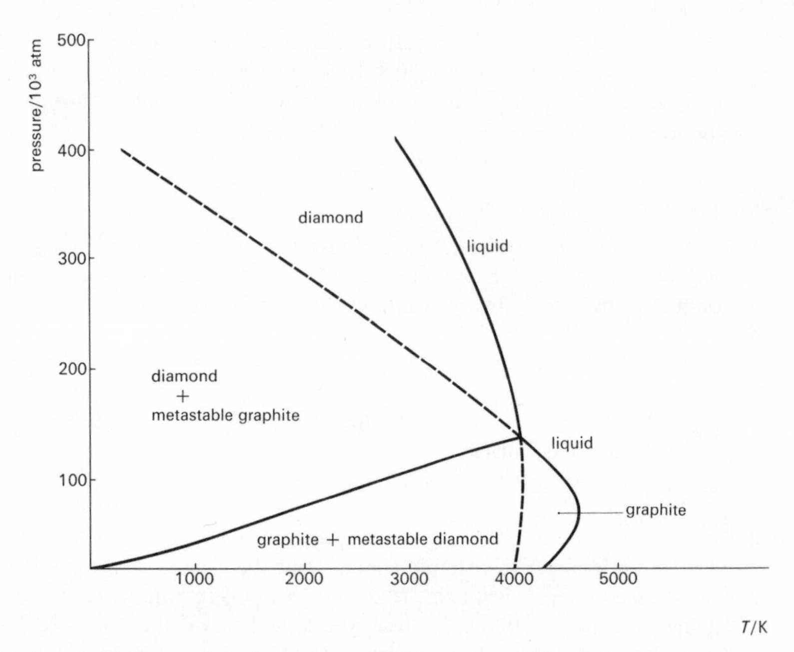

Figure 66 Part of the phase diagram of carbon. (Reproduced from F. P. Bundy, *Journal of Chemical Physics*, vol. 38, 1963, p. 631, with permission.)
Although the conversion graphite → diamond is favoured by high pressure and low temperature (400 000 atm would be required at room temperature), the rate of conversion is slow below about 3000 K. However, in the presence of a transition-metal catalyst (e.g. iron, cobalt, tantalum), diamond synthesis can be achieved using the relatively low temperature of 2000 K and about 70 000 atm pressure

Grey tin, having the more open diamond structure, has a lower density than either of the two metallic allotropes.

As already stated, only carbon of the group IV elements is able to form strong p_π–p_π bonds with itself and hence the graphite structure is unique to carbon. In this allotropic form the carbon atoms are linked in a planar hexagonal network and the separate layers stacked on each other at about the van der Waals distance (Figure 67). Hence there are no strong forces holding the layers in place, which gives rise to the lubricating properties of graphite; recent work indicates that adsorbed gases play some part in the lubrication mechanism as shown by the fact that graphite does not act as a lubricant under high vacuum conditions. Within the layers each sp^2 hybridized carbon atom forms σ-bonds with three neighbouring atoms in the same plane (see Figure 68). Every atom has a half-filled 2p orbital above and below this plane of atoms, which will interact to give a large number of molecular orbitals extending over the whole layer. The bond-

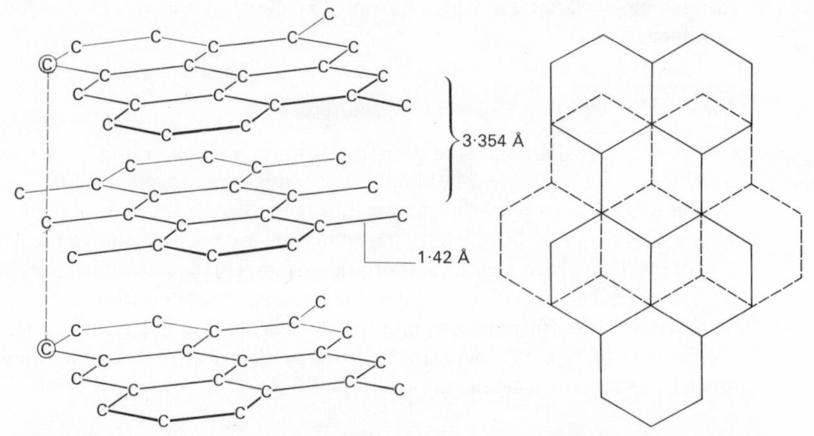

Figure 67 The structure of crystalline graphite. It is only alternate layers in graphite which are 'in phase'. Such an ababab . . . arrangement of the layers reduces the steric interaction between the carbon atoms in adjacent layers to a minimum

ing π molecular orbitals are full, the antibonding molecular orbitals are empty, but, because their energy separation is slight, excitation of electrons from the bonding to the antibonding orbitals occurs relatively easily (e.g. by absorption of light quanta or on the application of an electric field).

The high melting points of carbon (3930 °C), silicon (1420 °C) and germanium (959 °C) are due to the strong interatomic (covalent) forces which have to be broken in order to break up the lattice; the lowering of the melting points from carbon to germanium is due mainly to the decreasing strength of the M—M bonds down the group. The low melting points of tin (232 °C) and lead (327·5 °C) certainly indicate that these metals do not use all their outer electrons in binding the metallic lattice together, a property in keeping with the paucity of M^{4+} salts of tin and lead. (Compare the above figures to the higher melting points of

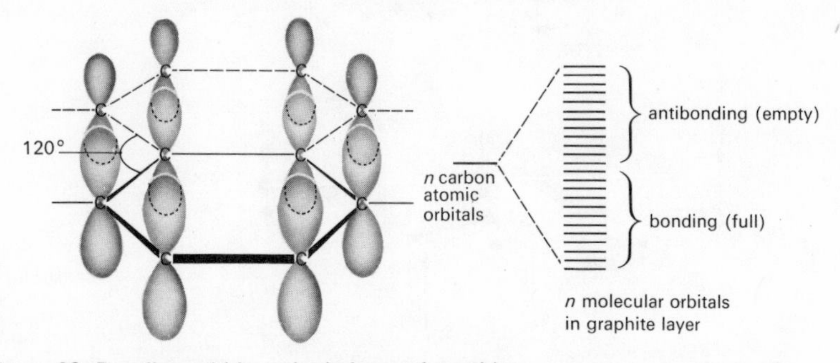

Figure 68 Bonding within a single layer of graphite

the alkaline-earth metals, which use only two electrons per atom in binding their metallic lattices.)

7.4.1 *Lamellar, or intercalation, compounds of graphite*

Compared with diamond, the graphitic form of carbon is quite reactive. The weak van der Waals forces holding the graphite layers together allow the penetration of a wide variety of reactants into the lattice giving, broadly, two types of product: (a) those in which the carbon layers are buckled, due to the destruction of the delocalized π molecular orbital system; (b) those in which the graphite layers retain their π-electron system, but become separated from each other by as much as 10 Å (in pure graphite the inter-layer distance is only 3·35 Å). In compounds of type (a) the graphite loses its electrical conductivity and often much of its colour, whereas those of type (b) retain the electrical conductivity.

Type (a). At 400 °C, pure graphite undergoes a quiet reaction with fluorine to give a product containing approximately equal amounts of carbon and fluorine $(CF_x)_n$; when x approaches 1·0 the colour is white. An X-ray examination of the product shows that the layers of carbon atoms have swelled apart to a distance of 6·6 Å. Figure 69 illustrates the structure commonly accepted for $(CF)_n$, the fluorines being covalently bound to the carbon atoms. The latter are now sp^3 hybridized and adopt the cyclohexane chair conformation; because the delocalized π molecular orbitals have been destroyed, the colour and electrical conductivity of the original graphite largely disappear.

Very strong oxidizing agents, such as mixtures of nitric acid and potassium chlorate, cause graphite to swell enormously during which the colour changes to brown. Analysis of the dried product indicates a carbon:oxygen ratio of

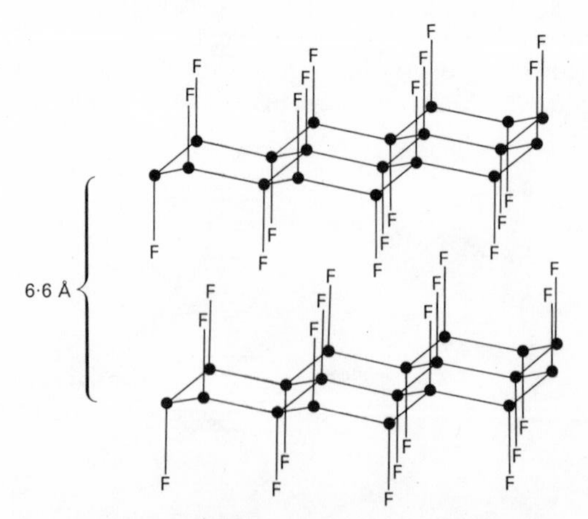

Figure 69 Suggested structure of 'graphite fluoride', showing the cyclohexane chair type of configuration adopted by the carbon atoms

about 2:1, but a small amount of combined hydrogen is always retained; the chemical properties of 'graphite oxide' suggest that at least three types of carbon–oxygen linkage are present:

(i) ether type (7.13), thought to bridge mainly the 1,3 positions as in (7.14);

C–O–C

(7.13)

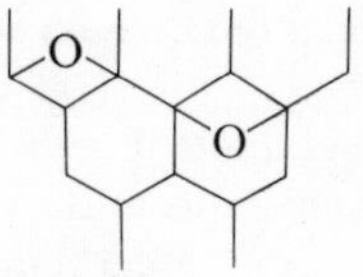

(7.14)

(ii) tertiary alcohols (7.15), can be methylated and esterified;

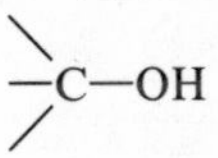

–C–OH

(7.15)

C=O

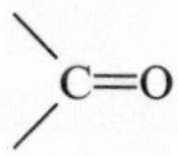

(7.16)

(iii) carbonyl groups (7.16), it has been suggested that keto–enol tautomerism is possible.

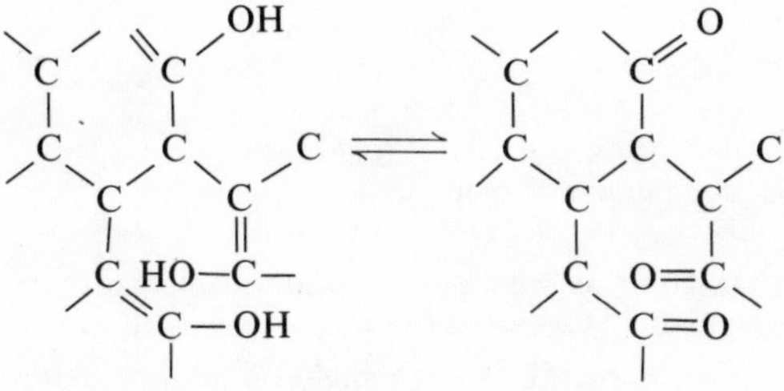

enol form ⇌ keto form

Type (*b*). Graphite readily absorbs the vapours of the heavy alkali metals to give highly coloured solids in which the original graphite structure is largely undisturbed except that the layers are farther apart to allow access of the metal atoms; on strong heating the alkali metal volatilizes unchanged out of the graphite lattice,

$$\text{e.g.}\quad C+K \rightarrow C_{60}K \xrightarrow{K} \underset{\text{deep blue}}{C_{48}K} \xrightarrow{K} C_{36}K \xrightarrow{K} \underset{\text{steel blue}}{C_{24}K} \xrightarrow{K} \underset{\text{bronze}}{C_8K}.$$

In the C_8M compounds there are alternate layers of carbon and metal atoms, the latter being arranged so that they lie above and below centre of two C_6 rings in the neighbouring planes (7.17). The stoichiometry of the several compounds arises from the number of carbon planes between each metal layer and the arrangement of the metal atoms within their layers as shown in Figure 70.

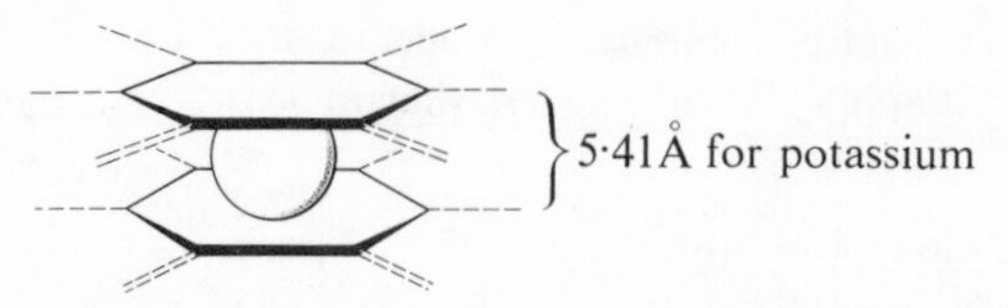

(7.17) each metal atom is equidistant from twelve carbon atoms

These metal–graphite intercalation compounds conduct electricity more readily than graphite due to the transference of electrons from the metal atoms into the π-orbitals on the carbon layers, where they are relatively 'free' to move on the application of an electric field.

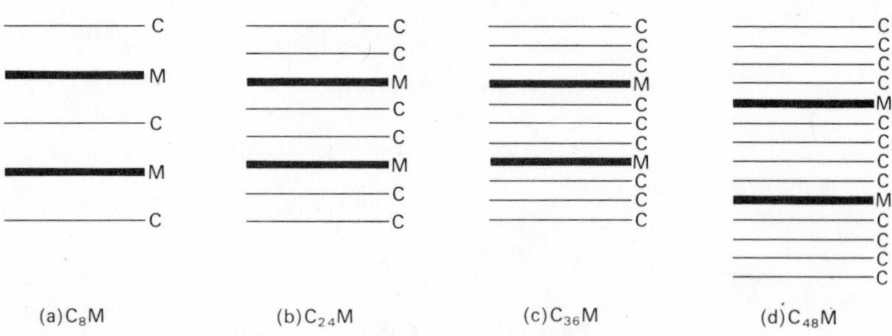

Figure 70 Type (b) lamellar compounds of graphite

The halogens, other than fluorine, react with graphite in a rather similar fashion to the alkali metals. It is also possible for a wide variety of metal halides (especially those of the transition metals in their higher oxidation states) to be absorbed into the host graphite. Only the compound formed between ferric chloride and graphite, which contains up to 56 per cent ferric chloride, has been studied structurally; it has alternate layers of carbon and ferric chloride, the latter layers consisting of Fe^{3+} and Cl^- ions in positions rather similar to those found in pure ferric chloride.

7.5 Hydrides of the group IV elements

The paraffin hydrocarbons belong to the realm of organic chemistry and will not be discussed here in any detail, although it should be remembered that they can be in the form of either straight chains or branched chains. This gives rise to the possibility of isomers, as shown for pentane C_5H_{12} (7.18–20).

The hydrides of silicon and germanium are similar to those of carbon, and hence the products arising from the acid hydrolysis of silicides and germanides (see p. 51) contain, not only mixtures of the hydrides having different numbers of silicon or germanium atoms, but also mixtures of isomers for those hydrides containing more than three group IV atoms; such mixtures have to be separated by

$$\underset{\text{H}_3\text{C}}{}\!\!\diagup^{\text{CH}_2}\!\!\diagdown_{\text{CH}_2}\!\!\diagup^{\text{CH}_2}\!\!\diagdown_{\text{CH}_3}$$

(7.18) n-pentane

$$(H_3C)_2CH-CH_2-CH_3$$

(7.19) iso-pentane (2-methylbutane)

$$C(CH_3)_4$$

(7.20) neo-pentane (2,2-dimethylpropane)

vapour-phase chromatography and the identity of each component checked by analysis and proton nuclear magnetic resonance spectroscopy. An intimate mixture of a silicide with a germanide on hydrolysis gives small amounts of hydrides which contain silicon and germanium atoms linked together (e.g. silylgermane $H_3Si—GeH_3$) in addition to silanes and germanes.

Tin forms only SnH_4 and Sn_2H_6, the latter decomposing very rapidly at room temperature. Plumbane PbH_4 is so unstable thermally that it can only be identified using a mass spectrometer.

7.6 Oxides of carbon

As the oxides of carbon differ markedly from those of silicon, germanium, tin and lead, they will be discussed separately. Unlike the other elements of group IV, carbon has three main oxides: the monoxide, the dioxide and the suboxide, all of which contain carbon–oxygen multiple bonds.

7.6.1 *Carbon monoxide* CO

From its mode of preparation via the dehydration of formic acid using concentrated sulphuric acid

$$HC(=O)OH \xrightarrow{-H_2O} CO$$

carbon monoxide could be considered as formic anhydride. However, it is very insoluble in water and only reacts with alkalis on heating

$$NaOH + CO \longrightarrow NaOOCH \quad \text{(sodium formate).}$$

The carbon monoxide molecule is iso-electronic to the nitrogen molecule and both have many physical properties in common; for example, they are colourless gases with very low melting points (nitrogen, -210 °C; carbon monoxide, -205 °C) and boiling points (nitrogen, -196 °C; carbon monoxide, -190 °C).

The carbon atom in carbon monoxide may be assumed to be sp hybridized:

one sp orbital containing a lone pair of electrons, the other making a σ-bond with an oxygen p-orbital; the remaining p-orbitals, two each on carbon and oxygen, overlap to form two π-bonds at right angles to each other. Therefore the electronic structure may be summarized as in (7.21). Oxygen has a higher electro-

$$\bigcirc C{\equiv}O$$

(7.21) carbon–oxygen distance is 1·13 Å

negativity than carbon and would therefore be expected to attract the greater share of the bonding electrons to itself, resulting in a substantial electric dipole moment for the carbon monoxide molecule

$$C^{\delta+}{\equiv}O^{\delta-}$$
$$\longmapsto$$

However, it is found that the dipole moment of carbon monoxide is very small (0·11 debye) due to the effect of the lone pair of electrons in the carbon sp hybrid orbital, which works in the opposite direction. The net effect is that these two 'internal' dipole moments almost cancel.

$$\bigcirc C{\equiv}O$$
$$\longleftarrow\!\!+$$

The sp lone pair of electrons confers on carbon monoxide the ability to act as a donor molecule. For example, diborane reacts with carbon monoxide under pressure to give borane–carbonyl

$$B_2H_6 + 2CO \longrightarrow 2H_3B \leftarrow CO.$$

Much more stable complexes are formed between carbon monoxide and the transition metals,

e.g. $Ni + 4CO \xrightarrow{50\,^{\circ}C} Ni(CO)_4$ (colourless, volatile liquid).

More generally, the finely divided transition metal is generated *in situ* under a high pressure of carbon monoxide and the mixture heated in an autoclave,

e.g. $MCl_x + xNa \xrightarrow[200\ \text{atm CO}]{\text{tetrahydrofuran}} M(CO)_y + xNaCl.$

The oxidation state of the transition metal in these metal carbonyls is zero, and in all but a very few cases the metal surrounds itself by sufficient carbon monoxide groups (each donating two electrons) to attain the same number of electrons as the next inert gas and hence fill up all the available low-energy orbitals:

Ni^0 $1s^2 2s^2 2p^6 3s^2 3p^6 3d^8 4s^2$.

Nickel(0) requires eight electrons to attain the same number of electrons as the next inert gas, krypton; this is the number supplied by four carbon monoxide ligands. The nickel atom uses sp^3 atomic orbitals to accommodate the electrons from carbon monoxide, hence the molecule is tetrahedral (7.22).

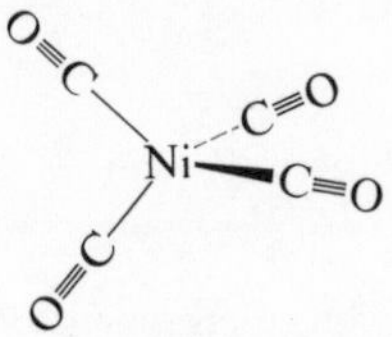

(7.22) nickel tetracarbonyl

In certain other metal carbonyls the carbon monoxide is also able to form 'bridges' between metal atoms; in such cases the carbon atom is better regarded as being sp^2 hybridized, two of the sp^2 orbitals (each containing one electron) are used in the formation of M—C σ-bonds (7.23).

Fe^0 $1s^2 2s^2 2p^6 3s^2 3p^6 3d^6 4s^2$.

Each iron(0) requires ten electrons to attain the electron count of krypton; this is achieved in $Fe_2(CO)_9$ as follows: six electrons from the three terminal carbon monoxide molecules, three electrons from the three bridge carbon monoxide molecules, one electron from the iron–iron bond.

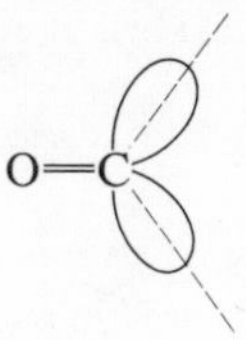

(7.23a)

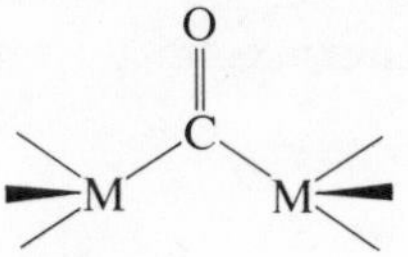

(7.23b)

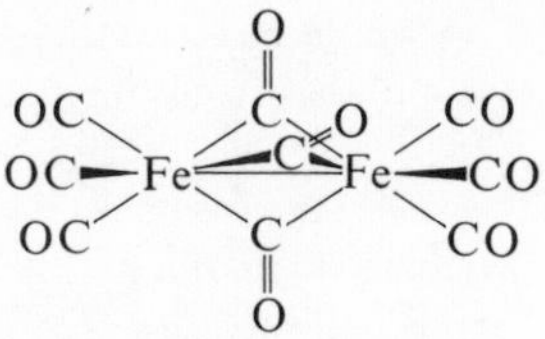

(7.24) di-iron enneacarbonyl $Fe_2(CO)_9$

The high thermal stabilities of the transition-metal carbonyl compounds compared with those formed by elements like boron are probably due to a secondary bonding effect in which filled metal d-orbitals interact with the empty antibonding π-orbitals on the carbon monoxide ligand (Figure 71). The σ-electron donation from the carbon monoxide results in a build-up of negative charge on the metal which has been shown by theoretical arguments to be a highly undesirable condition for a metallic element; the 'back-bonding' from metal to carbon monoxide tends to relieve this charge build-up on the metal.

It has been stated that the whole of the modern metallurgical industry depends entirely upon the existence and thermodynamic stability of carbon monoxide. Most metals are extracted from their oxides which either occur naturally or can be obtained by roasting an ore containing, say, the sulphide or carbonate. Hence for a reduction process based on carbon (and essentially all elements can be isolated by such an oxide reduction providing the temperature is sufficiently

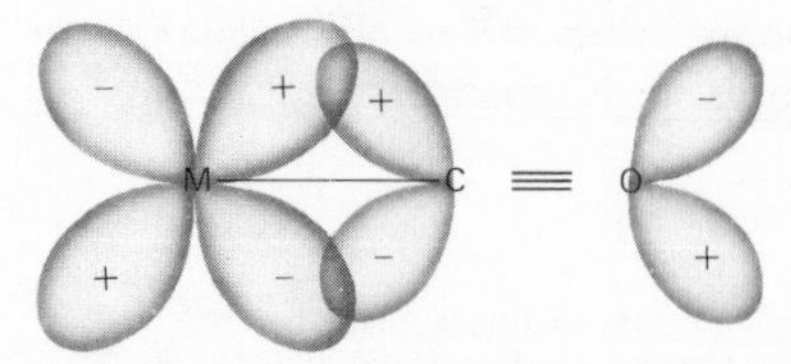

Figure 71 Back-bonding in the transition-metal carbonyls.
Filled metal d-orbitals of the correct symmetry to interact with empty antibonding π-orbitals on the carbon monoxide ligands. The placing of electron density in these antibonding orbitals causes a decrease in the carbon–oxygen bond order which is reflected by an increase in the carbon–oxygen distance as compared with the free carbon monoxide molecule

high) a knowledge is required of the thermodynamic functions relating to processes such as

$$2M+O_2 \longrightarrow 2MO,$$
$$2C+O_2 \longrightarrow 2CO,$$
$$C+O_2 \longrightarrow CO_2.$$

The approximately linear plots of ΔG° (the Gibbs free energy change) against temperature for these reactions are perhaps the most useful thermodynamic quantities in predicting possible reduction processes. The slope of such a plot would be equal to $-\Delta S$, the change in entropy for the reaction. This, of course, follows from the relation

$$\Delta G^\circ = \Delta H^\circ - T\Delta S^\circ,$$

the slope of a graph of ΔG° against T being $\partial(\Delta G^\circ)/\partial T$, that is, $-\Delta S$.

All such $\Delta G^\circ/T$ plots for the formation of element oxides (Figure 72) rise with increasing temperature except that representing the formation of carbon monoxide from carbon, which has a negative slope; this *positive* entropy change is due to the net generation of one mole of gas during the oxidation of one mole of carbon to carbon monoxide. At the temperature where the carbon monoxide plot intersects another oxide plot, the equilibrium constant for the reaction

$$MO+C \rightleftharpoons CO+M$$

is unity and above that temperature the reduction of MO will be increasingly in thermodynamic (and, equally important, kinetic) favour of metal formation. A point to note is that the $\Delta G^\circ/T$ plot for carbon dioxide follows the same pattern as the oxides of the other elements, emphasizing that the versatility of carbon as a reducing agent depends solely on the existence of the lower oxide, carbon monoxide.

Carbon monoxide is a reasonably reactive compound and, for example, combines readily with the halogens (except iodine), cyanogen, oxygen and sulphur,

$$CO+S \longrightarrow COS,$$
$$CO+X_2 \longrightarrow COX_2 \qquad (X = F, Cl, Br).$$

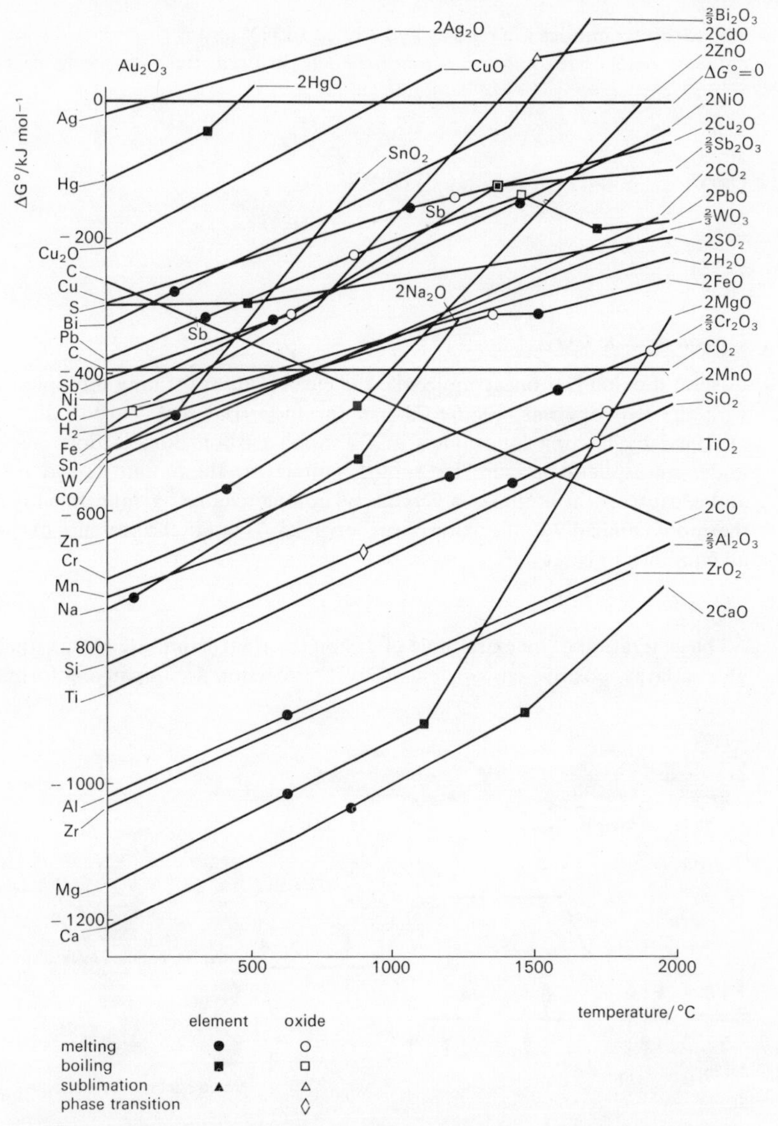

Figure 72 Plot of ΔG° against temperature for a variety of oxides (Reproduced from D. J. G. Ives, *Principles of the Extraction of Metals*, Royal Institute of Chemistry, 1960, p. 21, with permission).

The reaction $C + O_2 \rightarrow CO_2$ involves no change in the number of molecules of gaseous products or reactants, so there is only a small entropy change ($\Delta S^\circ = 2{\cdot}93\ \text{J K}^{-1}\ \text{mol}^{-1}$) and hence ΔG° is essentially independent of temperature as shown by the practically horizontal $\Delta G^\circ/T$ graph. On the other hand, the reaction $2C + O_2 \rightarrow 2CO$ has $\Delta S^\circ = 179{\cdot}5\ \text{J K}^{-1}\ \text{mol}^{-1}$, because one mole of gaseous reactant produces two moles of gaseous product; hence the stability of carbon monoxide increases with temperature (downward slope of $\Delta G^\circ/T$ graph)

The carbonyl halides are readily hydrolysed to HX and carbon dioxide, and the chloride reacts with gaseous ammonia to form urea, the diamide of carbonic acid (7.25).

$$O{=}C(Cl)_2 + 2NH_3 \longrightarrow 2HCl + O{=}C(NH_2)_2$$

(7.25)

7.6.2 *Carbon dioxide* CO_2

Carbon dioxide is a linear molecule, the carbon atom forming multiple bonds with the two oxygens, Figure 73. On the industrial scale carbon dioxide is obtained by heating limestone $CaCO_3$. Solid carbon dioxide does not melt under atmospheric pressure, but sublimes directly to the vapour state at -78 °C and because of this it makes a very useful commercial refrigerant. It is the most thermodynamically stable oxide of carbon at 25 °C but in the presence of carbon an equilibrium is set up,

$$CO_2 + C \rightleftharpoons 2CO.$$

The generation of an extra mole of gas on the right-hand side of this equation gives a large, positive entropy change to the reaction accompanying formation

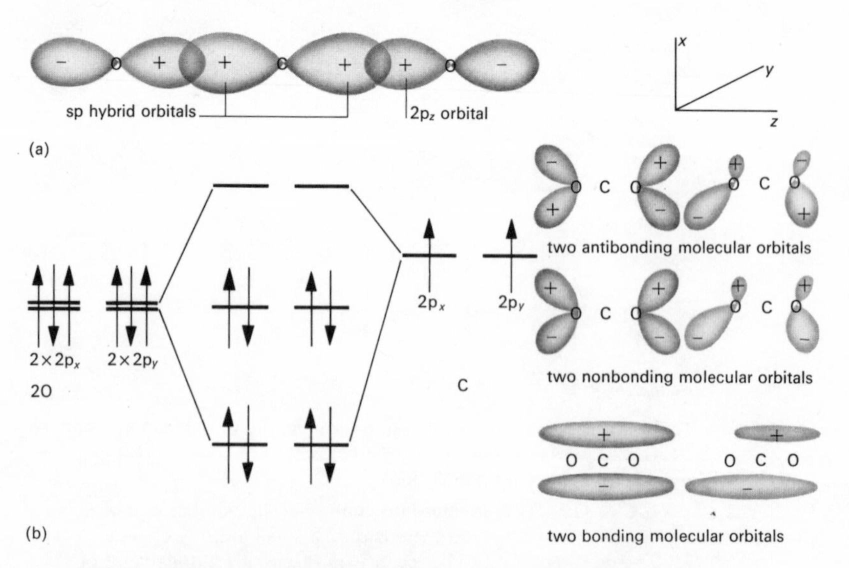

Figure 73 Bonding in the carbon dioxide molecule. (a) σ-Bonding interaction. (b) π-Bonding interaction and π energy-level diagram

of carbon monoxide (ΔS at 25 °C is +42·9 entropy units), thus ensuring that the formation of carbon monoxide is progressively favoured by increasing temperature. The reaction can be a nuisance in certain cases: the use of carbon dioxide gas as a heat exchanger in nuclear reactors constructed of graphite is hampered by the take-up of carbon in the hot zones and the deposition of carbon in the colder pipelines when the carbon monoxide disproportionates back to carbon and carbon dioxide.

Although carbon dioxide is the anhydride of carbonic acid, its aqueous solution contains few bicarbonate or carbonate ions, the bulk of the carbon dioxide remaining essentially unchanged

$$CO_2(g) \rightleftharpoons CO_2(aq) \overset{H_2O}{\rightleftharpoons} H_2CO_3 \rightleftharpoons H^+(aq)+HCO_3^-$$
$$\rightleftharpoons H^+(aq)+CO_3^{2-}$$

When finely divided sodium hydrogen carbonate is suspended in ether at −30 °C and treated with dry hydrogen chloride, no evolution of carbon dioxide occurs and on cooling the solution to −78 °C, snow-white crystals of $H_2CO_3.OEt_2$ are precipitated; this 'etherate of carbonic acid' is unstable and rapidly decomposes to carbon dioxide and water at −10 °C.

The carbonate ion CO_3^{2-} is planar, the carbon atom being sp^2 hybridized; the remaining carbon p-orbital interacts with similar orbitals on the oxygen atoms to give a π molecular orbital extending over the whole ion (Figure 74) so that the carbon–oxygen bond length is shorter than that expected for a carbon–oxygen single bond.

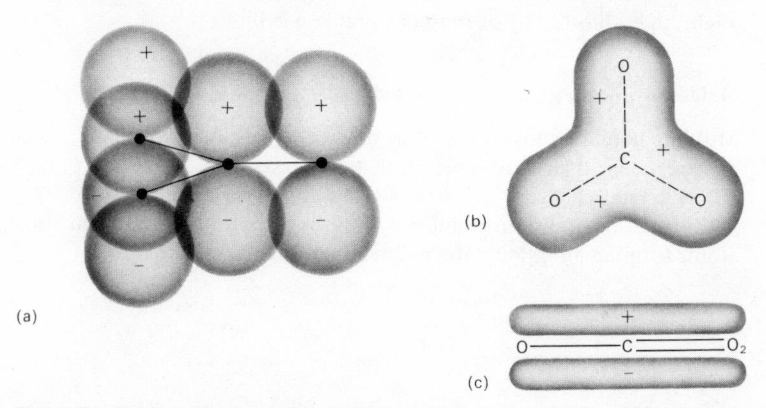

Figure 74 The p-orbitals used to form the bonding π molecular orbital in the carbonate ion. The carbon–oxygen bond length is 1·29 Å (the theoretical single-bond length is about 1·43 Å). (b) Plan and (c) side elevation of the bonding π molecular orbital in the carbonate ion

In the solid alkali-metal hydrogen carbonates (bicarbonates) the planar CO_3^{2-} ions are linked together into polymeric chains by hydrogen bonding involving the 'acid' hydrogen atoms (7.26).

(7.26)

7.6.3 *Carbon suboxide* C_3O_2

This gas, boiling point 6 °C, is prepared by dehydrating malonic acid or its esters with phosphorus pentoxide at 300 °C. In its reactions with water, ammonia and hydrogen chloride it behaves as the anhydride of malonic acid by giving the acid, the amide and the chloride respectively.

$$CH_2(COOH)_2 \xrightarrow{P_2O_5} C_3O_2 \begin{cases} \xrightarrow{H_2O} CH_2(COOH)_2 \\ \xrightarrow{NH_3} CH_2(CONH_2)_2 \\ \xrightarrow{HCl} CH_2(COCl)_2 \end{cases}$$

Electron diffraction and spectroscopic techniques show that C_3O_2 is a linear molecule. The carbon atoms are presumably sp hybridized, using the orbitals to form σ-bonds to neighbouring carbon atoms or oxygen atoms; two π molecular orbitals delocalized over the whole molecule can then be formed at right angles to each other by overlap of suitable p-orbitals.

7.7 Oxides of silicon, germanium, tin and lead

Multiple bonding of the p_π–p_π type does not occur between these elements and oxygen, hence their oxides bear little or no resemblance to those of carbon.

The crystal chemistry of silica SiO_2 is complex, but in all forms the silicon atom is tetrahedrally surrounded by four oxygen atoms, the silicon and oxygen atoms forming an infinite, three-dimensional network (7.27).

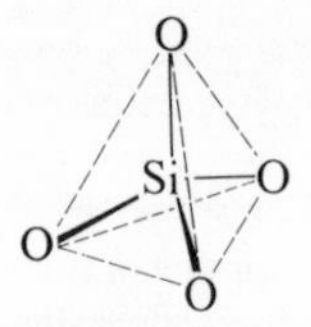

(7.27) the tetrahedral SiO_4 unit which is the basic 'building brick' in the various crystalline forms of silica

The stable form of silica at normal temperatures is α-quartz, the other crystal modifications having the following transition temperatures:

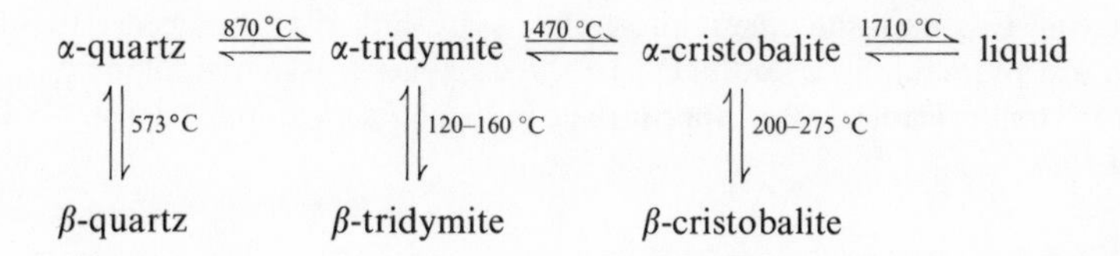

The structures of the quartz, tridymite and cristobalite forms of silica differ in the way in which the SiO_4 tetrahedra are linked together; in all three forms each oxygen atom is shared between two tetrahedra (i.e. the crystals have 4:2 co-ordination) so that the overall composition is SiO_2. The interconversion of these three crystalline types of silica requires the rupture of some silicon–oxygen bonds and the re-formation of others, so that the transitions occur only very slowly. Hence α-tridymite and α-cristobalite are found in nature as metastable crystals. The difference between the α and β forms of quartz, tridymite and cristobalite is slight and only involves the rotation of the SiO_4 tetrahedra relative to each other; this is reflected in the rapidity of the α–β transitions.

The crystallization of molten silica is very difficult to achieve and even on slow cooling a vitreous, non-crystalline solid is normally formed. This amorphous silica softens at about 1500 °C when it becomes plastic and can be worked and blown in either an oxy-hydrogen or oxy-acetylene flame to give transparent apparatus which is highly insensitive to thermal shock, due to the low coefficient of thermal expansion of silica. Quite complex pieces of blown-ware may be heated to redness and plunged into cold water without harm. Silica is also transparent in the ultraviolet region of the spectrum, which makes it useful in the construction of sample cells, prisms and lenses for use with ultraviolet light.

The dioxides of germanium, tin and lead normally adopt the rutile structure (6:3 coordination) although germanium dioxide transforms to the α-quartz structure above 1033 °C and also readily forms an amorphous glass like silica.

Divalent oxides of tin and lead, MO, are well known, but they do not adopt any of the expected ionic lattices. Instead they have an unusual layer structure in which each metal is bonded to four oxygens arranged in a square to one side of it. Possibly the lone pair of electrons present in the M^{II} state is stereochemically responsible for this metal coordination (7.28).

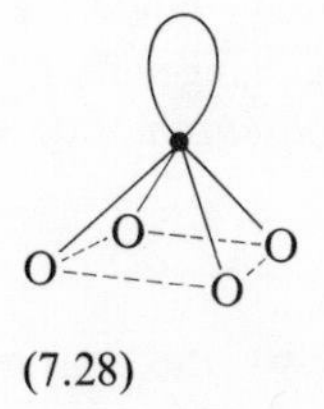

(7.28)

The oxide Pb_3O_4 contains lead in two oxidation states, $Pb^{II}_2Pb^{IV}O_4$. The lead(IV) atoms are at the centres of PbO_6 octahedra which are linked into chains by sharing two opposite edges; these chains are linked by the lead(II) atoms which are pyramidally coordinated to three oxygens. Again the lone pair of electrons on the lead(II) atom appear to be influencing the stereochemistry of the metal.

7.8 Silicates

The dioxides of silicon, germanium, tin and lead have acidic properties and give salts with a variety of metallic ions. The stannates and plumbates, which contain $M(OH)_6^{2-}$ octahedral anions, have been comparatively little studied, whereas the silicates occur widely, due mainly to both the preponderance of silicon and oxygen in nature (together they account for more than 74 per cent of the Earth's crust) and the strength of the silicon–oxygen bond. No silicon analogue of the carbonate ion is known because silicon and oxygen are unable to form strong p_π–p_π bonds with each other; all the silicon–oxygen bonds are single, and hence the simplest possible silicate ion is the tetrahedral orthosilicate SiO_4^{4-}, derived from the hypothetical orthosilicic acid $Si(OH)_4$.

In all the known silicates each silicon is tetrahedrally attached to four oxygen atoms. The wide variety of structures adopted by the silicates arise solely from the different ways in which the SiO_4 tetrahedra are linked together by sharing one, two or three oxygen atoms: this is illustrated for a few simple silicate ions in Figure 75. More complex silicates have the SiO_4 tetrahedra linked into sheets, or sometimes even into three-dimensional polymeric networks.

Electrical neutrality in the silicates is attained by the presence of a suitable number of cations, which may be monovalent, divalent or trivalent or mixtures of differently charged ions. In many silicates some of the silicon is replaced by aluminium (also surrounded tetrahedrally by four oxygen atoms), which means a reduction both in the negative charge carried by the silicate ions and in the number of accompanying cations, but which involves little change in the basic silicate structure. These aluminosilicates are found in micas, clays and the zeolites. Zeolites possess an open, three-dimensional structure and are able to absorb a number of small gaseous molecules, such as water, ammonia, alcohol and even mercury, into vacant cavities within the network. The cations are held more strongly in the zeolite lattice the higher their charge; that is, $M^{3+} > M^{2+} > M^{+}$. Hence on passing a solution of calcium ions (e.g. hard water) over a sodium zeolite, an exchange of sodium ions for calcium ions occurs within the zeolite structure (i.e. giving soft water). The calcium zeolite may be changed back to the sodium form by treatment with a concentrated solution of sodium chloride, when the law of mass action for the reaction

$$CaZ + 2Na^{+} \rightleftharpoons Na_2Z + Ca^{2+}$$

ensures the exchange of calcium ions for sodium ions. This is the basis of the Permutit water softeners.

The giant polymeric silicate ions are stable only in crystals; dissolution in strong aqueous base or fusion with alkali breaks down the solids into smaller, more soluble ions.

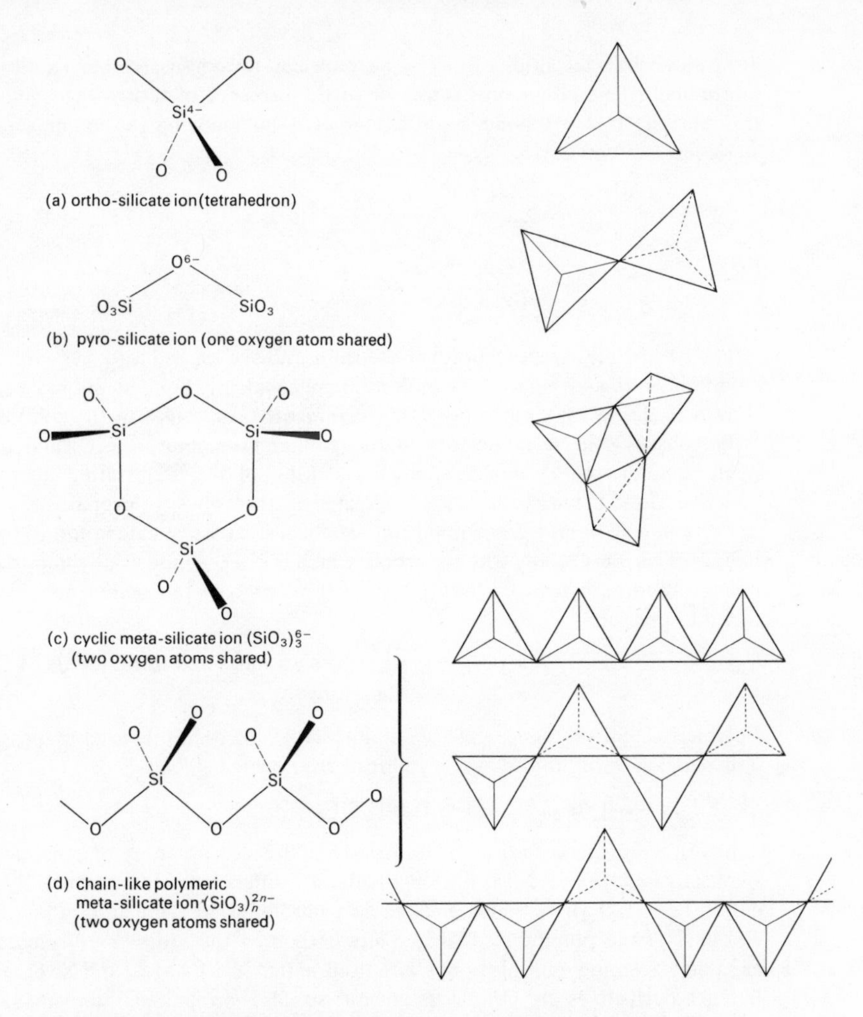

Figure 75 Some possible variations in simple silicate-ion structures using the basic building brick of the silicates, the SiO_4 tetrahedron.

7.9 Halides of the group IV elements

All the tetrahalides except lead tetraiodide are known; the strength of the M—X bond decreases with increasing atomic number of M (Table 21), so that at lead the lead–iodine bond is too weak to support the promotion of lead to lead(IV); even lead tetrachloride and tetrabromide are very unstable thermally. Although carbon–halogen bond strengths are high, the steric crowding round the carbon atom in carbon tetraiodide makes the molecule decompose rather readily to tetraiodoethylene either on heating or in the presence of ultraviolet light; in this way the I—C—I bond angle is increased, so reducing the iodine–iodine steric interaction.

$$CI_4 \ (109{\cdot}5^\circ) \longrightarrow I_2C{=}CI_2 \ (\sim 120^\circ) + I_2$$

The tetrahalides are, except for the tetrafluorides of tin and lead, very volatile compounds, suggesting that the molecules are covalently bonded and have only weak van der Waals forces operating *between* the molecules in the solid and liquid phases. The tetrafluorides of tin (sublimation point 705 °C) and lead (melting point 600 °C) are essentially ionic in the solid state, resulting in strong cohesive forces between the ions as indicated by their physical properties.

Carbon, silicon and germanium are capable of extensive catenation in their halides. This is especially true of carbon, which forms the industrially important polytetrafluoroethylene ('Fluon', 'Teflon') when tetrafluoroethylene is subjected to pressure

$C_2F_4 \longrightarrow —CF_2CF_2CF_2CF_2—$ (chain length measured in hundreds of carbon atoms).

Polymeric $—SiF_2—$ and $—SiCl_2—$ chains have also been prepared by passing either silicon tetrafluoride or tetrachloride over heated silicon

$$Si + SiF_4 \longrightarrow [SiF_2] \longrightarrow —SiF_2SiF_2SiF_2—.$$

When this polymeric fluoride is destructively distilled a variety of polyfluorosilanes Si_nF_{2n+2} (n = 1–14) are obtained; this differs from the thermal degradation of $—(C_2F_4)_n—$, which gives mainly tetrafluoroethylene and only minor amounts of the polyfluoroalkanes. Only carbon of the group IV elements is capable of forming multiple bonds with itself so that $X_2M{=}MX_2$ and $XM{\equiv}MX$ halogen derivatives are unique to carbon; so, also, are halogen derivatives of aromatic compounds containing delocalized π-bonds. A new branch of organic chemistry has grown up in recent years based on highly fluorinated aromatic molecules such as hexafluorobenzene (7.29).

$$C_6H_6 \xrightarrow{Cl_2} C_6Cl_6 \xrightarrow[\text{heat}]{KF} C_6F_6$$

hexachlorobenzene (7.29) hexafluorobenzene

The van der Waals radius of a fluorine atom bound to carbon is only slightly larger than that of hydrogen and hence virtually all the known aromatic derivatives are capable of having their fluorinated analogues; their chemistry is, however, considerably modified due to presence of the highly electronegative fluorine atoms.

The carbon atom in a CX_4 molecule has its maximum coordination number, whereas silicon, germanium, tin and lead are able to increase their coordination numbers to five or six, possibly by using their *nd* orbitals:

e.g. $GeF_4 + N(CH_3)_3 \longrightarrow F{-}Ge(F)_2(F){-}N(CH_3)_3$

trigonal-bipyramidal coordination round Ge atom

$SiF_4 + 2L \longrightarrow$ SiF_4L_2 (*trans*) or SiF_4L_2 (*cis*)

octahedral coordination round Si atom; both *cis* and *trans* complexes known

Dihalides of the elements can be made in the vapour state (see p. 146) but, on cooling, those of carbon and silicon polymerize to M^{IV} compounds. Only lead difluoride of the solid dihalides can safely be considered to be ionic, containing the Pb^{2+} ion (which is slightly larger than Sr^{2+}); it has the rutile structure, Figure 54. Hydrolysis of the dihalides occurs in aqueous solution although plumbous halides do give some Pb^{2+} ions on dissolution; halogens oxidize the dihalides to the tetrahalides in all cases except lead tetraiodide. Germanium dichloride adds gaseous hydrogen chloride to give the rather unstable germanium analogue of chloroform

$$GeCl_2 + HCl \xrightarrow{40\,^{\circ}C} Cl_3GeH.$$

The highly coloured, polymeric compounds $(SiCl)_n$ and $(GeCl)_n$ have been described, but their structure is unknown. They disproportionate on heating

$$4(MCl)_n \longrightarrow 3nM + nMX_4.$$

7.10 Carbides

Many elements form carbides M_xC_y when the free element is heated with either carbon or a hydrocarbon; in some cases the element may be made *in situ* by the carbon reduction of an oxide or some other suitable compound. The carbides of non-metals such as boron (B_4C) or silicon (SiC) contain covalently bonded lattices and are both hard and chemically inert. Silicon carbide, more commonly known as the abrasive carborundum, has a diamond-like structure in which alternate carbon atoms have been replaced by silicon. Many of the transition metals form 'interstitial' carbides in which carbon atoms occupy holes within a (usually) close-packed metallic lattice; they are high melting, extremely hard and chemically inert.

The carbides of the metals in groups I, II and III are more salt-like and contain essentially C^{4-} and C_2^{2-} anions; carbides containing the former ions (Na_4C, Be_2C, Al_4C_3) liberate methane on hydrolysis whereas those containing C_2^{2-} ions (MgC_2, CaC_2) form mainly acetylene when treated with water or dilute acids. The crystal structure of calcium carbide is very similar to that of sodium chloride, except that all the $[C\equiv C]^{2-}$ ions are lined up in parallel to each other, resulting in one crystal axis being slightly longer than the other two, Figure 51; the carbon–carbon distance of 1·20 Å is that expected for two triply bonded carbon atoms.

Magnesium carbide MgC_2 loses some carbon on heating and gives Mg_2C_3; the fact that this new carbide liberates methylacetylene $CH_3C\equiv CH$ on hydrolysis suggests the possible presence of C_3^{4-} ions in the crystal.

7.11 Organo-derivatives of silicon, germanium, tin and lead

The tetra-organo-derivatives are prepared in the usual way by treating halides with an excess of Grignard or lithium reagents

$$SiCl_4 + 4LiR \longrightarrow SiR_4,$$
$$2PbCl_2 + 4RMgX \longrightarrow [2PbR_2] \longrightarrow Pb + PbR_4.$$

When the tetrahalide is in excess the products are the alkyl, or aryl, halides,

e.g. $GeCl_4 + nLiR \longrightarrow R_nGeCl_{4-n}$,

although these halides are more easily prepared by heating stoichiometric amounts of MR_4 with the corresponding tetrahalide

$$GeBr_4 + GeR_4 \longrightarrow 2R_2GeBr_2,$$
$$GeBr_4 + 3GeR_4 \longrightarrow 4R_3GeBr.$$

The alkylsilicon chlorides required by the silicone industry are made on a large scale by 'direct synthesis' using silicon and an alkyl chloride

$$Si + CH_3Cl \xrightarrow{Cu} CH_3SiCl_3 + (CH_3)_2SiCl_2 + (CH_3)_3SiCl.$$

The most commercially important of the group IV organo-derivatives is tetraethyllead $(C_2H_5)_4Pb$, which is used widely as an 'anti-knock' in petrol. It is made on a large scale by treating ethyl chloride with a sodium–lead alloy

$$Na/Pb + C_2H_5Cl \longrightarrow NaCl + (C_2H_5)_4Pb.$$

The products are treated with water to remove excess of sodium and the tetraethyllead purified by steam distillation.

Attempts to prepare stable M^{II} organo-derivatives of these elements have so far failed; compounds having the empirical formula R_2M are always found to be cyclic,

$$\text{e.g.}\quad R_2GeCl_2 \xrightarrow{Na} \begin{array}{c} R_2Ge\text{—}GeR_2 \\ |\qquad\quad| \\ R_2Ge\text{—}GeR_2 \end{array}$$

(puckered ring)

By carrying out the above Würtz-type coupling reaction in the presence of R_3MCl derivatives it has proved possible, in the case of silicon for example, to couple up to twelve silicon atoms together in a chain

$$R_3SiCl + R_2SiCl_2 \xrightarrow{Na} R_3Si(SiR_2)_nSiR_3 \quad (n = 0\text{–}10).$$

7.12 Semiconductor properties of the group IV elements

Silicon and germanium are being increasingly used in semiconductor devices, and for this application they have to be obtained in a state of high purity. This is relatively easily accomplished using the technique of zone-refining (Figure 76).

Very pure silicon and germanium are not good conductors of electricity (Figure 77a); for example, the room temperature resistivity of germanium is $4700\ \Omega\ m^{-1}$. With increasing temperature their resistivity decreases rapidly due to electrons being thermally excited into an empty conduction band; trace impurities also cause a very marked increase in electrical conductivity and as little as 10^{-6} per cent of arsenic atoms will decrease the room temperature resistivity of germanium to $400\ \Omega\ m^{-1}$. Thus, to control the electrical properties of silicon and germanium to any useful degree, their initial purity must be controlled to a staggering degree and the element handled under 'sterile' conditions in glove-boxes. The zone-refining method is indicated in Figure 76; a rod of reasonably pure silicon or germanium has a small electrically heated ring oven placed round it at one end and a narrow strip of the rod melted; the ring oven then slowly traverses along the rod moving the molten zone along with it. The impurities (usually) concentrate in the molten phase and are gradually moved towards the

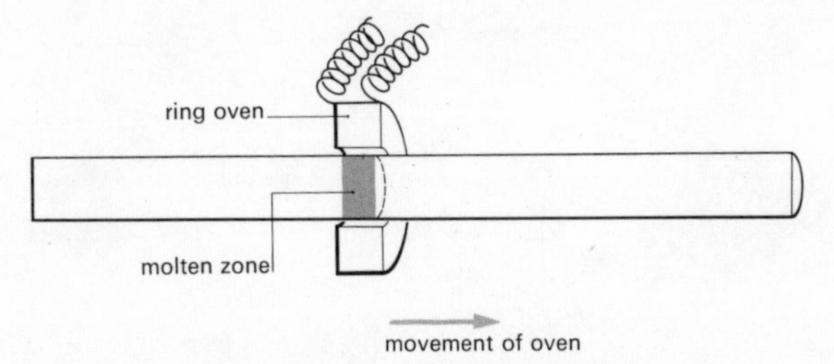

Figure 76 Zone refining; a section through the apparatus

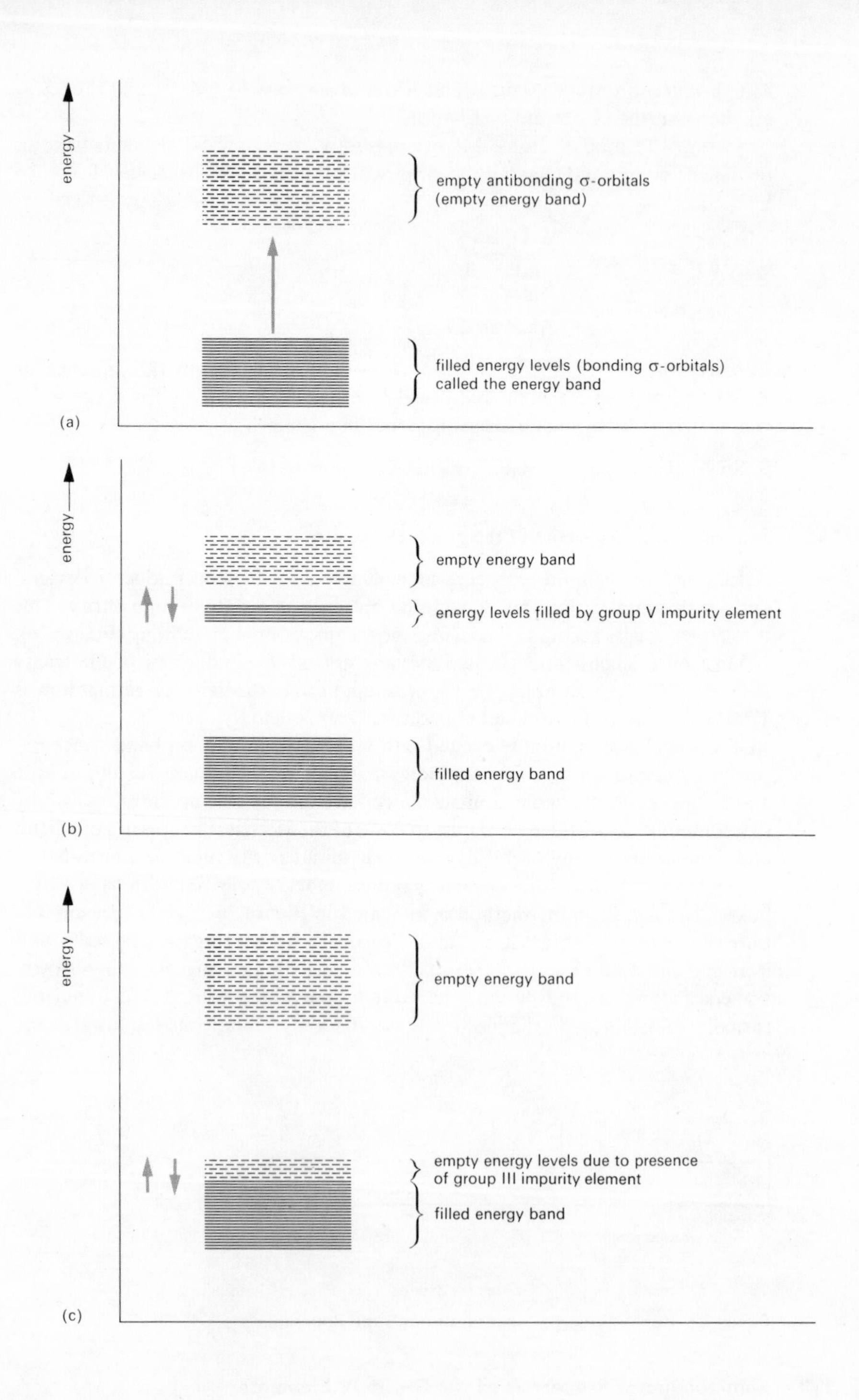

energy
empty antibonding σ-orbitals
(empty energy band)
filled energy levels (bonding σ-orbitals)
called the energy band
(a)
energy
empty energy band
energy levels filled by group V impurity element
filled energy band
(b)
energy
empty energy band
empty energy levels due to presence
of group III impurity element
filled energy band
(c)

far end of the rod when the oven completes its journey. The oven is placed at the beginning of the rod again and the process repeated several times. By using an inert atmosphere the concentration of impurity elements in germanium, for example, can be made as low as 10^{-7} parts per million each (except for oxygen and hydrogen).

The addition of a few parts per million of a group V element to the silicon or germanium lattice gives rise to a 'surplus' of electrons (one for each group V atom), which are accommodated in energy levels close to an empty conduction band of the host material, giving rise to an *n*-type semiconductor (*n* for negative electrons, which are the current carriers), see Figure 77(b). Conversely a few parts per million of a group III element (e.g. indium) results in a 'deficiency' of electrons within the host lattice and not all the available energy levels in the (normally filled) energy band are full; this gives a *p*-type semiconductor (*p* for positive holes acting as the current carriers), see Figure 77(c).

Figure 77 (a) A schematic representation of the energy levels in pure germanium. There is a relatively large gap between the completely filled energy levels and the vacant energy levels in the crystal ($\sim$0·72 eV) and few electrons are excited to the empty levels at room temperature. The lack of these mobile electrons results in a high resistance.

(b) Energy levels in germanium containing a group V element as an impurity. The addition of the group V element results in more electrons than are necessary to fill the bonding energy levels in the crystal. These extra electrons are accommodated in the normally empty, high-energy levels, resulting in a mobile electron system able to carry the electric current.

(c) Energy levels in germanium containing a group III element as an impurity. A group III element as an impurity in the germanium crystal results in too few electrons being available to fill all the normal energy levels. This gives mobility to the electrons when under the influence of an electric field. Basically, it appears that these empty levels ('holes') are causing electrical conduction, but in actual fact the holes only confer mobility upon the electrons which, of course, carry the electric current

Chapter 8
Group V. Nitrogen, Phosphorus, Arsenic, Antimony and Bismuth

8.1 Introduction

The immense amount of energy required to produce M^{5+} ions precludes their formation under normal conditions; the first, second and third ionization energies of antimony and bismuth are just low enough to allow them to form a few compounds containing M^{3+} ions. Thus, bismuth trifluoride, bismuth triiodide and, possibly, antimony trifluoride appear to be ionic solids and both elements in their M^{III} state form hydrated sulphates and nitrates typical of metallic elements; however, most compounds of antimony(III) and bismuth(III) tend to be hydrolysed very readily and their aqueous solutions contain mainly the SbO^+ or BiO^+ ions with only minor amounts of $M^{3+}(aq)$ being present. Bismuth reacts with bismuth trichloride under suitable conditions to form rather unusual polyatomic ions such as Bi_5^{3+}, Bi_8^{2+} and Bi_9^{5+}; the latter is the most well-characterized of these and has been shown by X-ray diffraction methods to contain the nine bismuth atoms arranged in the form of a tri-capped trigonal prism (Figure 78).

It is possible for the group V elements to achieve an inert-gas electron configuration by forming M^{3-} ions, but this is not energetically a very favourable process. For example, the production of N^{3-} from nitrogen atoms requires the absorption of energy equivalent to about 2130 kJ mol^{-1} and hence ionic nitrides are formed only by elements which have low ionization energies, form small ions and give nitrides having high lattice energies (e.g. lithium and magnesium).

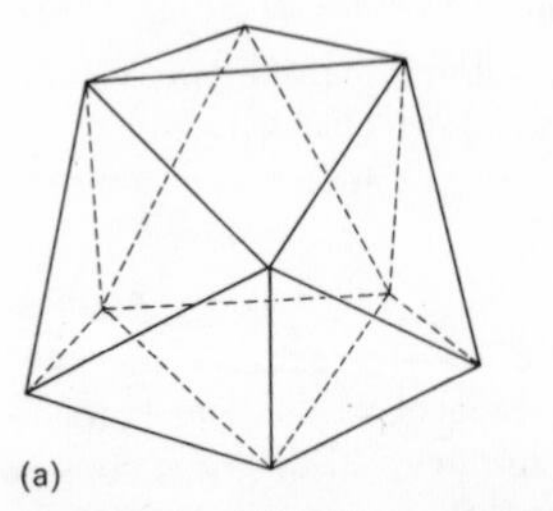
(a)

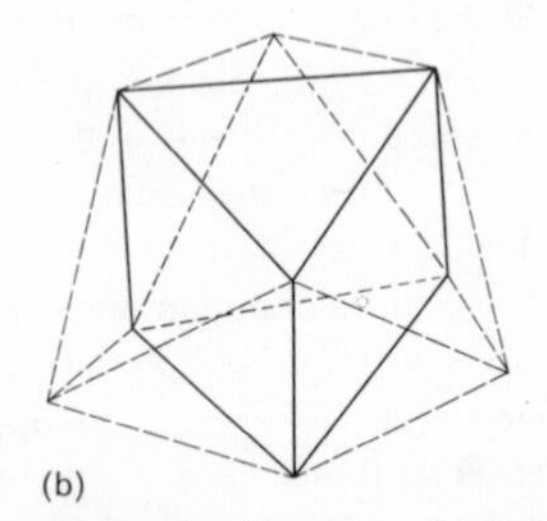
(b)

Figure 78 (a) The structure of the polyatomic Bi_9^{5+} cation. (b) A trigonal prism capped on each of its three square faces

Table 22 The Group V Elements

Element		1st IE / kJ mol^{-1}	2nd IE / kJ mol^{-1}	3rd IE / kJ mol^{-1}	Covalent radius/Å
Nitrogen N	$1s^2 2s^2 2p^3$	1402	2857	4578	0·70
Phosphorus P	$[Ne]3s^2 3p^3$	1063	1897	2910	1·10
Arsenic As	$[Ar]3d^{10} 4s^2 4p^3$	945·6	1950	2732	1·21
Antimony Sb	$[Kr]4d^{10} 5s^2 5p^3$	833·5	1590	2440	1·41
Bismuth Bi	$[Xe]4f^{14} 5d^{10} 6s^2 6p^3$	703·3	1610	2467	1·52

On the other hand, if a nitrogen atom were to lose an electron to become N^+ then it would be iso-electronic with carbon and thus capable of achieving the inert-gas structure of neon by forming four covalent bonds using sp^3 hybridization. This is apparently energetically feasible, because a wide variety of salts are known to contain the ammonium ion NH_4^+, which, like methane, has a tetrahedral structure. One or more of the hydrogen atoms in NH_4^+ may be substituted by alkyl (and sometimes aryl) to give, ultimately, NR_4^+ salts

$$NR_3 + RI \longrightarrow NR_4^+I^-.$$

The other members of group V form similar MR_4^+ ions but do so less readily than nitrogen. For a reaction such as

$$MH_3 + HX \longrightarrow MH_4^+X^- \quad (X = \text{halogen}),$$

there are several energy terms which must be considered, some of which are:

(a) the strength of the H—X bond,
(b) the energy required to promote M^+ to the valence state,
(c) the strength of the M—X bonds,
(d) the lattice energy of $MH_4^+X^-$.

The balance of these terms appears to occur at M = As and X = Br or I, since it is just possible to prepare $AsH_4^+Br^-$ and $AsH_4^+I^-$ but the H—Cl and H—F bond strengths do not allow the formation of arsonium chloride or fluoride. For antimony and bismuth, the M—H bonds are so weak (even BiH_3 is barely stable at 20 °C) that no SbH_4^+ or BiH_4^+ salts can be synthesized.

Most of the derivatives of these elements are covalent, the outer quantum shell being filled by electron sharing with three other atoms or groups of atoms. In this respect all the group V elements behave in a similar manner, but there are two main differences between the chemistry of nitrogen on the one hand and that of phosphorus, arsenic, antimony and bismuth on the other: (a) nitrogen is capable of forming strong p_π–p_π bonds whereas the other elements are not and (b) nitrogen is unable to coordinate more than four atoms or groups round itself whereas phosphorus, arsenic, antimony and bismuth are able to increase their coordination number to five and, sometimes, even to six.

The reasons for the weakness of p_π–p_π multiple-bond formation by the heavier group V elements are the same as those outlined previously for silicon (see p. 150), viz. the presence of nodes in np orbitals for $n > 2$ leads to less effective overlap than when 2p orbitals are used for π-bonding, and the inner electron shells cause strong repulsion forces as the elements attempt to achieve favourable np-orbital overlap by approaching each other more closely. As a consequence, the oxides of nitrogen, nitrites, nitrates, nitro, nitroso, azo, diazo, azides, nitriles and imino-derivatives (all of which have electronic structures involving p_π–p_π multiple bonds) have no phosphorus, arsenic, antimony or bismuth counterparts; similarly the elemental form of nitrogen is the diatomic gas, N≡N, whereas phosphorus, arsenic and antimony form single M—M bonds in their normal elemental states (even when several allotropic forms are known) and bismuth has a metallic lattice. When no other more suitable bonding is possible, as in the free gaseous M_2 molecules at high temperatures, it is conceivable that np_π–np_π bonding does occur even when $n > 2$. For example, below about 800 °C white phosphorus is present in the gas phase as tetrahedral P_4 molecules, the atoms of the tetrahedron being held by σ-bonds derived from pure 3p atomic orbitals (see Figure 79); however, as the temperature is increased the effect of entropy ensures that the tetramers dissociate and at 1700 °C it is found that the vapour consists of approximately equal amounts of P_4 and P_2 molecules. The phosphorus–phosphorus distance in P_2 has been calculated to be 1·89 Å from spectroscopic data, which is considerably less than the accepted value of 2·20 Å for a phosphorus–phosphorus single bond. It has therefore been suggested that the bonding in the P_2 molecule is similar to that which occurs in N_2 (Figure 80), but the dissociation energy of P≡P (489·6 kJ mol^{-1}) is considerably less than that of N≡N (941·4 kJ mol^{-1}) owing to both the relatively poor 3p–3p overlap and inner-shell repulsions which occur in the former molecule.

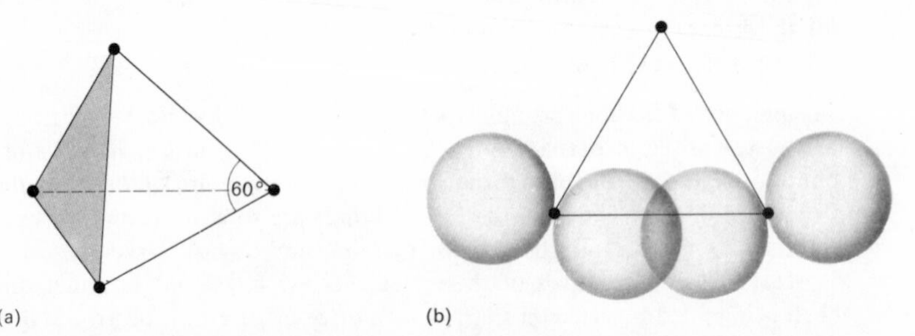

Figure 79 (a) The shape of the P_4 molecule. (b) The bonding between two phosphorus atoms in one edge of the P_4 tetrahedron.

Each phosphorus atom is assumed to use pure 3p orbitals for making σ-bonds to its three neighbouring phosphorus atoms. However, since the P—P—P bond angle is only 60°, the p-orbitals do not point directly at each other as in normal σ-bonds and maximum overlap cannot be achieved. A compromise situation occurs, giving a 'bent' bond somewhat similar to those found in cyclopropane

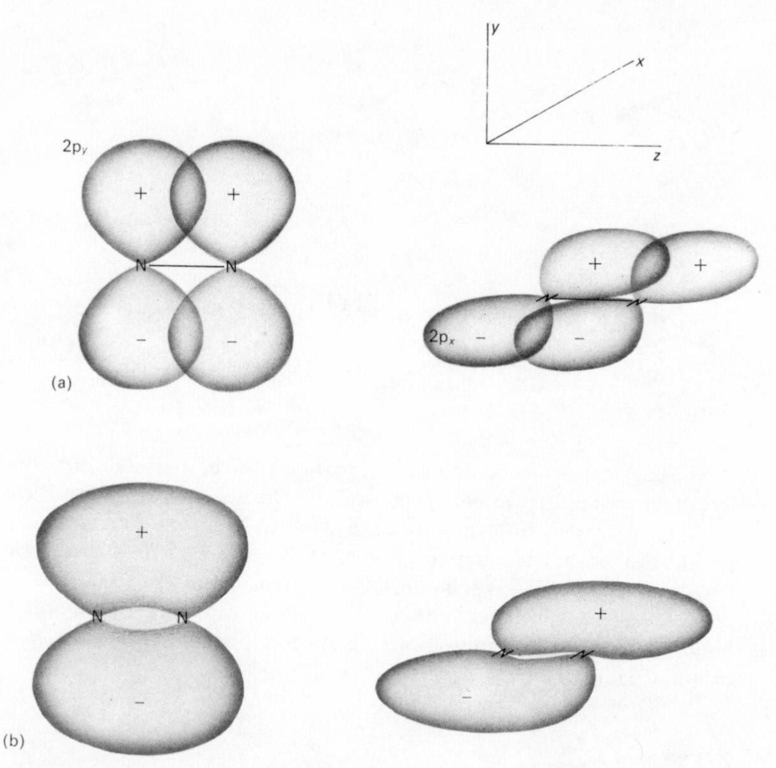

Figure 80 Bonding in the nitrogen molecule.

(a) Overlap of the $2p_x$ and $2p_y$ orbitals gives rise to two strong π-bonds in addition to the σ-bond formed by overlap of sp hybrid orbitals. Hence the molecule is held together by a triple bond N≡N.

(b) The bonding molecular orbitals formed by overlap of the $2p_x$ and $2p_y$ atomic orbitals

The expansion of the coordination numbers of phosphorus, arsenic, antimony and bismuth to five (8.1a, b, c) and six (8.1d, e, f) is often considered to involve the empty *n*d atomic orbitals of these elements giving sp^3d and sp^3d^2 hybridization respectively. As can be seen from Figure 16 (p. 33), the 3d orbitals of an isolated phosphorus atom, for example, are very much larger than the 3s and 3p orbitals, so it might have been expected that suitable sp^3d and sp^3d^2 hybrid orbitals could not be made by promoting electrons into the 3d orbitals; however, calculations suggest that the 3d orbitals contract to about the same size as 3s and 3p orbitals when electronegative ligands are placed around the phosphorus atom. At present it is not clear whether these atoms achieve their higher coordination numbers by using their *n*d orbitals or by forming 3c, 4e bonds to some of the ligands (see Figure 19).

(8.1a) (8.1b) $R = CH_3, C_6H_5$ (8.1c)

(8.1d) (8.1e) (8.1f)

Although experimental proof of the participation of d-orbitals in σ-bonding is lacking there appears to be a great deal of evidence that the group V elements use their d-orbitals to form both d_π–d_π and d_π–p_π bonds. For example, in the tetrahedral oxyhalides of phosphorus, OPX_3, the phosphorus–oxygen bond length of about 1·45 Å is somewhat shorter than would be expected for a single phosphorus–oxygen bond (1·61 Å). This can be explained by assuming d_π–p_π bonding between the empty d-orbitals on phosphorus and a filled 2p oxygen orbital (Figure 81).

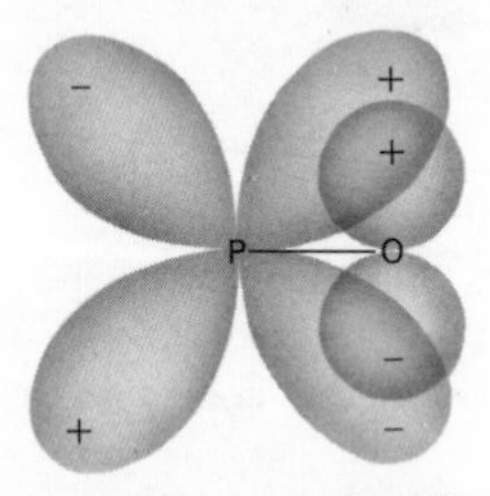

Figure 81 d_π–p_π Bonding between a phosphorus and an oxygen atom

The trihalides of phosphorus, as well as the trialkyls of phosphorus and arsenic, form many complexes with the transition metals, the initial σ-donation to the transition metal involving the sp^3 lone pair of electrons present in these molecules. There are strong reasons for believing that this σ-bonding is augmented by π-bonding from filled $(n-1)$d orbitals on the transition metal into the empty *n*d orbitals of phosphorus and arsenic (so-called 'back bonding', Figure 82).

The π-bonding which occurs in the tetrahedral phosphate ion PO_4^{3-} is similar to that occurring in the phosphorus oxyhalides, but is more difficult to visualize. Figure 83 illustrates the interaction between the $3d_{x^2-y^2}$ orbital on the phosphorus atom and the four 2p orbitals on the oxygen atoms, the large size of the 3d orbitals in this case facilitates simultaneous overlap with the four 2p orbitals.

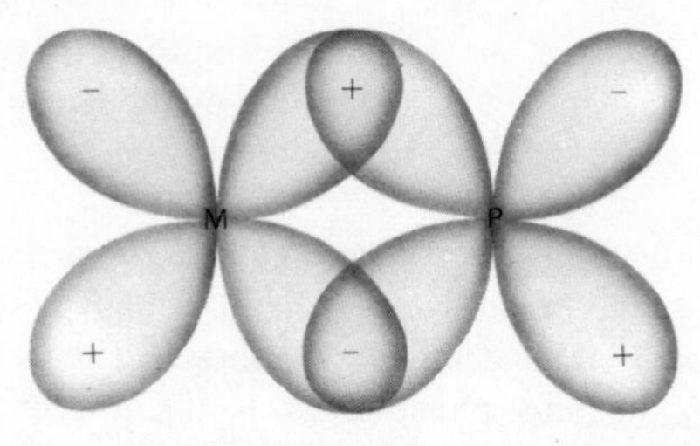

Figure 82 'Back bonding' between a transition-metal atom and a phosphorus atom

Catenation is not very extensive among the group V elements. For nitrogen, the maximum chain length is three, as found in the linear azide ion N_3^- and in hydrazoic acid HNNN. The longest open chains formed by the other elements contain only two atoms (as in the $R_2M—MR_2$ derivatives), although four (8.2), five (8.3) and six (8.4) atoms of phosphorus or arsenic can be joined together in puckered rings.

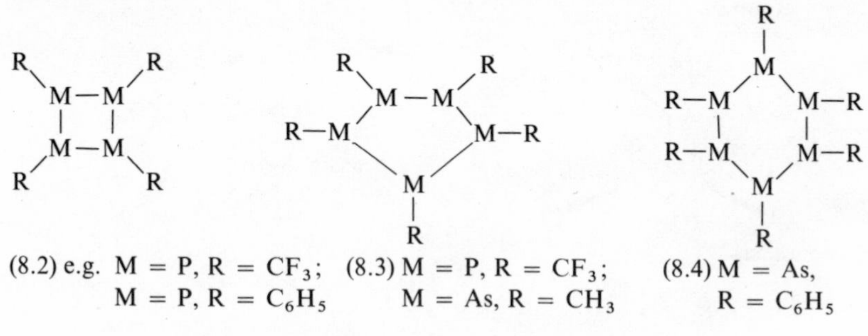

(8.2) e.g. $M = P$, $R = CF_3$; $M = P$, $R = C_6H_5$ (8.3) $M = P$, $R = CF_3$; $M = As$, $R = CH_3$ (8.4) $M = As$, $R = C_6H_5$

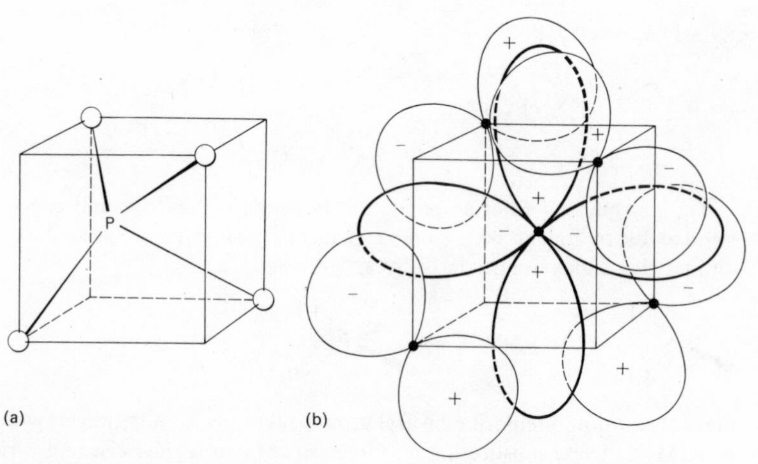

Figure 83 d_π–p_π Bonding in the PO_4^{3-} ion: (a) the tetrahedral PO_4^{3-} ion, (b) π-orbital overlap in the PO_4^{3-} ion

A rather unusual feature of phosphorus, arsenic and bismuth is that their atoms congregate readily into clusters. This is well illustrated by the tetrahedral molecules P_4, As_4 and Sb_4, and the rather peculiar multi-atomic ions formed by bismuth, viz. Bi_5^{3+}, Bi_8^{2+}, Bi_9^{5+}.

Although nitrogen and phosphorus themselves do not catenate appreciably, together they form an interesting series of polymers called the phosphazenes (still sometimes known by their old name 'phosphonitrilics'). The chlorophosphazenes are synthesized by heating phosphorus pentachloride with ammonium chloride, and can take the form of both rings $(Cl_2PN)_n$ and open chains $Cl_4P(NPCl_2)_nNPCl_3$. The rings usually contain either six or eight atoms, the six-membered rings being planar (8.5) and the eight-membered ones assuming both 'boat' and 'chair' conformations (8.6). In both types of ring all the phosphorus–nitrogen bond lengths are equal (1·56–1·59 Å) and are considerably shorter than the accepted bond length of 1·77 Å for a phosphorus–nitrogen single bond. This suggests that delocalized (i.e. 'aromatic') π-bonding occurs between the filled nitrogen 2p orbitals and empty 3d orbitals on the phosphorus atoms.

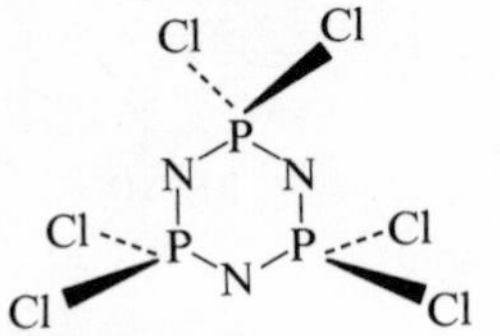

(8.5a)

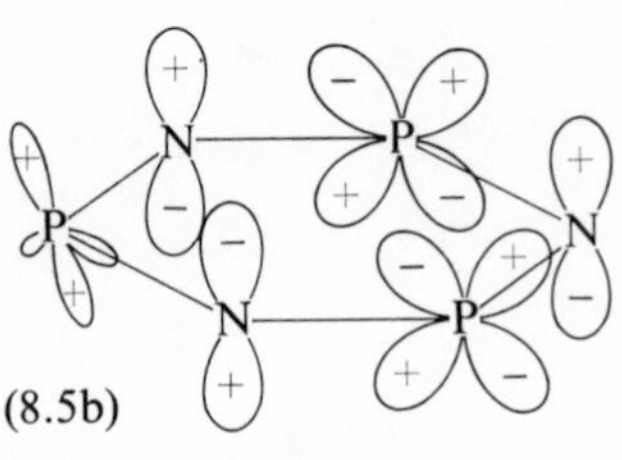

(8.5b)

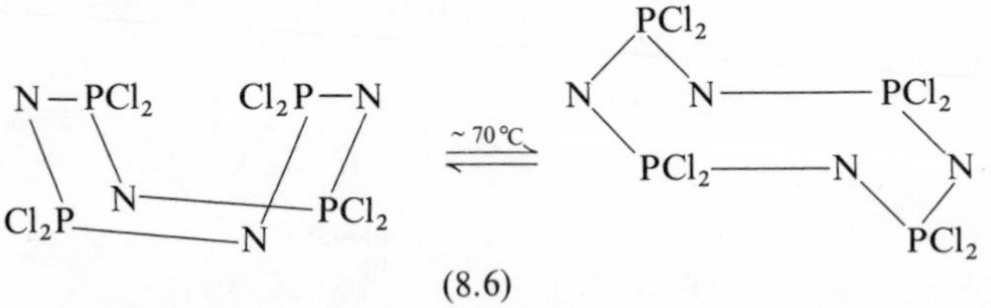

(8.6)

The phosphorus–chlorine bonds on the rings are reactive and can be substituted by treatment with suitable reagents. For example, secondary alkylamines give dialkylamino-derivatives,

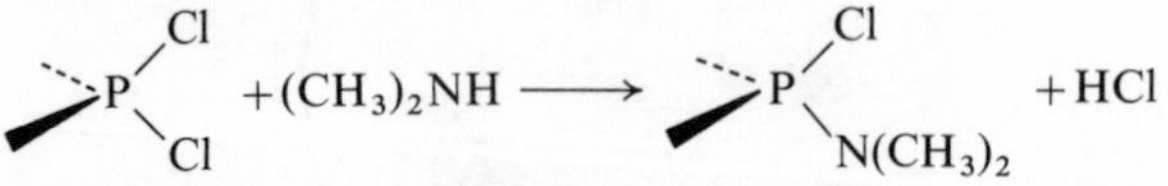

$$>PCl_2 + (CH_3)_2NH \longrightarrow >P(Cl)N(CH_3)_2 + HCl$$

and substitution of chlorine by aryl groups takes place on treatment with LiAr or ArMgX. These substitution reactions can, of course, give rise to a variety of isomers dependent both on the position of the substituted phosphorus atoms relative to each other in the ring and on the spatial arrangement of the appended groups. For a six-membered cyclic chlorophosphazene in which two of the chlorine atoms have been substituted by a group R, the three possible isomers are (8.7–9). The rubber-like chain polymers probably have the structure (8.10).

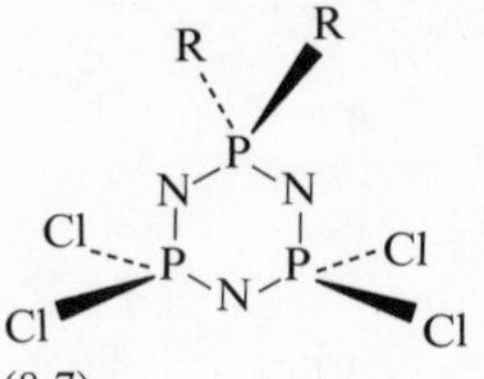

(8.7)

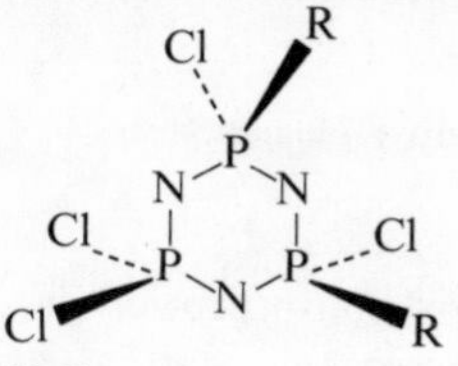

(8.8)

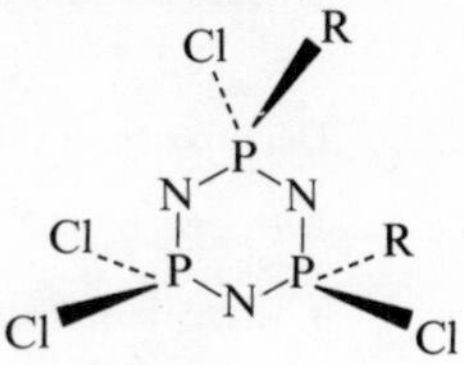

(8.9)

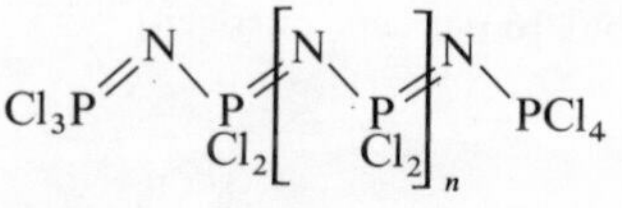

$$Cl_3P{=}N{-}\left[P(Cl_2){=}N{-}P(Cl_2){=}N\right]_n{-}PCl_4$$

(8.10)

An active field of current interest is that of 'nitrogen fixation'. It is hoped that this research will lead to the discovery of transition-metal catalytic systems which will combine nitrogen and hydrogen at room temperature and pressure to give ammonia for use in agriculture. At present ammonia is made in large quantities by the well-known Haber process but as this requires high pressures and temperatures it would obviously be cheaper to manufacture ammonia under less extreme conditions. Although no such catalyst has yet been synthesized it has recently proved possible to prepare metal complexes which readily absorb nitrogen gas at room temperature,

e.g. $$(Ru(NH_3)_5H_2O)^{2+} \xrightarrow[\text{aqueous solution}]{N_2\text{ bubbled through}} (Ru(NH_3)_5N_2)^{2+}$$
~ 50% yield

An X-ray structure determination of this ion shows that the M—N—N system is strictly linear, strongly suggesting the bonding is similar to that found in the transition-metal carbonyls and involves both the σ-donation of a nitrogen sp lone pair of electrons to the metal and back-donation from filled metal d-orbitals into the empty antibonding π-orbitals on the N_2 ligands (8.11). Other nitrogen complexes of the same type include $[(C_6H_5)_3P]_3CoN_2$ and $[(C_6H_5)_3P]_2Ir(N_2)Cl$.

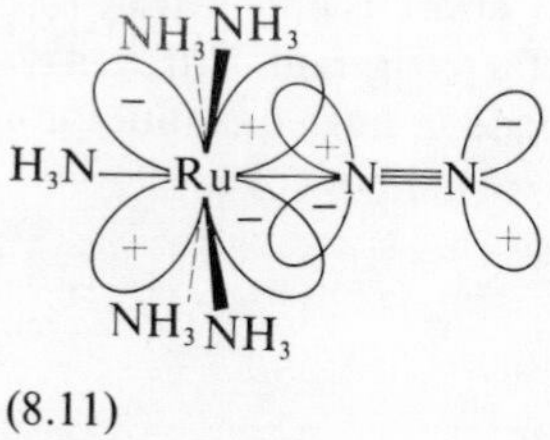

(8.11)

It is, therefore, conceivable that now the nitrogen molecule has been placed on a metal in this fashion, a system will eventually be found which can be reduced by hydrogen gas back to the original metal complex and ammonia. A continuous process for ammonia production would then be possible.

8.2 Occurrence of the group V elements

8.2.1 *Nitrogen*

The gas is the principal constituent of air, constituting about 78·1 per cent by volume. Combined nitrogen occurs as ammonia, nitrites and nitrates (e.g. 'Chile saltpetre' $NaNO_3$); as an essential element to life it occurs in organic matter (e.g. in proteins, which average about 16 per cent nitrogen).

8.2.2 *Phosphorus*

About 0·1 per cent in the lithosphere, almost exclusively as phosphates; principal ores are phosphorite $Ca_3(PO_4)_2$ and apatite $3Ca_3(PO_4)_2 . Ca(F, Cl)_2$.

8.2.3 *Arsenic*

About 5 p.p.m. of the Earth's crust; the minerals realgar As_2S_2 and orpiment As_2S_3 were known to the Ancients. Principal present source is the flue gases obtained in the extraction of nickel, iron and cobalt.

8.2.4 *Antimony*

0·5 p.p.m. of the Earth's crust; most important ore is stibnite Sb_2S_3.

8.2.5 *Bismuth*

0·1 p.p.m. abundance; normally extracted from the flue gases obtained during the roasting of lead sulphide ores.

8.3 Extraction

8.3.1 *Nitrogen*

Pure liquid nitrogen, boiling point −196 °C (1 atm.), is obtained by fractional distillation of liquid air; liquid nitrogen is a useful refrigerant. Nitrogen required for ammonia and cyanamide production is made in huge quantities using the producer-gas reaction, in which air is blown over red-hot coke

$$C + O_2 + N_2 \longrightarrow 2CO + N_2.$$

8.3.2 *Phosphorus*

Calcium phosphate is heated to 1300 °C with silica and carbon in an electric furnace; the calcium silicate $CaSiO_3$ formed under these conditions remains behind in the furnace while the phosphorus distils away and may be condensed in suitable receivers as white phosphorus:

$$2Ca_3(PO_4)_2 + 6SiO_2 \longrightarrow 6CaSiO_3 + P_4O_{10},$$
$$P_4O_{10} + 10C \longrightarrow P_4 + 10CO.$$

8.3.3 *Arsenic and bismuth*

The oxides obtained from flue gas deposits ('flue dusts') are reduced by heating with carbon.

8.3.4 *Antimony*

The sulphide ore is roasted in air to give the oxide which is reduced by carbon monoxide in a blast furnace.

8.4 The elements

Nitrogen is the only member of the group which solidifies (m.p. −210 °C) to a molecular lattice composed of diatomic molecules; under normal conditions of temperature and pressure it exists as a rather inert diatomic gas constituting slightly more than 78 per cent by volume of the Earth's atmosphere. The inert nature of nitrogen gas is undoubtedly due to the strong bonding in the N_2 molecule which requires the input of some 941·4 kJ before it can be broken up into atoms and undergo reaction; as a result of this only lithium of all the elements combines directly with nitrogen at room temperature, and then only slowly. By passing nitrogen through an electrical discharge at low pressure it can be dissociated into atoms which are much more reactive and combine directly at room temperature with such elements as mercury, sulphur, iodine, phosphorus and arsenic. The high dissociation energy of the N_2 molecule (which is the standard state of nitrogen at 25 °C) makes very many compounds of nitrogen endothermic and hence potentially unstable; for example, all the nitrogen oxides – except dinitrogen pentoxide, which is just exothermic – cyanogen and nitrogen trichloride are endothermic. This often means that a nitrogen derivative is thermally more unstable than its phosphorus analogue

(e.g. NCl_3, $\Delta H_f = +230$ kJ mol^{-1}; PCl_3, $\Delta H_f = -259$ kJ mol^{-1}).

Phosphorus, arsenic and antimony have several allotropes, the most stable form of all three (and for bismuth the only form) being that of a layer structure made up of puckered six-membered rings in the 'chair' conformation (see Figure 84 for the structure of black phosphorus which is of this type). The layer

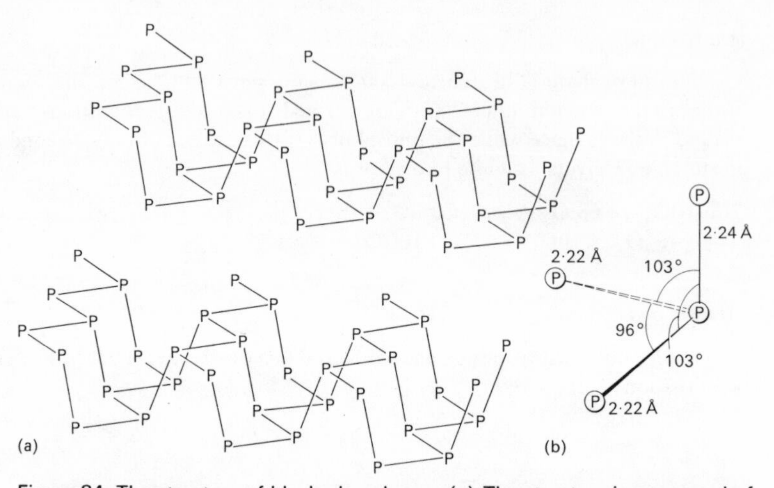

Figure 84 The structure of black phosphorus. (a) The structure is composed of layers of puckered six-membered rings – only two partial layers are shown.
(b) Each phosphorus atom is pyramidally coordinated to its three nearest neighbours

forms of arsenic, antimony and bismuth differ slightly from black phosphorus in that a degree of metallic bonding helps to hold the layers together (instead of purely van der Waals forces as in black phosphorus) allowing the solids to conduct both heat and electricity. This metallic interaction increases from arsenic to bismuth as shown by the fact that the three next-nearest neighbours from other layers become structurally significant and relatively more close (Table 23).

Table 23 Interatomic Distances in Group V Elements

	Three nearest neighbours at	*Three next-nearest neighbours at*
Arsenic	2·51 Å	3·15 Å
Antimony	2·87 Å	3·37 Å
Bismuth	3·10 Å	3·47 Å

Two other important forms of phosphorus are the white and the red modifications. Pyrophoric white phosphorus results whenever phosphorus vapour below about 1000 °C is condensed; it consists of a molecular crystal produced by the packing of P_4 tetrahedra. The action of light or heat converts the white form into the red. The latter is considerably less volatile and less soluble in organic solvents than white phosphorus and undoubtably contains larger units than P_4, but its structure has not yet been determined. The highly unstable yellow forms of arsenic and antimony, prepared in a similar fashion to white phosphorus, appear to have a structure containing M_4 molecules like that of white phosphorus.

The vapours of phosphorus, arsenic and antimony contain M_4 tetrahedra until quite high temperatures are reached; it is only at about 1000 °C that the P_4 molecules begin to dissociate slightly into P_2, and even at 1700 °C about 50 per cent of the tetrahedra remain intact. At very high temperatures the elements are present, at least partly, as atoms. Bismuth vapour contains only an equilibrium mixture of Bi_2 molecules and bismuth atoms, although Bi_2 molecules persist at high temperatures and can still be detected at 2000 °C.

8.5 Hydrides of the group V elements

Nitrogen forms four hydrides: hydrazoic acid N_3H, di-imide N_2H_2, hydrazine N_2H_4 and ammonia NH_3, of which the first two contain p_π–p_π multiple bonds, and hence have no analogues among the other elements of the group.

8.5.1 MH_3 *derivatives*

The small 1s orbital of hydrogen becomes less compatible with the orbitals of the group V elements as their atomic number increases. Hence the strength of the M—H bond decreases down the group, making the hydrides more thermally unstable: ammonia is thermally very stable and only begins to break down appreciably into its elements at temperatures around 2000 °C, whereas bismuthine BiH_3 has a half-life of only a few minutes at room temperature.

Table 24 Bond angles in MH_3 derivatives

NH_3	PH_3	AsH_3	SbH_3	BiH_3
106·6°	93·5°	91·8°	91·3°	—

The bond angles in the MH_3 derivatives (Table 24) strongly suggest that in ammonia the nitrogen atom is sp^3 hybridized, whereas the other elements use essentially pure p-orbitals to form the M—H bonds. Steric factors can account for the slight deviations from the expected H—M—H bond angles of 109·5° and 90°. The lone pair of electrons on nitrogen occupy a rather bulky sp^3 hybrid orbital which requires more space than the molecular orbitals because the lone-pair electrons, unlike the bonding electrons, are confined to only one nucleus (8.12). The lone-pair electrons in the other hydrides are in a spherical *ns* atomic orbital and can play no part in determining the molecular shape; the slight opening of the H—M—H bond angle from 90° is due to the mutual repulsion of the M—H bonds (8.13).

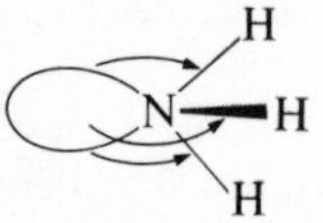

(8.12) 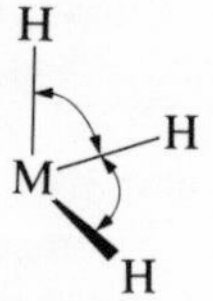 (8.13)

8.5.2 M_2H_4 *derivatives*

Nitrogen, phosphorus and arsenic form a lower hydride M_2H_4 in which there is an M—M single bond. In hydrazine N_2H_4 each nitrogen atom carries a lone pair of electrons in an sp^3 hybrid orbital, which dictate in part the 'gauche' shape which the molecule adopts in the gas phase (8.14).

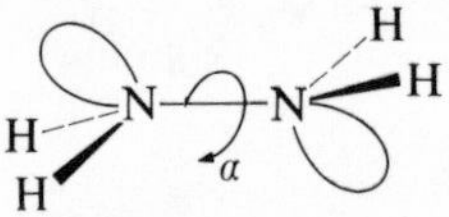

(8.14) angle of twist α from the *cis* configuration is about 90°

Commercially, hydrazine is manufactured by the Raschig process, in which ammonia is oxidized by sodium hypochlorite in aqueous solution

$$NH_3 \text{ (large excess)} + NaOCl \longrightarrow \underset{\text{(chloramine)}}{NH_2Cl} + NaOH,$$

$$NH_2Cl + 2NH_3 \longrightarrow N_2H_4 + NH_4Cl.$$

Glue or gelatine is added to the reaction mixture to inhibit the reaction

$$N_2H_4 + 2NH_2Cl \longrightarrow N_2 + 2NH_4Cl.$$

It is thought that the glue complexes with traces of heavy-metal ions present in the solution which catalyse this loss of hydrazine. A dilute ($\sim$ 2 per cent) solution of hydrazine is thus obtained, which can be concentrated by distillation to hydrazine hydrate $N_2H_4 . H_2O$; azeotropic distillation with aniline gives anhydrous hydrazine. It is a weaker base than ammonia, but does form a series of salts $N_2H_5^+X^-$.

8.5.3 M_2H_2 *derivatives*

A hydride of this formulation requires a double bond between the two M atoms, which is only possible when M is nitrogen. Di-imide N_2H_2 is the hydrogen analogue of the organic azo-derivatives RN=NR; it is highly unstable and cannot be isolated at room temperature, but is considered to be an intermediate in a few reactions. The infrared spectrum of di-imide, prepared by passing an electrical discharge through hydrazine and freezing out the di-imide at liquid nitrogen temperatures to arrest the decomposition, suggests that the molecule probably has the *cis* configuration as in (8.15).

8.5.4 *Hydrazoic acid* HN_3

Hydrazoic acid is interesting because it is one of the very few molecules in which more than two nitrogen atoms are linked directly together; the three nitrogens being collinear (8.16).

H H
N=N

(8.15)

1.34Å 1.34Å
N—N—N
111°
H

(8.16)

Hydrazoic acid is only slightly dissociated in aqueous solution ($pK_a \sim 5$); electropositive metals dissolve to give azides, but hydrogen is not released because it reduces part of the remaining hydrazoic acid to ammonia and nitrogen

$$4Li + 6HN_3 \longrightarrow 4LiN_3 + 2NH_3 + 2N_2,$$
$$Zn + 3HN_3 \longrightarrow Zn(N_3)_2 + NH_3 + N_2.$$

The azides of group I and group II elements are salts and contain the symmetrically linear azide N_3^- ion (8.17). The central nitrogen atom in the azide

$$N \overset{1{\cdot}15\,Å}{—} N \overset{1{\cdot}15\,Å}{—} N$$

(8.17)

ion can be assumed to be sp hybridized and makes a normal σ-bond to the other two nitrogens (which probably use pure p-orbitals for this interaction); the remaining two p-orbitals on each of the three nitrogen atoms overlap to form two π-bonds, which are mutually at right angles to each other (Figure 85).

8.5.5 MH_5

No binary hydrides have yet been isolated in which the group V element is pentacovalent, probably because the contraction of the d-orbitals required for bonding only occurs to a reasonable extent when there are very electronegative ligands in the axial positions. By placing fluorine atoms in the two axial positions it has recently proved possible to isolate PHF_4 (8.18) and PH_2F_3 (8.19).

F
H—P (F, F)
F

(8.18)

F
H—P (F, H)
F

(8.19)

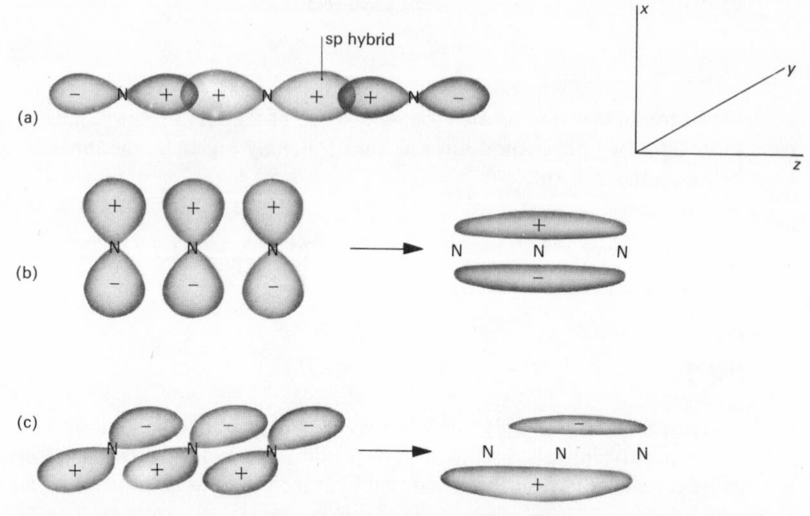

Figure 85 Bonding in the azide ion: (a) σ-interaction between sp orbitals on the central nitrogen atom and the $2p_z$ orbitals of the outer nitrogen atoms; (b) interaction of $2p_x$ orbitals on all three nitrogen atoms to produce a three-centre molecular orbital in the *xz* plane; (c) interaction of $2p_y$ orbitals on all three nitrogen atoms to produce a three-centre molecular orbital in the *yz* plane

8.5.6 *Hydroxylamine* $HONH_2$

This molecule is thought to have a structure in which the lone-pair electrons on nitrogen and oxygen are *trans* to each other (8.20).

(8.20)

Commercially, hydroxylamine is isolated from its bisulphate made by adding concentrated sulphuric acid to boiling nitromethane

$$CH_3NO_2 + H_2SO_4 \longrightarrow NH_3OH^+HSO_4^- + CO\uparrow.$$

The anhydrous compound is unstable and decomposes without melting into water, ammonia, dinitrogen monoxide and nitrogen monoxide; like hydrazine, it is a very weak base.

8.6 Oxides and oxy-acids of the group V elements

8.6.1 *Oxides of nitrogen*

As previously mentioned, in all the oxides of nitrogen there is p_π–p_π multiple bonding between the nitrogen and oxygen atoms, and consequently the oxides have no phosphorus, arsenic, antimony or bismuth analogues; therefore, it is convenient to discuss the oxides of nitrogen separately.

Dinitrogen monoxide N_2O. This colourless and rather inert gas may be prepared by heating ammonium nitrate; it has an asymmetric, linear structure (8.21). The central nitrogen atom can be regarded as being sp hybridized and forms normal

$$N \overset{1{\cdot}13\,Å}{\text{—}} N \overset{1{\cdot}19\,Å}{\text{—}} O$$

(8.21)

σ-bonds with p-orbitals on the oxygen and outer nitrogen atoms; the σ-bonding is augmented by π-bonding involving the available p-orbitals (two on each nitrogen atom and one on the oxygen atom; Figure 86).

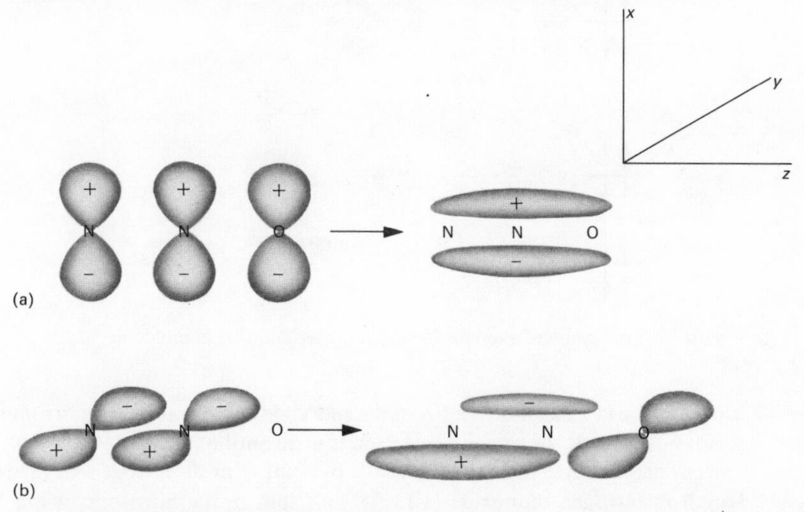

Figure 86 Bonding in the dinitrogen monoxide molecule: (a) interaction of $2p_x$ orbitals to form a three-centre molecular orbital in the *xz* plane; (b) interaction of the two $2p_y$ orbitals of the nitrogen atoms to form a π molecular orbital in the *yz* plane, leaving a $2p_y$ lone pair of electrons on the oxygen atom

Nitrogen monoxide NO. Nitrogen monoxide is a rather unusual molecule in that it contains one unpaired electron and yet shows no signs of dimerization until it is cooled to very low temperatures. On dimerization, nitrogen monoxide becomes diamagnetic, showing that all the electrons become paired, but the association forces are much weaker than normal single bonds, as shown by the low heat of

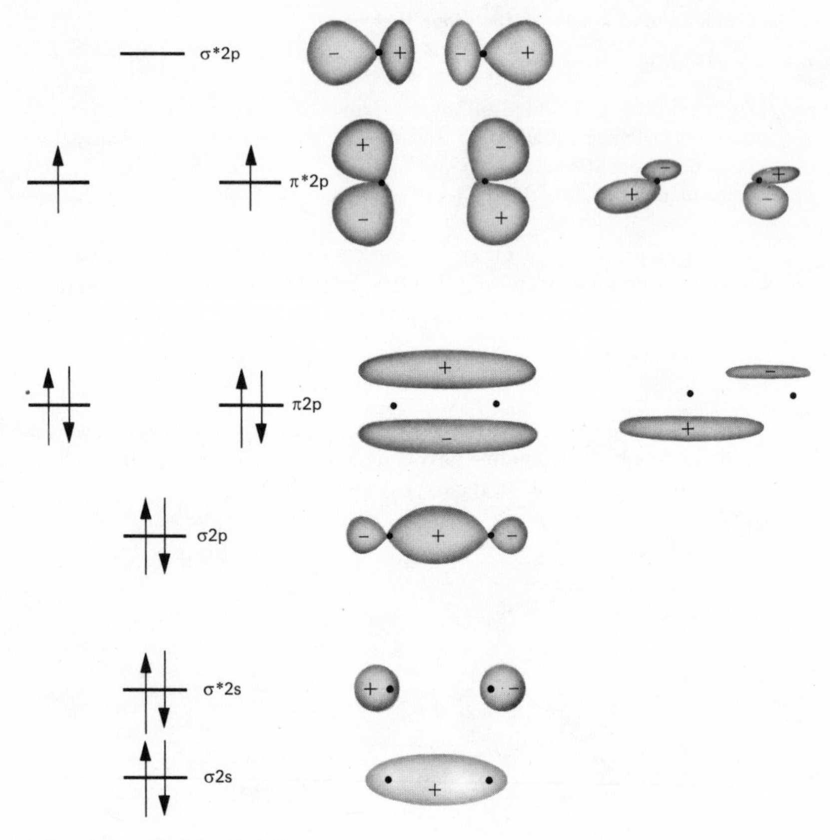

Figure 87 Energy-level diagram for the nitrogen monoxide molecule

dimerization (15·5 kJ mol^{-1}). From the energy-level diagram shown in Figure 87, it can be seen that the unpaired electron is in an antibonding orbital (π*2p). This can be demonstrated rather dramatically by comparing the nitrogen–oxygen bond length in nitrogen monoxide (1·15 Å) with that in the nitrosonium ion NO^+ (1·06 Å). To form the nitrosonium ion, an electron must be removed from the highest-energy orbital in nitrogen monoxide which, as seen from Figure 87, is the antibonding π*2p; this will have the effect of binding the nitrogen and oxygen atoms more strongly together, so causing a shortening of the nitrogen–oxygen bond in the nitrosonium ion relative to that in nitrogen monoxide. The infrared stretching frequency (which is some measure of a bond's strength) of the nitrogen–oxygen bond will therefore increase from 1877 cm^{-1} for nitrogen monoxide to about 2320 cm^{-1} for the nitrosonium ion.

Nitrogen monoxide forms a rather large number of compounds, which range from the covalent nitrosyl derivatives through transition-metal complexes to nitrosonium salts. Nitrosyl halides (except the iodide, which is unknown) are prepared simply by allowing nitrogen monoxide to react with the respective halogen,

e.g. $2NO + Cl_2 \xrightarrow{-80\,°C} 2Cl{-}NO.$

Their stability decreases with increasing atomic weight of the halogen, so that nitrosyl fluoride is stable at 20 °C, whereas nitrosyl chloride ClNO (0·5 per cent) and bromide BrNO (7 per cent) are partially dissociated into nitrogen monoxide and halogen under the same conditions. The molecules are bent (8.22).

X–N–O (bond angle α at N)

(8.22: X = F: $\alpha = 110°$; X = Cl: $\alpha = 116°$; X = Br: $\alpha = 117°$)

Many of the transition-metal complexes of nitrogen monoxide can be envisaged as arising from the donation of three electrons from nitrogen monoxide to the metal. As back-donation from metal d-orbitals to antibonding orbitals on the nitrogen monoxide molecule apparently contributes towards the stability of these metal nitrosyls, the transition metals are often in low oxidation states, so that the maximum electron density is present in the d-orbitals to facilitate such back-bonding. The effective atomic number rule (see under metal carbonyls, p. 160) is normally found to apply to these complexes, e.g. for iron:

$$Fe_3(CO)_{12} + NO \longrightarrow Fe(NO)_2(CO)_2 \quad \text{(iso-electronic with } Fe(CO)_5\text{)}$$

$$Fe(NO)_2(CO)_2 \xrightarrow{I_2} Fe(NO)_2I$$

$$FeI_2 + NO \xrightarrow{100\,°C} Fe(NO)_2I,$$

$$Fe(CO)_5 + NO \xrightarrow[\text{pressure}]{50\,°C} Fe(NO)_4.$$

The structures of these products are probably (8.23–25).

(8.23) tetrahedral: $(OC)_2Fe(NO)_2$

(8.24) iron–iron bond: $(ON)_2Fe(\mu\text{-}I)_2Fe(NO)_2$

(8.25) ionic: $NO^+[Fe(NO)_3]^-$

Although the π^*2p orbital in nitrogen monoxide is antibonding relative to the free nitrogen and oxygen atoms, it still requires about 912 kJ mol^{-1} to remove an electron from that orbital. The simplest method of synthesizing compounds containing the nitrosonium ion is to treat a nitrosyl halide with a halide of an element which is capable of forming a complex halo-anion,

e.g. $NOCl + AlCl_3 \longrightarrow NO^+AlCl_4^-,$

$NOF + PF_5 \longrightarrow NO^+PF_6^-.$

Many of these nitrosonium salts have been found to be isomorphous with the corresponding ammonium salts, the ammonium ion being only slightly larger than the nitrosonium ion.

Dinitrogen trioxide N_2O_3. An equimolar mixture of nitrogen monoxide NO and nitrogen dioxide NO_2, when condensed at low temperatures, gives dinitrogen trioxide N_2O_3 as a blue solid. Dinitrogen trioxide melts to a blue liquid, but in the gas phase it completely dissociates back into nitrogen monoxide and nitrogen dioxide. Recent spectroscopic data suggest that the molecule has a planar structure (8.26) with non-equivalent nitrogen atoms.

O
N—O—N
O O
(8.26)

N
O 134° O
(8.27)

Dinitrogen tetroxide N_2O_4. On the laboratory scale, the thermal decomposition of lead nitrate is a satisfactory source of dinitrogen tetroxide N_2O_4; the gas is also formed when certain metals (e.g. copper) dissolve in concentrated nitric acid.

Solid dinitrogen tetroxide is both colourless and diamagnetic; it melts at −10 °C to a pale yellow liquid and boils at 21 °C, the vapour at the boiling point containing about 16 per cent of the dark brown nitrogen dioxide NO_2. At 150 °C the vapour contains only paramagnetic nitrogen dioxide NO_2 molecules and is almost black. Dinitrogen tetroxide is a planar molecule and has a rather unexpectedly long nitrogen–nitrogen bond of 1·64 Å (cf. the nitrogen–nitrogen bond length of 1·47 Å in hydrazine); the question as to why this bond is so long and so weak has not yet been satisfactorily answered.

Nitrogen dioxide NO_2 is an angular molecule (8.27). A satisfactory approximation to the bonding may be obtained by assuming that the nitrogen atom is originally sp^2 hybridized, one sp^2 hybrid containing the unpaired electron and the other two making σ-bonds with p-orbitals on the oxygen atoms; the remaining nitrogen p-orbital combines with similar oxygen orbitals to make a delocalized π-type molecular-orbital system extending over the whole molecule (Figure 88).

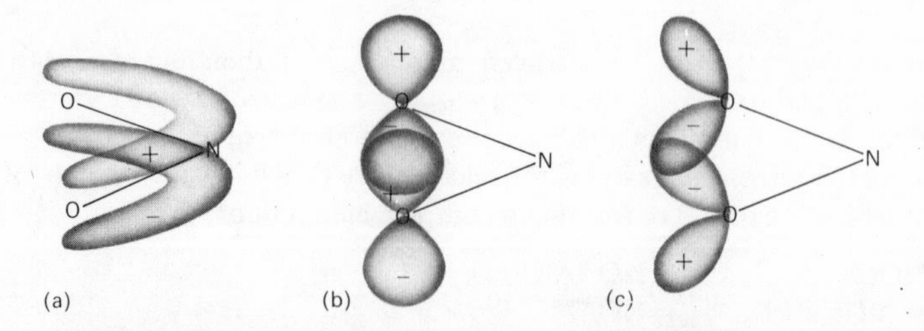

Figure 88 Molecular orbitals in the nitrogen dioxide molecule: (a) bonding molecular orbital containing two electrons, (b) nonbonding molecular orbital containing two electrons, (c) empty antibonding molecular orbital

On this idealized model the O—N—O bond angle would, of course, be 120°; because only one electron is in an sp^2 orbital the mutual repulsion between electrons in the two nitrogen–oxygen bonds will force the O—N—O angle to open up to more than 120°. Treatment of nitrogen dioxide with fluorine or chlorine results in the formation of nitryl halides XNO_2, the previously unpaired nitrogen dioxide electron now being used in bonding to the halogen atom; nitryl bromide and iodide appear to be unknown.

If the unpaired electron were to be removed from nitrogen dioxide, the repulsion between the nitrogen–oxygen bonds would cause the O—N—O bond angle to increase to 180° (when the nitrogen atom could be described as being sp hybridized); this is the shape found experimentally for the nitronium ion NO_2^+ (Figure 89).

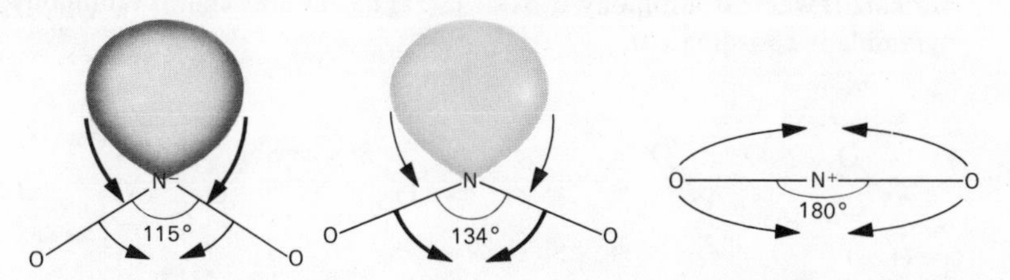

Figure 89 The change in O—N—O bond angle in the series, nitrite ion NO_2^-, nitrogen dioxide NO_2 and nitronium ion NO_2^+. (a) Nitrite ion; two sp^2 electrons. (b) Nitrogen dioxide; one sp^2 electron. (c) Nitronium ion; no sp^2 electrons. (The repulsion effects are shown by means of arrows, the heavy arrows being the dominant ones)

The nitronium ion is the active electrophilic species present in the mixtures of concentrated sulphuric and nitric acids used as 'nitrating mixtures' in organic chemistry

$$HNO_3 + 2H_2SO_4 \longrightarrow NO_2^+ + H_3O^+ + 2HSO_4^-.$$

Dinitrogen pentoxide N_2O_5. In the crystalline state this oxide is a salt, nitronium nitrate ($NO_2^+NO_3^-$), but the vapour contains *covalent* N_2O_5 molecules, the structure of which has not definitely been established.

8.6.2 *Oxides of phosphorus, arsenic, antimony and bismuth*

Trioxides. All four elements react directly with oxygen to form the trioxides, but a limited supply of oxygen must be used with phosphorus to stop the formation of phosphorus pentoxide. In the solid state, phosphorus trioxide and the high-temperature modifications of arsenic trioxide and antimony trioxide contain M_4O_6 molecules, the structures of which are not entirely unrelated to the M_4 tetrahedral molecules found in some allotropes of these elements, the tetrahedra being somewhat expanded and having M—O—M bridges linking the group V atoms (8.28–30). The trioxides sublime directly as M_4O_6 molecules,

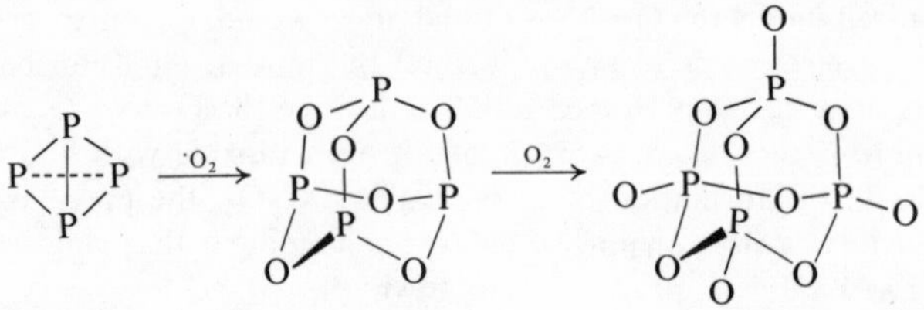

(8.28) P_4 (8.29) P_4O_6 (8.30) P_4O_{10}

although at high temperatures they begin to dissociate into pyramidal M_2O_3 molecules. Arsenic and antimony trioxides also exist in low-temperature forms; that of arsenic trioxide (8.31) contains puckered layers (arsenic pyramidally coordinated) whereas antimony trioxide (8.32) has double chains (antimony also pyramidally coordinated).

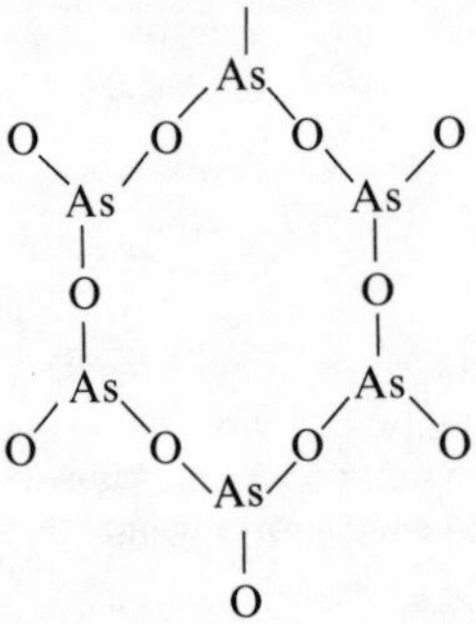

(8.31) 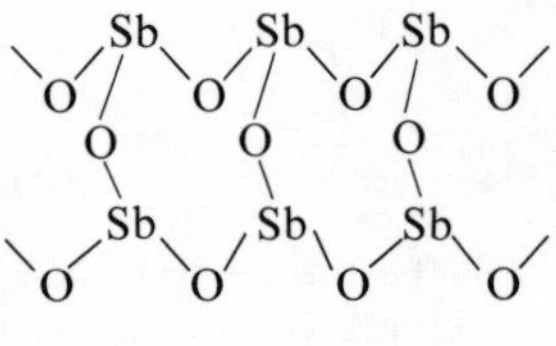 (8.32)

The structural chemistry of the various polymorphs of bismuth trioxide is rather complex and will not be discussed.

Pentoxides. Phosphorus trioxide takes up further oxygen on heating to give the pentoxide, which is present in the vapour state as dimeric P_4O_{10} molecules, the structure of which is shown above (8.30). Condensation of the vapour produces phosphorus pentoxide as a white solid containing discrete P_4O_{10} molecules. This particular form of phosphorus pentoxide is actually metastable and two other modifications have been described in which PO_4 tetrahedra link up by sharing oxygens to give polymeric structures; these polymeric forms are much less volatile and less reactive than the metastable P_4O_{10}.

The pentoxides of arsenic, antimony and bismuth are less well defined and have not been studied structurally. They cannot be made by heating the trioxides in oxygen; careful dehydration of arsenic acid H_3AsO_4 and the so-called 'antimonic acid' is said to give arsenic pentoxide and antimony pentoxide, respectively. Bismuth pentoxide is obtained as an ill-defined solid by oxidizing bismuth trioxide with a strong oxidizing agent such as chlorine, potassium chlorate or a persulphate.

8.6.3 *Oxy-acids of nitrogen*

The bonding in the oxy-acids of nitrogen involves p_π–p_π nitrogen–oxygen bonding and, since phosphorus, arsenic, antimony and bismuth are unable to form strong p_π–p_π bonds with oxygen, there can be no direct analogues for these elements.

Hyponitrous acid $H_2N_2O_2$. Although hyponitrous acid is decomposed by sulphuric acid to give dinitrogen monoxide, this oxide is not the acid anhydride;

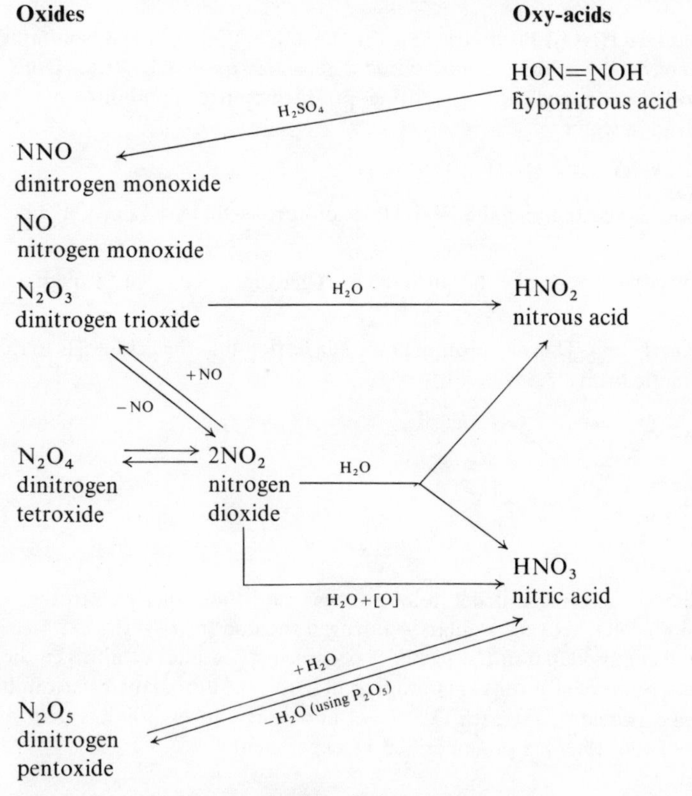

Scheme 3 Oxides and oxy-acids of nitrogen

dinitrogen monoxide dissolves quite readily in water, but gives a neutral solution. It is probable that hyponitrous acid has the *trans* configuration (8.33a) rather than *cis* (8.33b) because the hyponitrite ion $N_2O_2^{2-}$ has Raman and infrared spectra entirely consistent with the *trans* structure. (The *cis* hyponitrite ion is formed when nitrogen monoxide reacts with a solution of an alkali metal in liquid ammonia.)

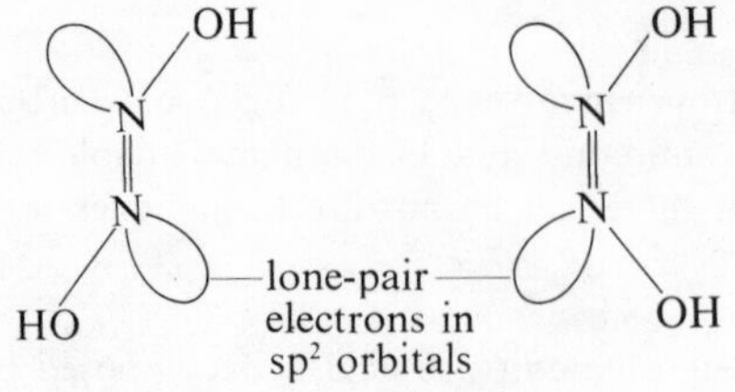

$$\text{HO}-\text{N}=\text{N}-\text{OH}$$

(8.33a) *trans* (8.33b) *cis*

lone-pair electrons in sp^2 orbitals

Nitrous acid HNO_2. The free acid is very unstable; dilute aqueous solutions may be prepared by adding a mineral acid to an alkali-metal nitrite. Dinitrogen tetroxide is a 'mixed' acid anhydride, producing nitrous and nitric acids when dissolved in water

$$N_2O_4 + H_2O \longrightarrow HNO_2 + HNO_3.$$

Organic and certain metallic derivatives of nitrous acid are known in which the NO_2 residue is bonded either through nitrogen, as $-N{\overset{O}{\underset{O}{\lessgtr}}}$, or through oxygen, as —O—N—O. This has prompted the suggestion that the acid exists in the two tautomeric forms (8.34a and b).

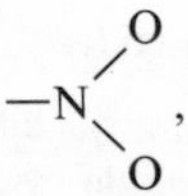

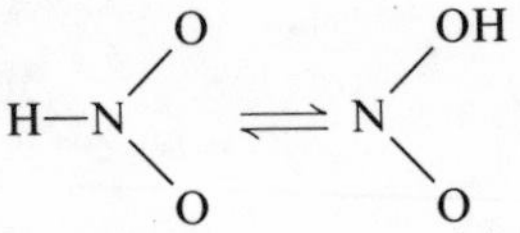

$$\text{H}-\text{N}{\overset{O}{\underset{O}{\lessgtr}}} \rightleftharpoons \text{N}{\overset{OH}{\underset{O}{\lessgtr}}}$$

(8.34a) (8.34b)

While the electronic structure of nitrous acid is not exactly clear, that of the nitrite ion NO_2^- is very similar to nitrogen dioxide (p. 194) the extra electron being accommodated in the partially occupied sp^2 orbital of nitrogen dioxide. Because a pair of electrons are now present in this sp^2 orbital their steric influence will be expected to close the O—N—O bond angle to slightly less than the sp^2 angle of 120° (8.35), a point verified by experiment.

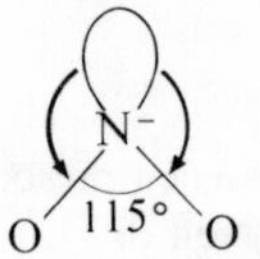

N^-, O, O, 115°

(8.35)

Nitric acid HNO_3. The anhydride of nitric acid is dinitrogen pentoxide

$$H_2O + N_2O_5 \longrightarrow 2HNO_3;$$

the reverse reaction, that of dehydration of nitric acid to give the oxide, may be accomplished using phosphorus pentoxide. On the industrial scale nitric acid is manufactured by the direct oxidation of ammonia using platinum as a catalyst.

The nitrate ion is planar with all the O—N—O bond angles equal to 120°, which suggests that the nitrogen atom uses sp^2 hybrid orbitals to form σ-bonds to each of the three oxygen atoms; this σ-interaction is augmented by π-bonding involving the remaining nitrogen p-orbital and similar orbitals on oxygen (Figure 90).

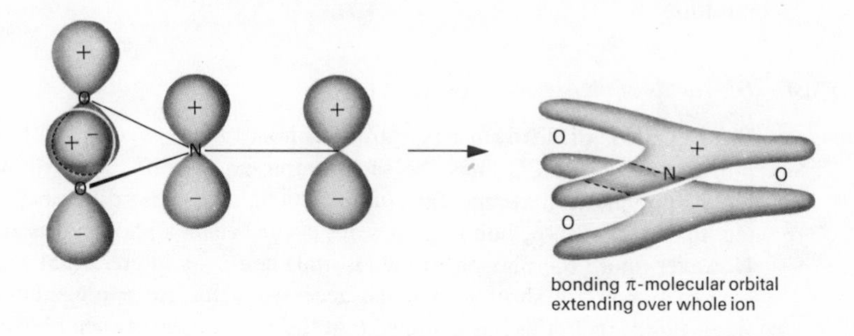

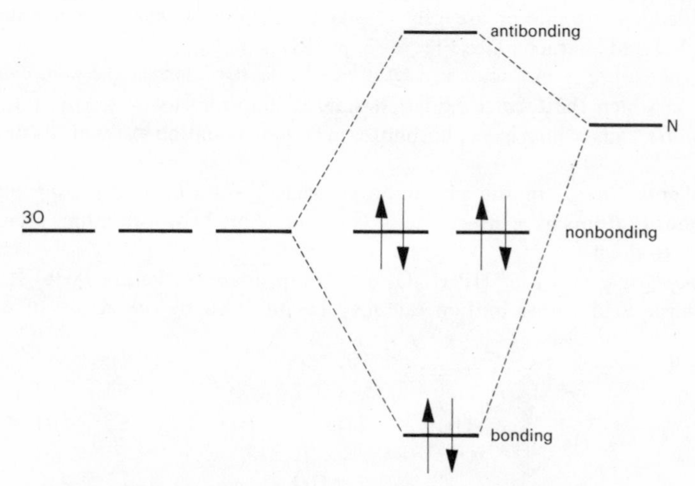

Figure 90 Energy-level diagram showing the electron population of the π molecular orbitals in the nitrate ion

Nitric acid dissolves in pure sulphuric acid to give the nitronium ion

$$HNO_3 + 2H_2SO_4 \longrightarrow NO_2^+ + H_3O^+ + 2HSO_4^-,$$

which is the active agent when this acid mixture is used for nitrating aromatic systems in organic chemistry

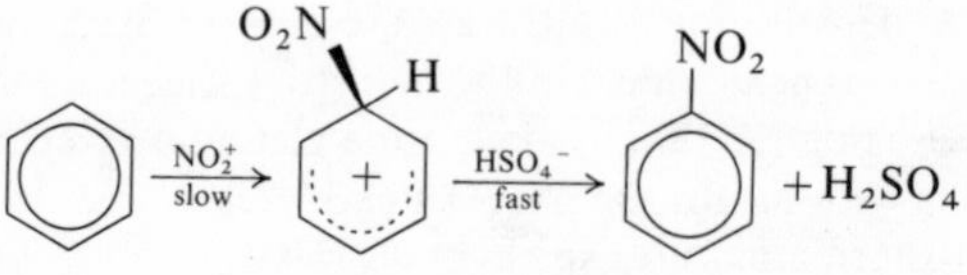

$$\text{C}_6\text{H}_6 \xrightarrow[\text{slow}]{NO_2^+} [\text{C}_6\text{H}_6\text{NO}_2]^+ \xrightarrow[\text{fast}]{HSO_4^-} \text{C}_6\text{H}_5\text{NO}_2 + H_2SO_4$$

The sulphuric acid is therefore a necessary constituent of 'nitrating mixture' because it ensures that all the nitric acid is converted into the active species of the slow step in the nitration mechanism (NO_2^+) thus maximizing the rate of nitration.

8.6.4 *Oxy-acids of phosphorus*

The oxy-acids of nitrogen have no phosphorus analogues and although metaphosphoric acid HPO_3 has the same empirical formula as nitric acid it is a highly polymeric material quite unlike nitric acid. The difference lies in the fact that strong p_π–p_π bonding does not occur between phosphorus and oxygen. However, in all the phosphorus acids and their salts the terminal phosphorus–oxygen bonds are shorter than the accepted value for a phosphorus–oxygen single bond; from this it is assumed that the σ-bonding between phosphorus and oxygen is augmented by d_π–p_π bonding between empty 3d orbitals on phosphorus and filled 2p orbitals on oxygen so giving the phosphorus–oxygen linkages multiple-bond character (see Figure 83, p. 181).

The phosphorus oxy-acids can be divided into two classes: the *phosphorous* acids, in which the formal oxidation state of phosphorus is +1 or +3, and *phosphoric* acids, which have phosphorus in formal oxidation states of +4 or +5.

Phosphorous acids. In the phosphorous acids the tetrahedrally coordinated phosphorus atoms have either one or two non-acidic hydrogen atoms attached directly to them.

Hypophosphorous acid $HP(H)_2O_2$ has the probable structure (8.36) It is a monobasic acid whose barium salt may be prepared by the action of white

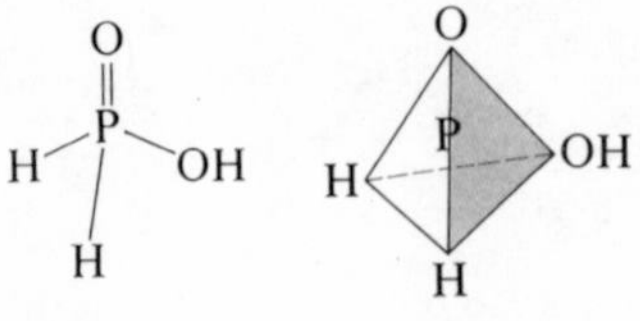

(8.36)

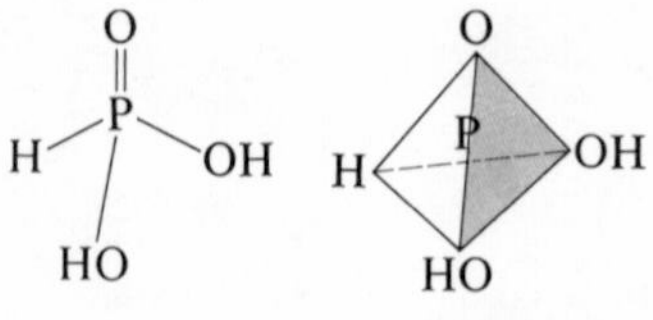

(8.37)

phosphorus on barium hydroxide; the free acid is obtained by the addition of sulphuric acid.

$$Ba(OH)_2 + P\ (\text{white}) \longrightarrow Ba(H_2PO_2)_2 \xrightarrow{H_2SO_4} HP(H)_2O_2$$

Orthophosphorous acid $H_2P(H)O_3$ is a dibasic acid with the structure (8.37). It may be prepared by the reaction between phosphorus trichloride and water at 0 °C.

$$PCl_3 + H_2O \xrightarrow{0\,^\circ C} H_2P(H)O_3$$

Pyrophosphorous acid $H_2P_2(H)_2O_5$ is dibasic. The free acid (8.38) has not

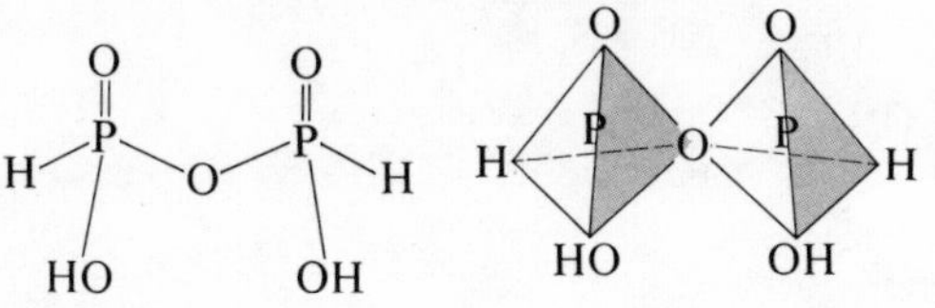

(8.38)

been isolated, but several of its salts are known; they can be prepared by heating monohydrogen orthophosphites.

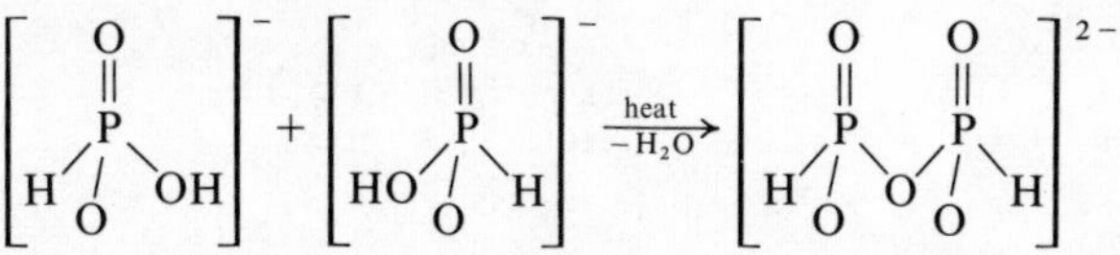

Phosphoric acids. In the phosphoric acids the phosphorus atoms are tetrahedrally surrounded by four oxygen atoms and the variations which occur within this class arise via the sharing of no corner, one corner or two corners of the PO_4 tetrahedra. If three corners are shared, then the entity which is formed is electrically neutral and has the composition P_2O_5. Thus it is not possible to obtain polymeric phosphates with layer structures similar to those found in some of the polysilicates.

Orthophosphoric acid H_3PO_4 is a tribasic acid of structure (8.39). It is prepared industrially by treating phosphate rock with sulphuric acid.

$$Ca_3(PO_4)_2 + H_2SO_4 \longrightarrow CaSO_4\downarrow + H_3PO_4$$

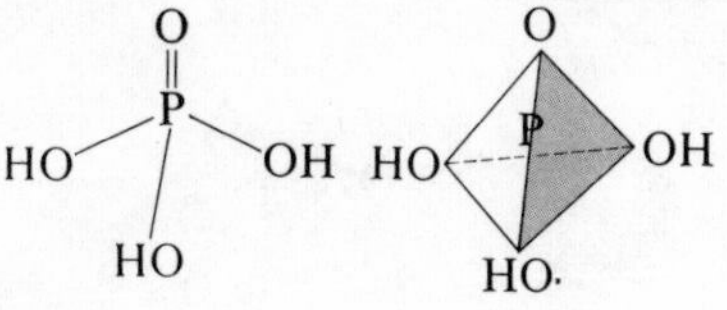

(8.39)

Pyrophosphoric acid $H_4P_2O_7$ has the structure (8.40) and is tetrabasic. It is prepared by dehydrating orthophosphoric acid between 213 and 316 °C. Pyrophosphates can be made by heating monohydrogen orthophosphates, when a molecule of water is lost between two orthophosphate ions.

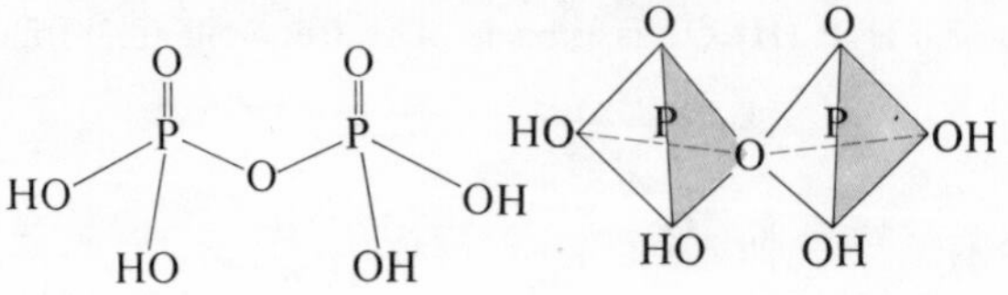

(8.40)

Triphosphoric acid $H_5P_3O_{10}$ (8.41, pentabasic); the free acid is unknown, but salts containing the triphosphate ion $P_3O_{10}^{5-}$ have been isolated.

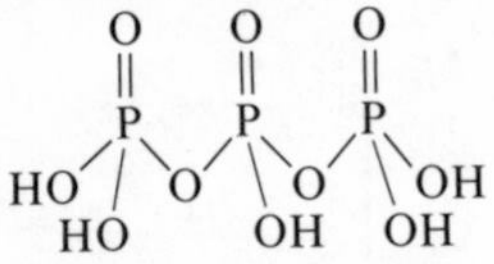

(8.41) 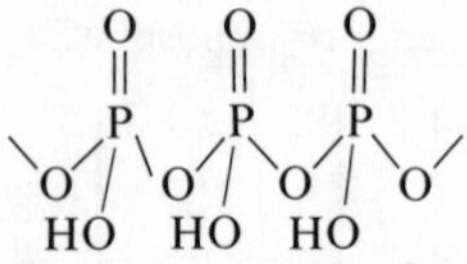 (8.42)

Metaphosphoric acid HPO_3 has the structure (8.42) and is monobasic per HPO_3 unit. This is a highly polymeric acid in which the PO_4 tetrahedra share two corners. Metaphosphoric acid is prepared by dehydrating orthophosphoric acid at 316 °C

$$H_3PO_4 \xrightarrow{316\,°C} HPO_3 + H_2O.$$

Several metaphosphates (and meta-arsenates) contain polymeric chains similar to those found in the acid, see Figure 91. However, some metaphosphates are known in which the PO_4 tetrahedra link up into either six-membered rings (trimetaphosphate $P_3O_9^{3-}$, 8.43) or eight-membered rings (tetrametaphosphate $P_4O_{12}^{4-}$, 8.44) having the general formula $(P_nO_{3n})^{n-}$. The eight-membered rings are known in both boat and chair conformations.

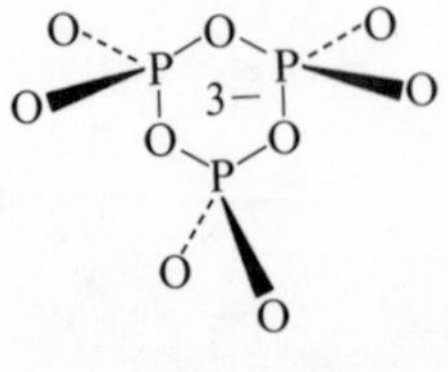

(8.43) 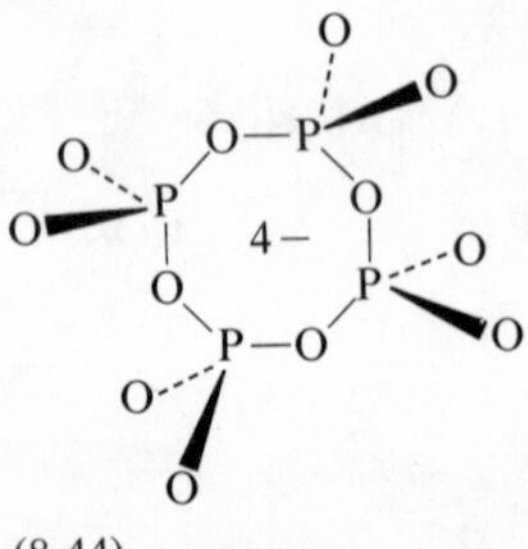 (8.44)

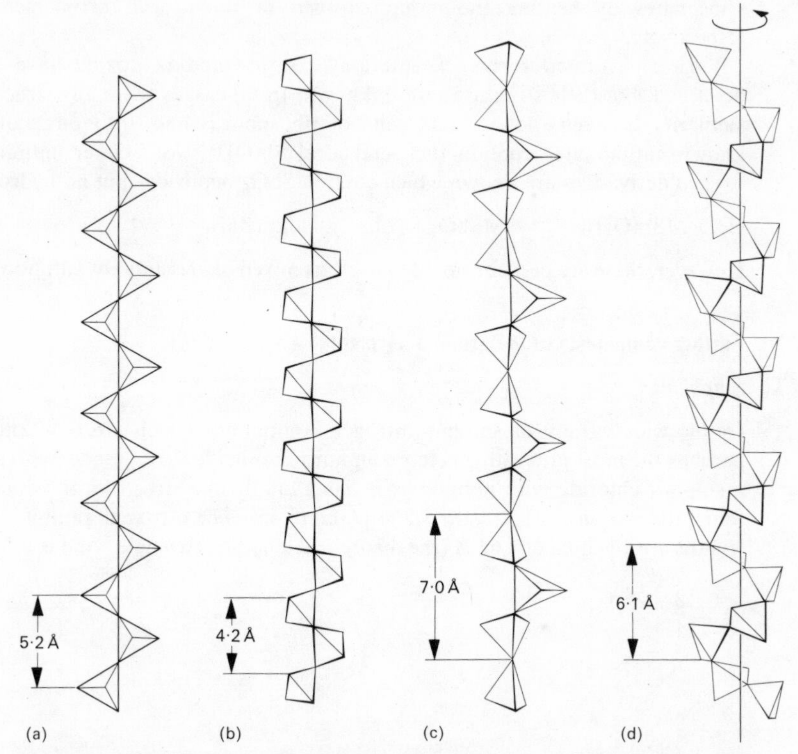

Figure 91 Various chain types in polymeric metaphosphates of the alkali metals: (a) lithium metaphosphate, (b) rubidium metaphosphate, (c) sodium metaphosphate (high-temperature form), (d) sodium metaphosphate. (Reproduced from H. J. Emeleus and A. G. Sharpe (eds.), *Advances in Inorganic Chemistry and Radiochemistry*, vol. 4, Academic Press, 1962, p. 54, with permission)

Hypophosphoric acid $H_4P_2O_6$ is rather out of place among the phosphoric acids. It is a tetrabasic acid which is unusual in having the two phosphorus atoms linked to each other directly and not bridged by an oxygen atom (8.45).

$$(HO)_2P(=O)\text{—}P(=O)(OH)_2$$

(8.45)

8.6.5 *Oxy-acids of arsenic and antimony*

The arsenates are similar to their phosphorus analogues except that the pyro- and meta-arsenic acids are not known as clearly defined compounds; their salts, however, can be made in a similar manner to pyrophosphates and metaphosphates by heating the monohydrogen or dihydrogen orthoarsenates, respectively.

Various hydrated forms of antimony trioxide and pentoxide have been described, but their structures are unknown. In no case is there any structural similarity between the oxy-acid salts of phosphorus and antimony; all the known antimonates contain the octahedral $Sb(OH)_6^-$ ion. Other antimony–oxygen derivatives are known which contain SbO_6 octahedra but no hydrogen,

e.g. $MSb(OH)_6 \xrightarrow[-H_2O]{heat} MSbO_3$ (M = alkali metal).

However, these are best considered purely as mixed oxides and not antimonates.

8.7 Further compounds of the group V elements

8.7.1 *Sulphides*

Of the wide variety of sulphur–nitrogen compounds which are now known, perhaps the most interesting is tetrasulphur tetranitride S_4N_4, made by treating a sulphur chloride with ammonia; it has a cradle-like structure, in which all four nitrogen atoms lie in the same plane (8.46). The nitrogen–sulphur bond lengths are all equal at 1·62 Å (the theoretical length for a single bond is 1·76 Å).

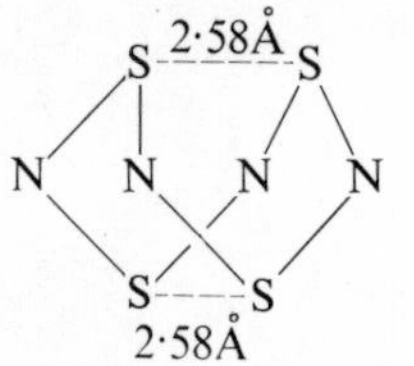

S 2·58Å S

N N N N

S S
2·58Å

(8.46)

The pairs of sulphur atoms above and below the plane of four nitrogen atoms are within 2·58 Å of each other (the sum of van der Waals radii for two sulphur atoms is 3·7 Å and the normal sulphur–sulphur single-bond length is 2·04 Å) which suggests that there is a weak sulphur–sulphur interaction. Delocalized π-bonding involving sulphur 3d and nitrogen 2p orbitals is thought to occur over the whole molecule, which would account for the rather short sulphur–nitrogen bonds.

Depending on the reagents employed, groups can be placed on either the nitrogen or sulphur atoms of tetrasulphur tetranitride, with resultant loss of delocalized π-bonding and a change in structure to a puckered eight-membered sulphur–nitrogen ring (e.g. 8.47 or 8.48).

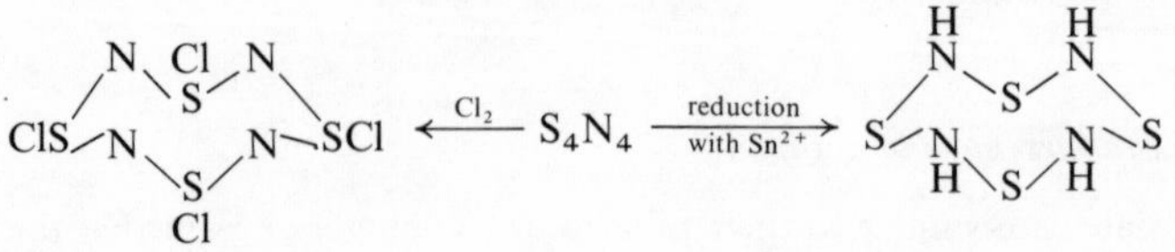

$$S_4N_4Cl_4 \xleftarrow{Cl_2} S_4N_4 \xrightarrow[\text{with } Sn^{2+}]{\text{reduction}} S_4(NH)_4$$

(8.47) $S_4N_4Cl_4$ (8.48) $S_4(NH)_4$

Tetrasulphur tetraimide $S_4(NH)_4$ (8.48) is one of the products arising from the reaction of ammonia with sulphur chlorides, and it can be considered as arising from a puckered S_8 ring by substitution of four sulphur atoms by nitrogen. All the sulphur imides in the series $S_{8-n}(NH)_n$, where $n = 1, 2, 3$ or 4, have been isolated and separated into their respective isomers (no isomers exist with two nitrogen atoms linked directly to each other in the ring). For example, $S_6(NH)_2$ has three isomers (8.49a–c). By treating the sulphur imides with S_2Cl_2 the rings can be coupled together via —S—S— bonds

$$2S_7NH + S_2Cl_2 \longrightarrow$$ N—S—S—N

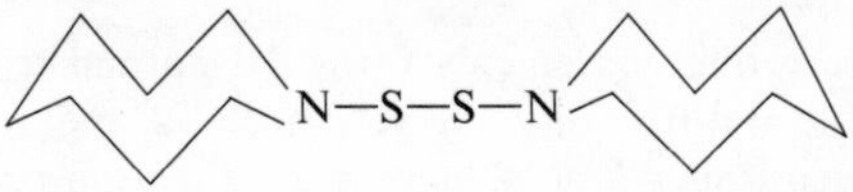

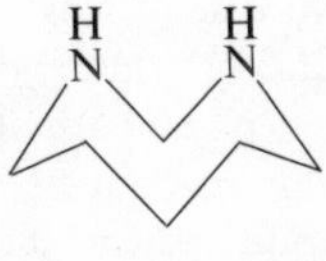

(8.49a) 1,3-isomer 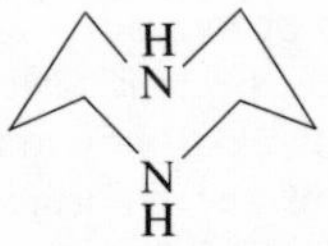 (8.49b) 1,5-isomer 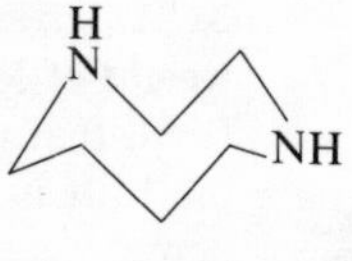 (8.49c) 1,4-isomer

As many as five rings have been coupled in this manner using the isomers of $S_6(NH)_2$ and S_2Cl_2.

The direct reaction of phosphorus and sulphur gives, under the appropriate conditions, four sulphides, of formulae P_4S_3 (8.50), P_4S_5 (8.51), P_4S_7 (8.52) and P_4S_{10} (8.53). The absence of P_4S_6 is surprising, because As_4S_6 has the same structure as P_4O_6, and P_4S_{10} has the P_4O_{10} structure. Although all these sulphides have a P_4S_x formula there is no P_4 unit common to their structures.

(8.50) P_4S_3 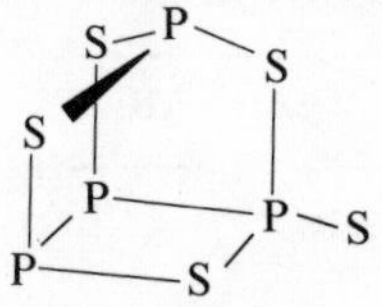 (8.51) P_4S_5 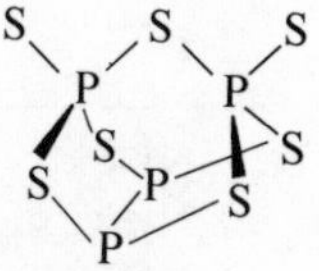 (8.52) P_4S_7

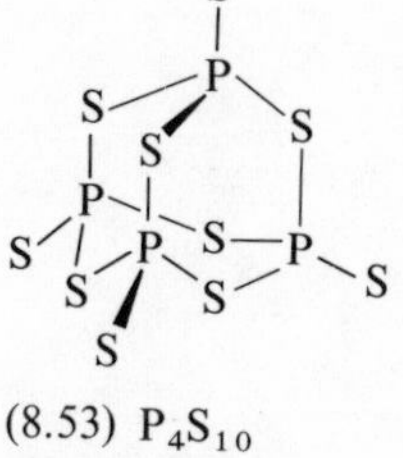

(8.53) P_4S_{10}

Arsenic is the only other group V element to react with sulphur to give a tetrasulphide. Arsenic tetrasulphide As_4S_4 has a cradle-like structure both in the solid and gaseous phases. Hydrogen sulphide passed into aqueous, acidified arsenites, acidified antimony salts or bismuth salts precipitates the corresponding trisulphide M_2S_3. On heating, arsenic trisulphide sublimes as the dimer As_4S_6, which has the same structure as P_4O_6.

8.7.2 *Halides*

All the group V elements form the normal trihalides, although nitrogen tribromide and tri-iodide are very unstable and are not known free of ammonia. Nitrogen is unable to achieve a coordination number of five, and hence has no MX_5 derivatives; the M—X bond strength decreases with increasing atomic weight of M (and X) and becomes less able to stabilize the M^V state, with the result that the pentahalides become fewer in number from phosphorus to bismuth. It is not completely clear why arsenic pentachloride has not been prepared; phase diagrams of the chlorine–arsenic trichloride system show no compound formation even at low temperatures. Attempts to stabilize the arsenic(V) state by partial replacement of chlorine in arsenic pentachloride by fluorine result in the formation of AsF_3Cl_2, but structural studies show that this is an ionic compound $AsCl_4^+ AsF_6^-$.

Only nitrogen is able to form strong p_π–p_π bonds and hence *cis* and *trans* difluorodiazine (8.54), which have a double bond between the nitrogen atoms, have no analogues among other elements in the group.

The trihalides of phosphorus, arsenic, antimony and bismuth can be made by

Table 25 Halides of the group V elements

	F	Cl	Br	I
N	NF_3	NCl_3	(NBr_3)	(NI_3)
	N_2F_4	—	—	—
	N_2F_2 (*cis* and *trans*)	—	—	—
P	PF_3	PCl_3	PBr_3	PI_3
	P_2F_4	P_2Cl_4	—	P_2I_4
	PF_5	PCl_5	PBr_5	—
As	AsF_3	$AsCl_3$	$AsBr_3$	AsI_3
	AsF_5	—	—	—
Sb	SbF_3	$SbCl_3$	$SbBr_3$	SbI_3
	SbF_5	$SbCl_5$	—	—
Bi	BiF_3	$BiCl_3$	$BiBr_3$	BiI_3
	BiF_5	—	—	—

N=N with F, F

(8.54a) *cis*

F, N=N, F

(8.54b) *trans*

direct reaction, whereas those of nitrogen are produced by treating ammonia or ammonium salts with the respective halogen. In the case of nitrogen trifluoride, the fluorine is produced *in situ* by electrolysing fused ammonium bifluoride

$$NH_4F.HF \xrightarrow{\text{electrolysis}} (F_2) \xrightarrow{NH_4^+} NF_3,$$

cis- and *trans*-difluorodiazine are minor products in this reaction. The instability of nitrogen trichloride, tribromide and tri-iodide is largely due to the high bond energy of the nitrogen N_2 molecule which makes these halides very endothermic compounds (e.g. ΔH_f for NCl_3 is $+230$ kJ mol^{-1}).

It is worth noting the different products formed on hydrolysis of nitrogen and phosphorus trichlorides

$$NCl_3 + 3OH^- \longrightarrow 2NH_3 + 3OCl^-,$$
$$PCl_3 + 3OH^- \longrightarrow H_2P(H)O_3 + 3Cl^-.$$

This has been attributed to the electronegativity sequence N > Cl > P, which results in differing polarities in the M—Cl bonds:

$$\overset{\delta-}{N}—\overset{\delta+}{Cl} \quad \text{and} \quad \overset{\delta+}{P}—\overset{\delta-}{Cl};$$

the negatively charged hydroxyl ion will attack the chlorine atoms in nitrogen trichloride and the phosphorus atom in phosphorus trichloride.

Although nitrogen has a high electronegativity, that of fluorine is even higher, which results in $\overset{\delta+}{N}—\overset{\delta-}{F}$ bond polarities in the nitrogen fluorides. The very small electric dipole moment of 0·2 debye possessed by nitrogen trifluoride (compare this to the dipole moment of ammonia, which is 1·44 debye) is thought to be due to the partial cancelling of the dipole moment due to the lone pair by the nitrogen–fluorine bond polarities, which act in the opposite direction (Figure 92).

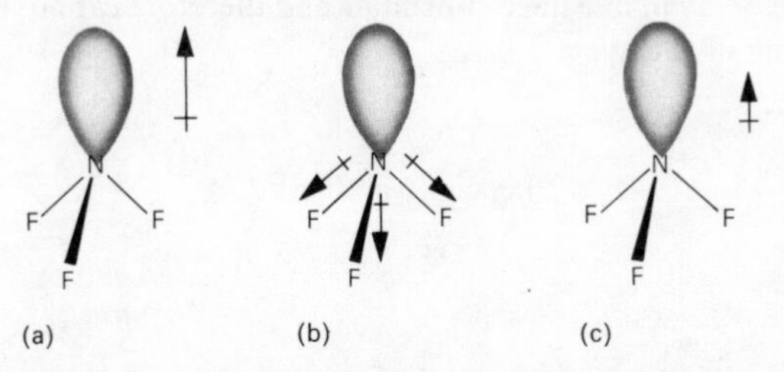

Figure 92 The dipole moment of nitrogen trifluoride: (a) lone-pair moment, (b) three nitrogen–fluorine bond moments, (c) small resultant moment

By passing nitrogen trifluoride or phosphorus trichloride through an electric discharge between mercury electrodes, one halogen atom is removed from each molecule by excited mercury atoms. The MX_2 fragments so formed couple together to give tetrafluorohydrazine $F_2N—NF_2$ or diphosphorus tetrachloride $Cl_2P—PCl_2$ respectively. In gaseous tetrafluorohydrazine at room temperature there are about equal amounts of the *trans* (8.55b) and *gauche* (8.55c) forms.

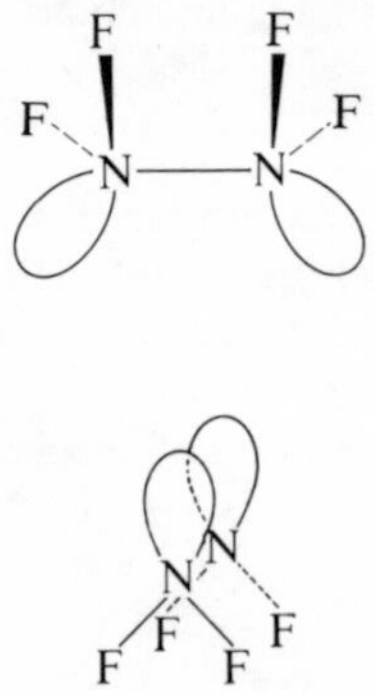

(8.55a) *cis* 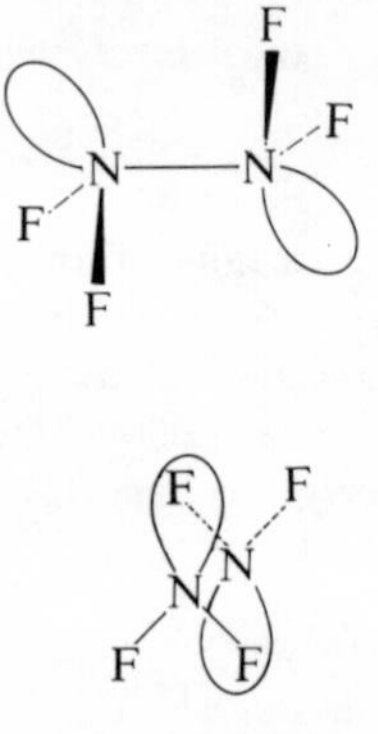 (8.55b) *trans* 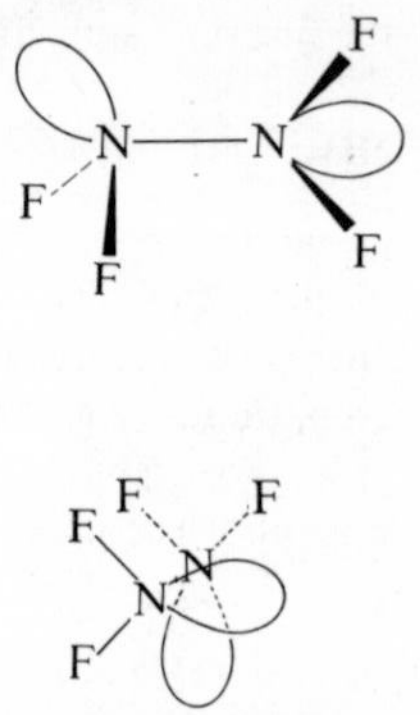 (8.55c) *gauche*

Phosphorus trifluoride and trichloride are very weak donor molecules towards non-transition elements such as boron because the highly electronegative fluorine and chlorine atoms reduce the effective availability of the lone-pair electrons by electron withdrawal. In sharp contrast these molecules form stable complexes with many transition metals,

e.g. $4CO + Ni(PCl_3)_4 \xleftarrow{PCl_3} Ni(CO)_4 \xrightarrow{PF_3} 4CO + Ni(PF_3)_4$.

This can be rationalized by assuming that, although the phosphorus-to-metal σ-donation is weakened by the presence of halogens on the phosphorus, overlap of filled metal d-orbitals and empty phosphorus 3d orbitals occurs, this π-interaction being strengthened by the electronegative halogens (8.56). The non-transition elements have no available filled d-orbitals and therefore cannot enter into secondary π-bonding of this type.

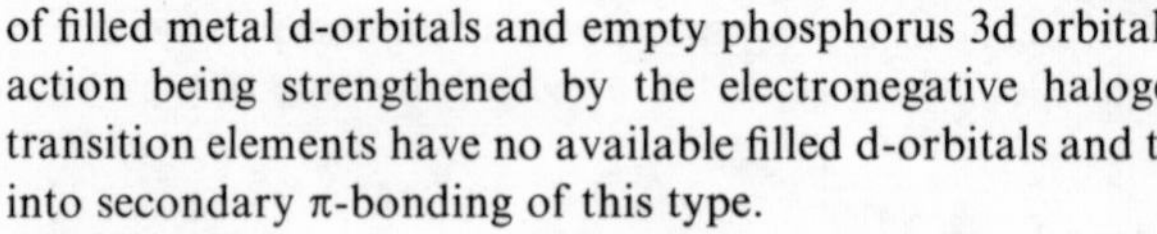

i.e. M—P—F

(8.56)

The structural variations found among the pentahalides deserve some comment. Phosphorus pentafluoride is found to have the trigonal-bipyramidal structure (8.57) both in the solid and vapour states. This is the structure adopted

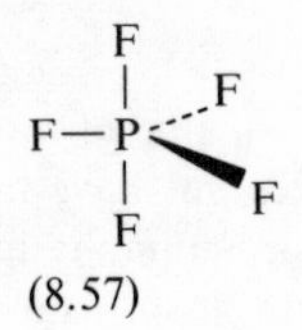

(8.57)

by *gaseous* phosphorus pentachloride, phosphorus pentabromide and antimony pentachloride, although they readily dissociate to MX_3 and free halogen. However, crystalline phosphorus pentachloride is ionic $PCl_4^+PCl_6^-$; this can be visualized as a salt in which phosphorus pentachloride has ionized to the tetrahedral PCl_4^+ cation and a chloride ion, the latter being 'solvated' by another phosphorus pentachloride molecule, giving the octahedral PCl_6^- anion. Phosphorus pentabromide on the other hand forms an ionic salt, $PBr_4^+Br^-$; apparently the energy released by forming the solvated ion PBr_6^- is not enough to compensate for the loss of lattice energy accompanying the increase in anionic radius as the bromide ion is changed into PBr_6^-. In some solvents phosphorus pentachloride ionizes to give PCl_4^+ and chloride ion, the latter now being solvated by the solvent and not with a further molecule of phosphorus pentachloride.

8.7.3 *Oxy-halides*

Phosphorus trihalides react with oxygen to produce the phosphoryl halides

$$2PX_3 + O_2 \longrightarrow 2OPX_3 \quad (\text{cf. } PX_3 + S \longrightarrow SPX_3),$$

which also result from the partial hydrolysis of the pentahalides

$$PX_5 + H_2O \xrightarrow{-2H_2O} OPX_3 \xrightarrow{3H_2O} H_3PO_4 + 3HX.$$

The phosphoryl halides possess a tetrahedral structure, the phosphorus–oxygen bond being considerably shortened by d_π–p_π bonding. They are powerful donor molecules and form complexes with compounds of both transition and non-transition elements, the *oxygen* atom being the donor site (8.58–61).

$Cl_3PO \rightarrow BCl_3$ (8.58) $\quad Cl_3PO \rightarrow SbCl_5$ (8.59) $\quad Cl_3PO \rightarrow FeCl_3$ (8.60) $\quad (Cl_3PO)_2 \rightarrow TiCl_4$. (8.61)

The controlled hydrolyses of the trihalides of antimony and bismuth produce the oxy-halides MOX; these have rather complex crystal structures and will not be discussed further.

8.7.4 *Salts of oxy-acids*

Only antimony and bismuth of the group V elements are capable of forming salts with the common oxy-acids, and even antimony does not have many; the rather unstable sulphate $Sb_2(SO_4)_3$, made by dissolving antimony in sulphuric acid, is perhaps the most well characterized. Bismuth dissolves in both nitric and sulphuric acids to give $Bi(NO_3)_3.5H_2O$ and $Bi_2(SO_4)_3$ respectively; these are hydrolysed quite readily to insoluble basic salts such as $BiO(NO_3)$ and $(BiO)_2SO_4$.

8.8 Organo-derivatives of the group V elements

A discussion of the amines R_nNH_{3-n} belongs to organic chemistry and will not be given here.

The halides of phosphorus, arsenic, antimony and bismuth react readily with an excess of either Grignard reagents or lithium reagents to form organo-derivatives,

$$\text{e.g.}\quad \begin{aligned} MCl_3+3RMgX &\longrightarrow MR_3,\\ OPCl_3+3LiR &\longrightarrow OPR_3,\\ MX_5+5LiR &\longrightarrow MR_5. \end{aligned}$$

A deficiency of the Grignard or lithium reagents normally leads to the formation of intermediate organo-halides, which are useful for synthetic work,

$$\text{e.g.}\quad \begin{aligned} PCl_3+LiR &\longrightarrow RPCl_2,\\ PCl_3+2LiR &\longrightarrow R_2PCl \xrightarrow{Na} R_2P{-}PR_2. \end{aligned}$$

Direct synthesis of the organo-derivatives by simply heating a group V element with an alkyl or aryl halide is sometimes possible, as in the case of the trifluoromethyls

$$CF_3I+P \xrightarrow{heat} P(CF_3)_3+IP(CF_3)_2+I_2PCF_3$$

$$IP(CF_3)_2 \xrightarrow[Hg]{heat} HgI_2+(CF_3)_2P{-}P(CF_3)_2 \qquad I_2PCF_3 \xrightarrow[Hg]{heat} (PCF_3)_n+HgI_2$$

$(n = 4,5$ or 6; cyclic derivatives)

The group V organo-derivatives are strong donor molecules, and for this reason alone they have been extensively studied in recent years. As mentioned on p. 180 the initial σ-donation to the central metal in a complex is thought to be augmented by d_π–p_π back-bonding from filled metal d-orbitals to empty d-orbitals on the group V atom.

Chapter 9
Group VI. Oxygen, Sulphur, Selenium, Tellurium and Polonium

9.1 Introduction

Energy considerations for these elements completely rule out the possibility of an inert-gas configuration produced by loss of all six outer electrons to give the M^{6+} ions, but Te^{4+} and Po^{4+} (both having an ns^2 electron configuration) appear to have been established in a few cases: thus tellurium dioxide TeO_2 and polonium dioxide PoO_2 are essentially ionic crystals and there is evidence of a hydrated polonium sulphate $Po(SO_4)_2$ and a nitrate $Po(NO_3)_4$. A few polyatomic cations have also been described,

e.g. $O_2 + PtF_6 \longrightarrow O_2^+PtF_6^-$,
$M + \text{pure } H_2SO_4 \longrightarrow M_4^{2+}$ (M = Se, Te).

The yellow Se_4^{2+} cation has a square planar atomic arrangement.

The most usual ways in which these elements attain a rare-gas electronic structure are by

(a) forming M^{2-} ions, as in the alkali-metal oxides, sulphides, selenides, tellurides and polonides;

(b) forming two covalent bonds as in H_2O and $(CH_3)_2S$;

(c) forming MR^- ions as in OH^- and SH^-.

Oxygen differs from sulphur, selenium, tellurium and polonium in being too small to accommodate more than four electron pairs round itself, and hence

Table 26 The Group VI Elements

Element		1st IE / kJ mol^{-1}	2nd IE / kJ mol^{-1}	3rd IE / kJ mol^{-1}	4th IE / kJ mol^{-1}	Covalent radius/Å
Oxygen O	$1s^22s^22p^4$	1314	3393	5301	7469	0·66
Sulphur S	$[Ne]3s^23p^4$	1000	2260	3379	4564	1·04
Selenium Se	$[Ar]3d^{10}4s^24p^4$	941·4	2075	3088	4138	1·17
Tellurium Te	$[Kr]4d^{10}5s^25p^4$	870·4	1795	3012	3683	1·37
Polonium Po	$[Xe]4f^{14}5d^{10}6s^26p^4$	811·7	—	—	—	—

there are no oxygen analogues of, for example, sulphur tetrafluoride (9.1) or sulphur hexafluoride (9.2).

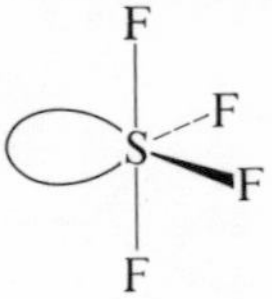

(9.1) trigonal-bipyramidal coordination

(9.2) octahedral coordination

The shielding effects of the s- and p-electrons ensure that the outer d-orbitals decrease in size across the periodic table so that, from group IV to group VII they become more and more spatially suitable for interaction with filled oxygen 2p orbitals, resulting in progressively stronger d_π–p_π bonds. This is the basic reason for the differences in behaviour of oxy-acids and oxy-salts derived from the elements silicon, phosphorus, sulphur and chlorine. In the silicates, silicon–oxygen π-bonding is relatively weak and polymerization to polysilicates via O—Si—O σ-bonding is very important. In the phosphates the effect is less marked, but a fairly wide variety of polyphosphates are known. At sulphur, π-bonding becomes more dominant and hence few oxy-acids or their salts containing more than one S—O—S bridge are known. No oxy-acid derivatives of chlorine have been prepared containing more than one chlorine atom; the chlorine–oxygen bond lengths in the chlorite, chlorate and perchlorate groups are very short and suggest substantial d_π–p_π bonding between the chlorine and oxygen atoms.

In passing down the group from oxygen to polonium there is a gradual change towards metallic character due to the increased shielding effect of the inner electrons on the outer *n*s and *n*p electrons as the quantum number *n* varies from 2 to 6. Chemically, the developing metallic character will manifest itself in various ways, such as an increasing tendency to form positive ions, a decreasing stability of the M^{2-} ions and a weakening of the M—X bonds. Physically, it is reflected both in the structures of the free elements, which change from a diatomic molecule (oxygen) through rings and chain molecules (sulphur, selenium and tellurium) to a simple metallic lattice (polonium), and in the electrical properties of the elements; oxygen and sulphur being insulators, selenium and tellurium being semiconductors, and polonium showing metallic conduction.

The electronegativity of oxygen is very high, being second only to that of fluorine. This leads to a greater degree of ionic character in M—O bonds, to the limit where many oxides actually have ionic structures. For the same reason the phenomenon of hydrogen bonding (section 2.8) within this group is almost completely restricted to hydroxy-compounds.

The oxygen molecule O_2 is rather unusual in that it is paramagnetic, with two unpaired electrons occupying the antibonding π^*2p_x and π^*2p_y orbitals (see Figure 93). Due to the antibonding nature of the π^*2p molecular orbitals, the

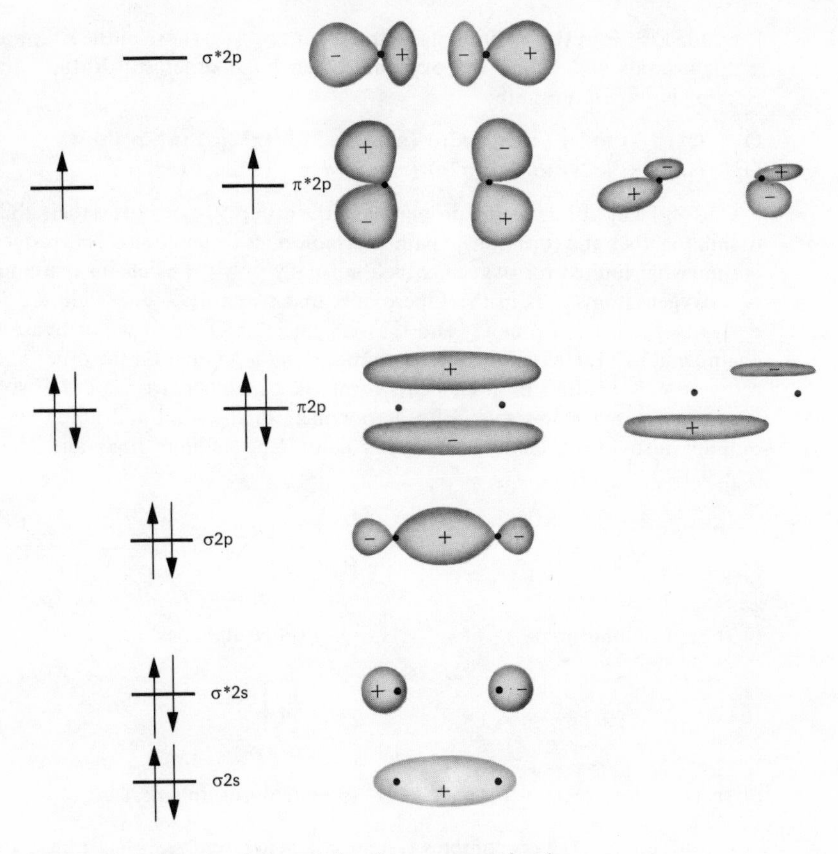

Figure 93 Energy-level diagram for the oxygen molecule

presence of the two unpaired electrons reduces the overall bond order in the oxygen molecule to two. This can be demonstrated by removing one of the unpaired electrons to give O_2^+, when the molecule actually becomes *more* stable even though it now has fewer electrons (Table 27). In the superoxide O_2^- and

Table 27

	Bond length/Å	*Dissociation energy*/kJ mol^{-1}	*Bond order*
O_2^+	1·12	623	2·5
O_2	1·21	490	2·0
O_2^-	1·28	—	1·5
O_2^{2-}	1·49	398	1·0

peroxide O_2^{2-} ions the additional electrons pair up with those in the π^*2p_x and π^*2p_y orbitals and reduce the oxygen–oxygen bond order still further; their electron configurations are:

$$O_2^- \quad (1s)^2(1s)^2(\sigma 2s)^2(\sigma^*2s)^2(\sigma 2p)^2(\pi 2p_x)^2(\pi 2p_y)^2(\pi^*2p_x)^2(\pi^*2p_y)^1,$$
$$O_2^{2-} \quad (1s)^2(1s)^2(\sigma 2s)^2(\sigma^*2s)^2(\sigma 2p)^2(\pi 2p_x)^2(\pi 2p_y)^2(\pi^*2p_x)^2(\pi^*2p_y)^2.$$

One of the most interesting properties of the group VI elements is their ability to link together and form long, chain-like molecules (catenation). This property is somewhat limited for oxygen, which normally only forms chains containing two oxygen atoms – as in the superoxides and peroxides – although in a few derivatives, such as ozone O_3 and the very unstable O_4F_2, it will increase the chain length to three or four atoms. Sulphur and selenium, on the other hand, catenate very readily; both elements form eight-membered rings in the solid, liquid and gaseous states, and linear polymers in the solid and liquid states. Sulphur also forms many compounds containing sulphur rings and chains (e.g. 9.3–6).

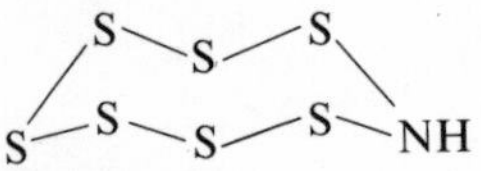

S–S–S–S–S–S–S–NH ring

(9.3) heptasulphurimide

H–S[S]$_n$S–H

(n = 0–6)

(9.4) (polysulphanes)

$M_2^+S_n^{2-}$

(n = 2–4)

(9.5) (polysulphides)

Cl–S[S]$_n$S–Cl

(n = 0–4)

(9.6) (chlorosulphanes)

Liquid sulphur is a very complex system, which in recent years has been studied using a wide variety of techniques. On melting at about 115 °C, rhombic sulphur (containing S_8 rings) gives a mobile yellow liquid which consists mainly of S_8 rings, but also probably contains a low percentage of S_8 chains

$$S_8(\text{rings}) \xrightleftharpoons{\text{m.p.}} S_8(\text{rings}) + S_8(\text{chains}).$$

On further heating, the viscosity of the liquid falls (as is typical of all normal liquids), until at 159 °C the viscosity begins to rise very sharply, reaching a maximum value at 170 °C (see Figure 94a). This increase in viscosity, which parallels a darkening of the colour to deep red, is due to the production of very-long-chain molecules in the liquid, the average chain containing about 10^6 atoms at 170 °C. At least some, if not all, of the chains are diradicals, and the liquid is consequently found to be slightly paramagnetic; the number of free radicals increases by nearly two-hundred-fold between 200 °C and 375 °C as the long chains break up into smaller ones. Other species, such as large rings, are also thought to be present (Figure 94b), but make up only a small proportion of the liquid. At the boiling point (444 °C) chains containing many thousands of sulphur atoms are still present in the liquid, but the vapour contains mainly S_7 (40 per cent), S_6 (30 per cent) and S_8 (20 per cent), with minor amounts of S_5, S_4 and S_2. Selenium melts at 220 °C and the viscosity of the liquid decreases rapidly with increasing temperature, indicating that the selenium chains originally present in the solid state are breaking up into smaller units, in contrast to the behaviour of sulphur. At the boiling point of selenium (685 °C) the vapour consists essentially of Se_6 and Se_2 molecules.

In recent years it has been found that certain transition-metal derivatives will form complexes which contain molecular oxygen apparently π-bonded to the metal (e.g. 9.7, 9.8).

$$Pt(P\phi_3)_4 + O_2 \longrightarrow 2P\phi_3O + (P\phi_3)_2PtO_2 \quad (9.7)$$

$$IrX(CO)(P\phi_3)_2 + O_2 \longrightarrow O_2IrX(CO)(P\phi_3)_2 \quad (9.8)$$

$$(\phi = C_6H_5;\ X = Cl, Br, I)$$

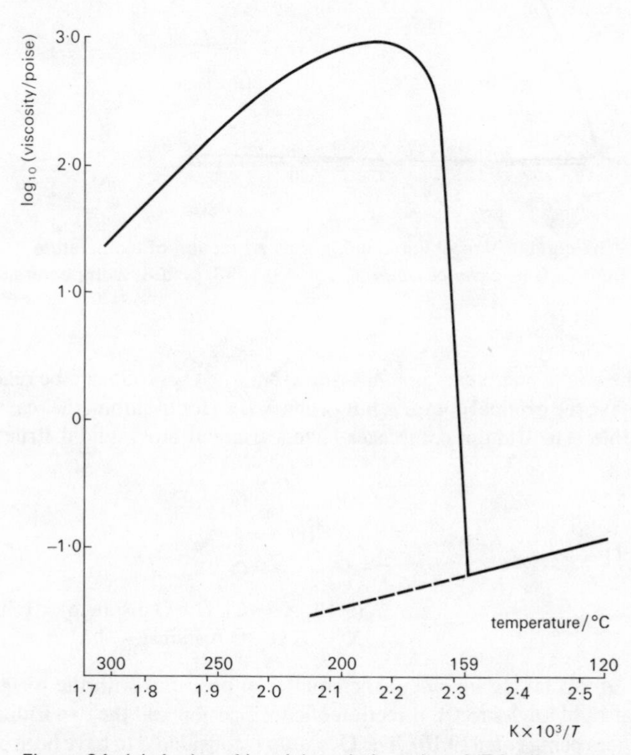

Figure 94 (a) A plot of log (viscosity) against $1/T$ for liquid sulphur. The dotted part of the curve is the extrapolated viscosity for S_8. (Reproduced from G. Gee, *Science Journal*, vol. 43, 1953, p. 204, with permission)

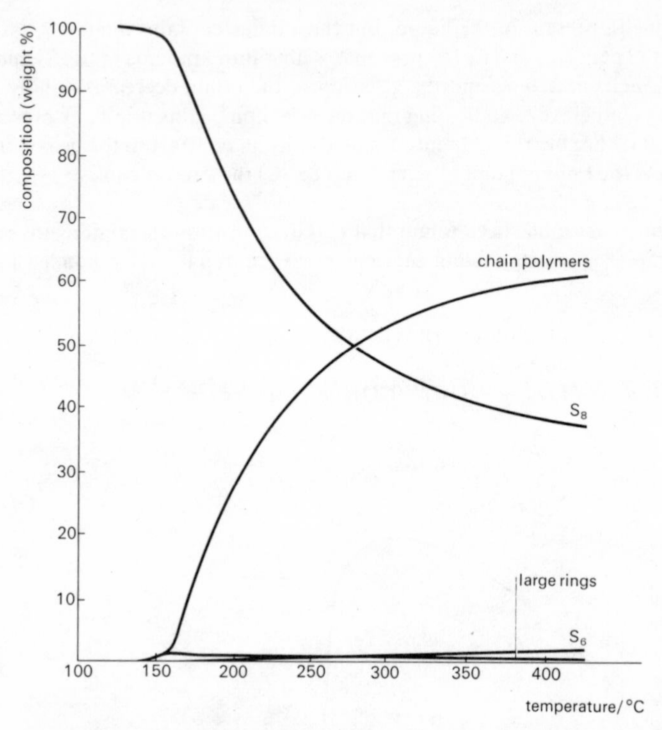

Figure 94 (b) The composition of liquid sulphur as a function of temperature. (Reproduced from G. Gee, *Science Journal*, vol. 43, 1953, p. 208, with permission)

When X in the above complexes is a chlorine atom, the O_2 group can be readily removed to leave the original complex but, when X is an iodine atom, the reaction is not reversible. The iridium complexes have a trigonal-bipyramidal structure

$P\phi_3$ X
OC—Ir O
$P\phi_3$ O

(9.9)

X O
90°
Ir
OC O

(9.10; X = Cl, O—O distance = 1·30 Å; X = I, O—O distance = 1·51 Å.)

(9.9), the O_2 group taking up *one* of the equatorial positions with the oxygen–oxygen axis at right angles to the direction of coordination and the two iridium–oxygen distances being equal (9.10). The O_2 group is considered to have both oxygen atoms sp^2 hybridized with *two* lone pairs of electrons on *each* oxygen and a double bond $(\sigma + \pi)$ between the atoms; this differs from the electronic structure of the free oxygen molecule in that it results in a diamagnetic state iso-electronic with ethylene. The iridium is dsp^3 hybridized and receives an electron pair from the π-bond in O_2; back-bonding from a filled d-orbital on iridium to the empty π^* (antibonding) orbital on the O_2 group probably also occurs (9.11). The extent of

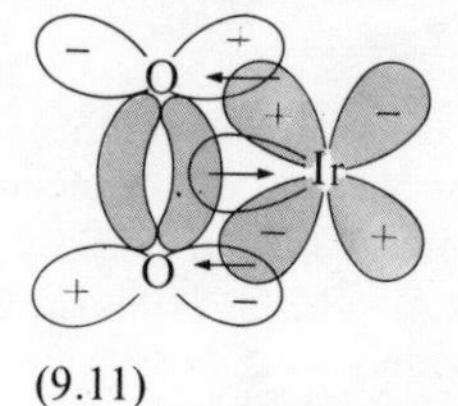

(9.11)

back-bonding will depend on the available electron density at the iridium atom and will be expected to be least when the most electronegative halogen is present in the complex, resulting in a weaker complex when X is a chlorine atom than when X is an iodine atom. This may explain why the formation of the complex $O_2IrX(CO)(P\phi_3)_2$ is reversible when X is a chlorine atom but not when it is an iodine atom. Furthermore, since the back-bonding involves the filling of an antibonding orbital on the O_2 group, the amount of back-donation will be least in the chloro-complex and the oxygen–oxygen distance will consequently be shorter (see data given above).

These molecular-oxygen complexes (and more especially the 'reversible complexes') are of great interest because they are closely related to the iron–oxygen complex which occurs in oxyhaemoglobin. This complex acts as an oxygen carrier in the blood, the formation of the iron–oxygen system being readily reversible. Haemoglobin is an iron-containing protein which has iron(II) protoporphyrin (9.12), called haem, as its prosthetic group (i.e. as its reactive site).

H_3C CH_2CH_2COOH

$H_2C{=}CH$ N CH_2CH_2COOH

N—Fe—N

H_3C N CH_3

$H_2C{=}CH$ CH_3

(9.12)

Blood haemoglobin has four such haem groups and has a molecular weight of about 64 500. In oxyhaemoglobin the O_2 ligand is thought to lie with the oxygen–oxygen axis parallel to the plane of haem and with the two iron–oxygen distances equal (9.13).

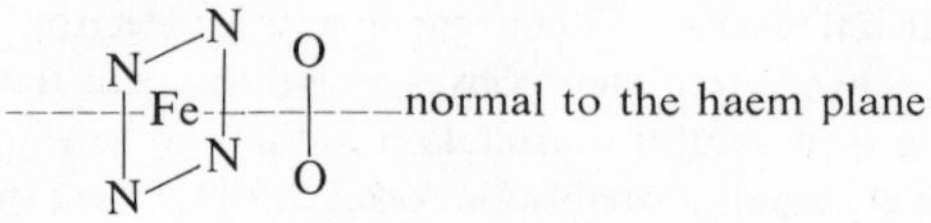

```
     N
  N⁄  |    O
-----|-Fe-|------|------normal to the haem plane
     |   N     O
  N⁄
```

(9.13)

All twenty-seven known isotopes of polonium are radioactive and of these only three have sufficiently long half-lives to allow a tracer study of the element's chemistry: polonium-210 (half-life 138·4 days), polonium-208 (half-life 2·9 years) and polonium-209 (half-life 42 years). Unfortunately, polonium-210 is the most accessible isotope of the three and hence is most widely used in tracer work, elaborate safety precautions having to be taken because of its intense α-activity (the maximum amount of polonium which is thought permissible in the human body is only about 10^{-11} g, or 10^{10} atoms). The work is normally carried out inside a glove box which is essentially a miniature laboratory to which the scientist's only access is via either thick rubber gloves efficiently sealed into one side of the box or remote-controlled robot arms. As well as hampering chemical study because of the danger to personnel, the intense α-activity causes fairly rapid decomposition of polonium compounds; for example, iodine is liberated from polonium iodate, organic acid derivatives of polonium char and ammines are reduced to free polonium metal (possibly by atomic hydrogen arising from α-bombardment of the ammonia ligands).

9.2 Occurrence of the group VI elements

9.2.1 *Oxygen*

Most abundant of all the elements, making up about 49·4 per cent of the Earth's crust and 23 per cent (by weight) of the atmosphere.

9.2.2 *Sulphur*

About 0·05 per cent abundance; occurs in the free state, as sulphides and as sulphates.

9.2.3 *Selenium*

0·09 p.p.m. of the Earth's crust; occurs as selenides of lead, copper, silver, mercury and nickel, and the main source is as a by-product from the treatment of the sulphide ores of these elements, especially those of copper.

9.2.4 *Tellurium*

At 0·002 p.p.m. abundance, it is about as rare as gold; found as metallic tellurides, often of silver and gold; main source is anode slime from electrolytic purification of copper.

9.2.5 *Polonium*

One of the rarest naturally occurring elements (about 10^{-10} p.p.m. of the Earth's crust); main natural source is uranium ores, which contain about 10^{-4} g of polonium per tonne. All isotopes are radioactive.

9.3 **Extraction**

9.3.1 *Oxygen*

The main industrial source of oxygen is from the fractional distillation of liquid air. The material may be transported either as the highly compressed gas or as the liquid in vacuum-jacketed tanks.

9.3.2 *Sulphur*

Native sulphur of high purity (about 99·5 per cent) is mined using the Frasch process in which a boring is made down to the deposit and lined with three concentric pipes (Figure 95). Superheated water, when pumped down the outer pipe, melts the sulphur which is then pumped to the surface by applying compressed air to the central tube.

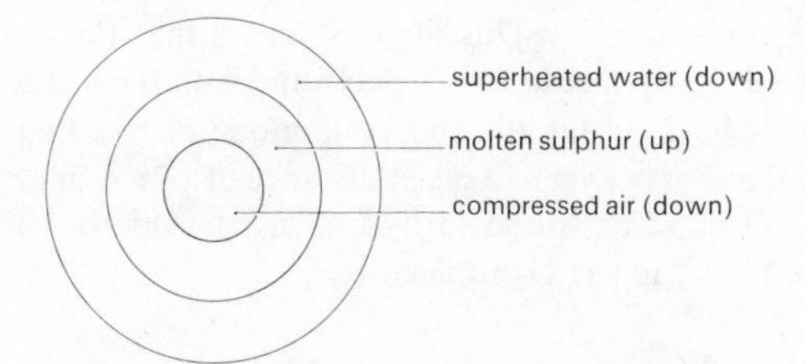

Figure 95 Cross-section through the concentric pipes of the Frasch process

9.3.3 *Selenium and tellurium*

The 'anode mud' resulting from the electrolytic purification of copper contains 3–28 per cent of selenium and about 8 per cent of tellurium. The dried mud is fused with soda ash and silica and blown with air to oxidize the selenium and tellurium to their dioxides. Addition of sodium hydroxide to the melt converts the dioxides to sodium selenite and sodium tellurite. After cooling and leaching with water, sulphuric acid is added to neutralize the solution, when tellurium dioxide is precipitated leaving selenious acid in solution. Tellurium dioxide is reduced to tellurium with carbon at red heat and selenium precipitated from the mother liquor by adding sulphur dioxide

$$H_2SeO_3 + 2SO_2 + H_2O \longrightarrow Se\downarrow + 2H_2SO_4.$$

9.3.4 *Polonium*

Natural sources are not normally used for the production of polonium because it occurs in such minute amounts. Polonium-210 is now made artificially via the neutron irradiation of natural bismuth (100 per cent bismuth-209)

$$^{209}_{83}Bi \xrightarrow{\text{n, } \gamma \text{ reaction}} {}^{210}_{83}Bi \xrightarrow[T_{\frac{1}{2}},\ 5 \text{ days}]{\beta^- \text{ decay}} {}^{210}_{84}Po.$$

9.4 The elements

The increasing metallic character down group VI finally manifests itself with the typically metallic lattices adopted by the two allotropic forms of elemental polonium. Both forms have a polonium coordination of six, the low-temperature α-polonium being cubic and the high-temperature β-polonium rhombohedral,

$$\alpha\text{-Po} \underset{}{\overset{\sim 36\,^\circ\text{C}}{\rightleftharpoons}} \beta\text{-Po}.$$

The vapour at the boiling point (962 °C) contains Po_2 molecules and presumably represents the maximum chain length attainable by polonium.

Oxygen normally exists as the diatomic molecule O_2, the electronic structure of which was discussed on p. 212. The action of a silent electric discharge converts the paramagnetic O_2 into diamagnetic ozone O_3, which is a non-linear molecule having an O—O—O bond angle of 117°. This angle suggests that the central oxygen atom is sp^2 hybridized (9.14). Ozone is iso-electronic with the nitrite ion NO_2^-, and the two molecules have essentially the same mode of bonding (see p. 198), which accounts for the short oxygen–oxygen distance of 1·28 Å in ozone. (The length of a single O—O bond as found in hydrogen peroxide is 1·49 Å, whilst the O=O distance is 1·21 Å in the O_2 molecule.)

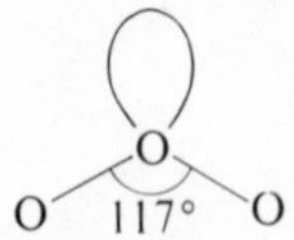

O

O 117° O

(9.14)

Sulphur and selenium are rather similar to one another in that both normally adopt either M_8 rings or polymeric chains in their various allotropic forms. (Sulphur has a complex allotropy and about thirty different modifications have been described, several arising from the different crystal packings of S_8 rings;

S_6, S_7, S_{10} and S_{12} ring derivatives can also be prepared.) The eight-membered rings of sulphur and selenium are puckered (9.15).

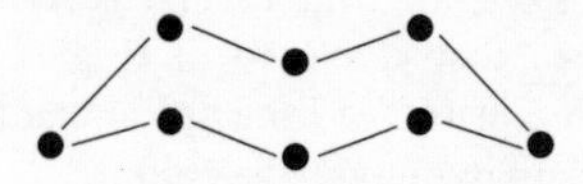

(9.15; S—S distance 2·04 Å, Se—Se distance 2·34 Å, angle S—S—S 108°, angle Se—Se—Se 106°.)

Table 28 Some Common Allotropic Forms of Sulphur and Selenium

Name of allotrope	*Basic unit of allotrope*	*Preparation*
Sulphur		
Rhombic sulphur (S_α)	S_8 rings	Recrystallize sulphur from carbon disulphide; all other forms revert to S_α on standing.
Monoclinic sulphur (S_β)	S_8 rings	Slow crystallization of hot solutions of sulphur in organic solvents; transition point for $S_\alpha \rightleftharpoons S_\beta$ is probably 96·5 °C but this has been disputed.
Plastic sulphur	Long chains of sulphur atoms	Fuse sulphur and rapidly chill by pouring into water.
Engel's sulphur	S_6 rings	Acidify sodium thiosulphate solution, extract the precipitated sulphur with toluene and crystallize.
Selenium		
Grey (or metallic) selenium	Long chains of selenium atoms	Slowly cool molten selenium or heat any other modification.
Red crystalline selenium (Se_α and Se_β)	Se_8 rings	Two crystalline modifications of red selenium are known; both obtained by crystallizing carbon disulphide solution of black selenium.
Black (vitreous) selenium	Probably very large rings of selenium atoms ($\sim Se_{1000}$)	Pour molten selenium into cold water.

With sulphur the eight-membered ring is the stable form at normal temperatures (as rhombic sulphur or S_α) and all other allotropes including the chain polymers (plastic sulphur) revert to S_α on standing; with selenium the reverse is true, the chain polymer found in grey selenium is the most stable form (Figure 96). Tellurium forms only one modification which consists of long spiral chains of tellurium atoms, very similar to the chains occurring in grey selenium.

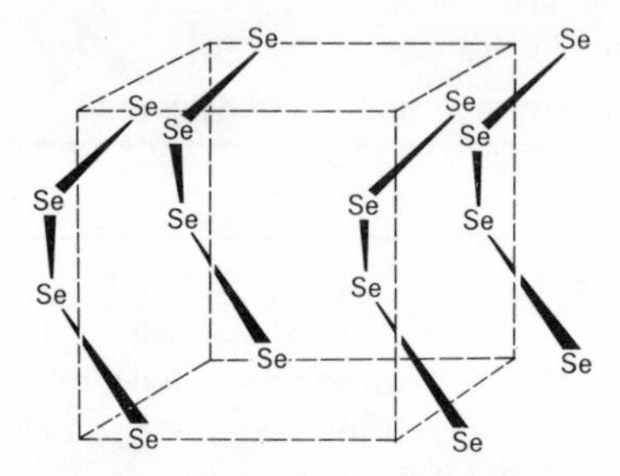

Figure 96 The crystal structure of grey selenium showing spiral chains of selenium atoms

To a certain extent it is possible to interchange the group VI elements in the eight-membered rings. For example, the vapour above heated mixtures of sulphur and selenium contains cyclic S_nSe_{8-n} (n = 1–7) species and crystalline S_4Se_4 has been described; heated mixtures of tellurium and sulphur produce only S_7Te in the gas phase.

9.5 Semiconductor properties of selenium

Grey selenium and tellurium have electrical conductivities intermediate between those of insulators and those of metals and which increase with rise in temperature. These semiconductor properties are explained by assuming that at low temperatures all the available low-energy orbitals are completely filled, so that no electrons are mobile (or 'free') to carry an electric current; under these conditions selenium and tellurium behave as insulators (Figure 97a). As the temperature rises, thermal energy promotes a small number of electrons to low-lying empty orbitals which allows some movement of electrons in the now incompletely filled orbitals when an electric field is applied (Figure 97b).
Illumination of selenium, especially with light having a wavelength of about 7000 Å, also induces promotion of electrons into the empty orbitals and allows the passage of an electric current. In this way the conductivity of selenium can be increased by as much as a thousandfold and the phenomenon forms the basis of selenium photocells used in the measurement of light intensity.

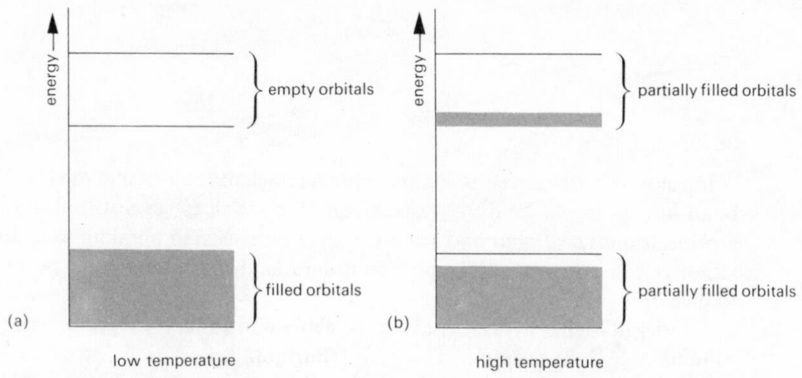

Figure 97 Energy-level diagrams of grey selenium: (a) low temperature, (b) high temperature

9.6 Hydrides of the group VI elements

All the MH_2 hydrides can be made directly from the elements, except for polonium hydride. A better synthesis for hydrogen sulphide, selenide or telluride is to treat a metal sulphide, selenide or telluride with aqueous acid. The small hydrogen 1s orbital becomes less and less compatible with the bonding orbitals of the heavier group VI elements, making the M—H bonds decrease in strength down the group. This is reflected in the heats of formation of the hydrides, listed in Table 29. Consequently hydrogen selenide and hydrogen telluride are relatively unstable and strong reducing agents.

Table 29 Heats of Formation of the Hydrides of the Group VI Elements

	Heat of formation/kJ mol^{-1}
Water H_2O	−241·8
Hydrogen sulphide H_2S	−20·1
Hydrogen selenide H_2Se	+85·8
Hydrogen telluride H_2Te	+154·4

The H—O—H bond angle of 105° in the isolated water molecule is generally assumed to involve sp^3 hybridization of the oxygen atom, the H—O—H angle being slightly less than the expected tetrahedral value of 109° 28′ due to steric influence of the lone-pair electrons. Since the lone pairs are confined to only one nucleus (oxygen), their orbitals are relatively large and require more space round the oxygen atom than the bond-pair electrons, which are attracted to two nuclei (oxygen and hydrogen) (9.16).

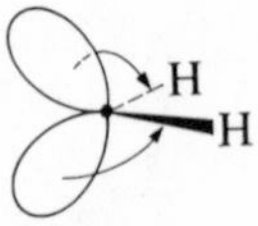

(9.16)

In gaseous hydrogen sulphide and hydrogen selenide molecules the H—M—H bond angles are 92° and 91° respectively; these values suggest that only pure p-orbitals on the sulphur and selenium atoms are used in bonding to hydrogen, although some theoretical chemists consider that d-orbitals may also be involved to give dp hybrids.

A series of higher hydrides having the general formula H_2M_x (oxygen, $x = 2$; sulphur, $x = 2$–8; selenium, $x = 2, 3$; tellurium, $x = 2$) are known. Except for hydrogen peroxide (see section 9.6.2), these higher hydrides are very unstable and readily deposit the free element

$$H_2M_x \longrightarrow (x-1)M + H_2M \quad \begin{cases} \Delta H_f^\circ,\ H_2S_2 = -8{\cdot}4\ \text{kJ mol}^{-1}, \\ \Delta H_f^\circ,\ H_2Se_2 = +105\ \text{kJ mol}^{-1}, \\ \Delta H_f^\circ,\ H_2Te_2 = +166\ \text{kJ mol}^{-1}. \end{cases}$$

A mixture of polysulphanes may be prepared by treating sodium polysulphide with hydrochloric acid, and it is separated into the various components ($x = 2$–8) by simultaneous distillation and thermal cracking; the purified sulphanes must be stored at low temperatures to prevent decomposition. Although the rapid decomposition of the sulphanes to give free sulphur might suggest otherwise, it has been established that the polysulphanes (e.g. 9.17) do not contain branched sulphur chains.

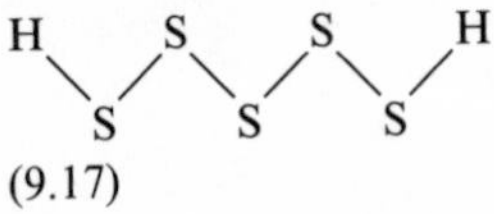

(9.17)

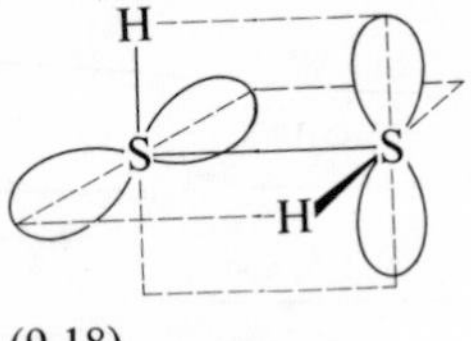

(9.18)

The molecular structure of disulphane (9.18) is similar to that of hydrogen peroxide, showing that the lone-pair electrons in the sulphur 3p orbitals exert considerable steric influence and cause the molecule to adopt a configuration in which the lone pairs are farthest apart and hence interact least with each other.

9.6.1 *Water* H_2O

A graphical plot of the boiling points of the group VI hydrides MH_2 against the atomic weight of M clearly shows that the boiling point of water is anomalously high (see Figure 37, p. 62); direct extrapolation of the graph from hydrogen sulphide to water gives the expected boiling point of water as about −80 °C, which is almost 200 °C too low. As discussed in section 2.8, this anomalous boiling point of water is due to hydrogen bonding which augments the normal intermolecular forces (van der Waals forces) within the liquid. The hydrogen bond occurs along the line of the sp^3 lone-pair electrons on the oxygen atoms (9.19).

(9.19)

The tetrahedral coordination forced upon the oxygen atoms gives water an open structure both in the solid and liquid states; despite the open structure of liquid water, the solubility of non-polar compounds is very low because of the solvent's strong intermolecular forces. The only substances soluble in water are those capable of opening up the hydrogen bonds: for example, many metal salts do this by forming strong ion–dipole interactions between the ions (both cations and anions) and neighbouring water molecules; furthermore, the anions can sometimes form hydrogen bonds with water molecules if they contain highly electronegative atoms (e.g. F^-, BF_4^-, SO_4^{2-}). The water solubility of alcohols, sugars, organic carboxylic acids and phenols is due largely to their being able to participate in hydrogen bonding with the solvent molecules. Even in the gas phase it is possible to detect hydrogen bonding between water molecules; for example, mass-spectral analyses of water vapour show the presence of clusters containing as many as ten H_2O molecules.

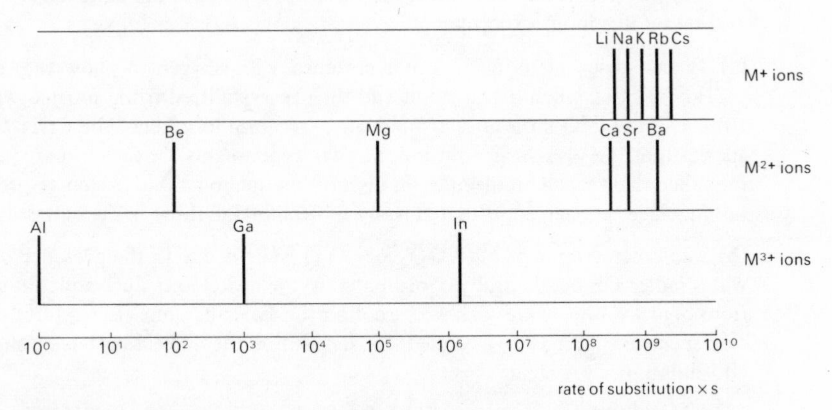

Figure 98 The rate of water substitution in hydrated M^+, M^{2+} and M^{3+} ions. The rate of water substitution is proportional to the strength of the water–cation interaction and hence, from the diagram, the interaction is clearly in the order $M^{3+} > M^{2+} > M^+$ for ions in the same quantum shell

That ions are hydrated in aqueous solution is amply demonstrated by the formation of crystalline hydrates when many salt solutions are carefully evaporated. Most of the 'water of crystallization' in such hydrates is coordinated around the cation, although some water molecules can either be hydrogen bonded to other groups in the crystal or simply be held in suitably sized cavities (as in the zeolites). Hydration of cations in crystals depends mainly on two factors: ionic charge and ionic size. Since the water molecules are held essentially by ion–dipole forces, the strength of the water–cation interaction will increase with increasing ionic charge (see Figure 98) and hence the extent of hydrate formation by salts containing cations of similar size will be in the order $M^{3+} > M^{2+} > M^{+}$. Ionic size simply restricts the number of water molecules which are able to coordinate round the metal ion, the usual coordination numbers being 4, 6 and 8. For example, in group II the effect of increasing ionic size down the group allows the maximum hydration number to increase from four at beryllium to eight at calcium; for example, $Be(H_2O)_4O_2$, $Mg(H_2O)_6SO_4$ and $Ca(H_2O)_8O_2$.

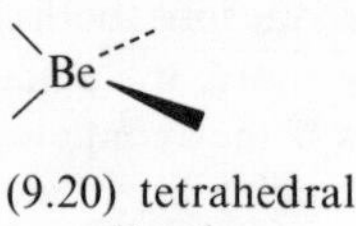

(9.20) tetrahedral coordination

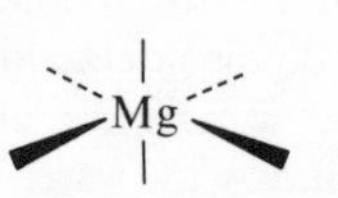

(9.21) octahedral coordination

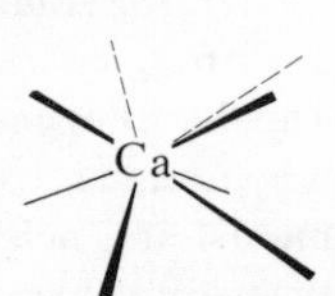

(9.22) square antiprismatic coordination

The hydration numbers of metal ions in crystals are relatively easy to determine using X-ray diffraction techniques, but it is much more difficult to ascertain the number of water molecules which interact with a given ion in aqueous solution and rather ingenious experiments have to be devised, for example:

(a) Tracer experiments using water enriched with oxygen-18 show that solid $Cr(H_2{}^{18}O)_6Cl_3$, when dissolved in and then recrystallized from, normal water, still has six $H_2{}^{18}O$ molecules around the chromium ion; hence the $Cr(H_2O)_6^{3+}$ species must be present in solution. Similar experiments show that many other ions shed their $H_2{}^{18}O$ molecules on dissolution due to a rapid exchange process taking place between coordinated water molecules and those in the bulk solvent.

(b) The green colour of $Ni(NO_3)_2.6H_2O$ crystals is due to the presence of six water molecules octahedrally coordinated to the nickel ion; since aqueous solutions of nickel nitrate are of the same colour (i.e. have the same electronic absorption spectrum), it may be assumed that in solution the nickel ions have a similar coordination.

(c) Oxygen-17 has a nuclear spin and, like the proton, may be used in nuclear magnetic resonance (n.m.r.) experiments. By observing the intensities of the two n.m.r. signals from $H_2{}^{17}O$ coordinated to the cation being investigated and $H_2{}^{17}O$ in the free solvent, it is possible to deduce the coordination number of the metal ion if the concentration of the salt solution is known. In this way the coordination numbers of Be^{2+} and Al^{3+} in aqueous solution have been established as four and six respectively.

(d) Freezing-point–composition diagrams for water and various salts being studied show that compounds of definite stoichiometry are formed as water is added to the salt; this is illustrated in Figure 99 for ferric chloride.

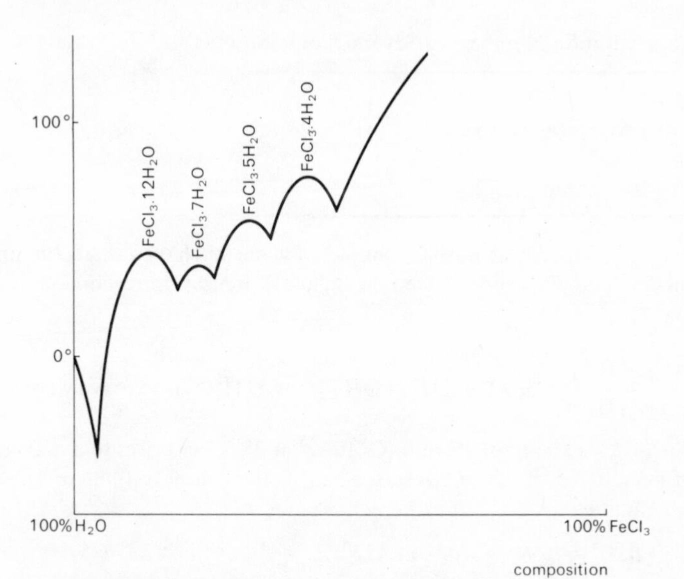

Figure 99 Melting point–composition diagram for the system ferric chloride–water showing hydrate formation

These methods, however, only work in favourable cases where the ion–dipole interactions are particularly strong (e.g. with multiply charged ions). Furthermore, they only give a measure of the number of water molecules in the immediate vicinity of the cation. These particular water molecules will be strongly polarized by the presence of the metal ion, and their protons will become positive relative to protons in the free solvent; hence it has to be expected that more water molecules will be hydrogen bonded to these relatively positive protons forming a second (and then a third, fourth, etc.) hydration layer round the cation. The influence of the cationic charge will, of course, decrease rapidly with each successive layer of water molecules so that the outer layers become more and more weakly bound relative to those in the innermost hydration layer. Therefore, in solution, the number of molecules of water associated with a particular ion is difficult to define since it is actually a matter of degree, and the value obtained depends markedly on the particular experiment used to determine it. Experiments in which a sugar is added to a salt solution prior to electrolysis in order to probe the amount of water associated with diffusing cations (assuming that the non-electrolyte, sugar, distributes evenly between bulk water molecules and water in the various hydration spheres of the cation) give much higher hydration numbers than the four experiments mentioned above. This is because the two types of experiment are measuring different aspects of water coordination; typical hydration numbers calculated from electrolysis and similar experiments are given in Table 30 and clearly demonstrate, again, the effect of ionic charge and size.

Table 30 Hydration Numbers of Several Common Ions

Li^+	5 ± 1	—		—	
Na^+	5 ± 1	Mg^{2+}	15 ± 2	Al^{3+}	26 ± 5
K^+	4 ± 2	—		—	
Rb^+	3 ± 1	Ba^{2+}	12 ± 4	—	

Even the most rigorously purified samples of water still have a small, but finite, electrical conductivity which is due to a slight self-ionization reaction

$$H_2O \rightleftharpoons H^+(aq) + OH^-,$$

$$K = \frac{[H^+][OH^-]}{[H_2O]}; \quad \text{or } K' = [H^+][OH^-], \text{ since } [H_2O] \text{ is effectively constant.}$$

The value of K' is about 10^{-15} at 0 °C, 10^{-14} at 25 °C and rises to a maximum value of $6{\cdot}5 \times 10^{-12}$ at 220 °C. Hence at 25 °C the concentrations of H^+ and OH^+ are minute,

$$[H^+] = [OH^-] = \sqrt{K'} = 10^{-7} \text{ mol l}^{-1}.$$

The bare proton H^+ does not exist as such in aqueous solution and behaves like any other cation by becoming hydrated. One water molecule is held particularly strongly to give the oxonium ion H_3O^+ found in the solid monohydrates of many acids; for example, $HCl.H_2O$ is $H_3O^+Cl^-$. The ion is pyramidal in shape with a lone pair of electrons on the oxygen atom (9.23). In solution the proton is thought to be quite closely associated with a further three water molecules, which are probably hydrogen bonded to the H_3O^+ ion (9.24).

The protons of water molecules in contact with metal ions in solution become positively charged relative to the free water molecules as a result of the ion–dipole

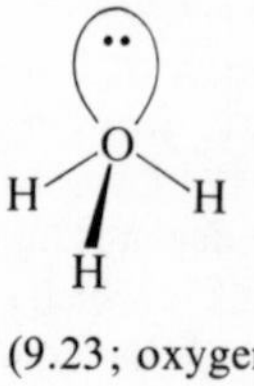

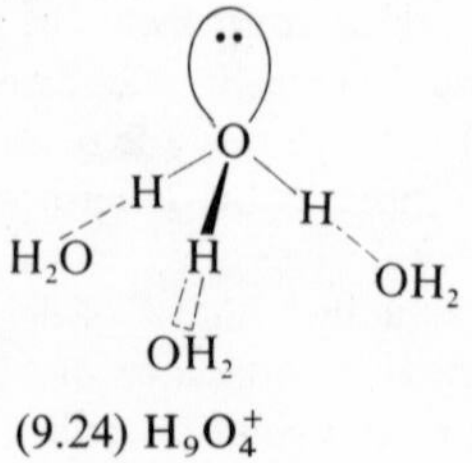

(9.23; oxygen atom is sp^3 hybridized) (9.24) $H_9O_4^+$

interaction. There is therefore a stronger tendency for such hydrogen atoms to ionize as protons than the hydrogen atoms in the bulk of the solvent,

e.g.
$$(H_2O)_nM^{3+}\text{—}O\begin{matrix}H\\ \\H\end{matrix} \rightleftharpoons (H_2O)_nM^{2+}\text{—}OH + H^+(aq)$$
$$\Updownarrow$$
$$H^+(aq) + M(OH)_3 \rightleftharpoons (H_2O)_nM^+(OH)_2 + H^+(aq)$$

This effect will be most marked for small, highly charged ions (e.g. Be^{2+}, Al^{3+}), since they will polarize the protons to the greatest extent. Such ions are readily hydrolysed, giving rise to either basic salts or a precipitate of the metal hydroxide, the hydrolysis being catalysed by alkali because the added hydroxyl ions disturb the above equilibria by removing the protons from solution. Conversely, singly charged cations do not polarize the protons very strongly and aqueous solutions of their salts are stable to hydrolysis (e.g. the alkali metals).

The hydrolysis of certain covalent compounds such as the silicon halides is thought to proceed via an S_N2 process involving hydroxyl ions arising from the self-ionization of water

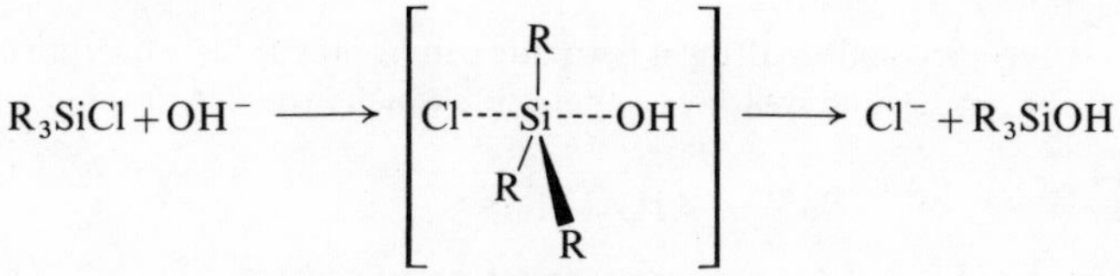

$$R_3SiCl + OH^- \longrightarrow \left[Cl\text{----}\underset{R\ \ R}{\overset{R}{Si}}\text{----}OH^- \right] \longrightarrow Cl^- + R_3SiOH$$

five-coordinate
reaction intermediate

9.6.2 *Hydrogen peroxide* H_2O_2

This pale blue, heavy liquid (m.p. $-0{\cdot}5$ °C) has a high Trouton constant indicating strong association between the molecules due to hydrogen bonding. X-ray

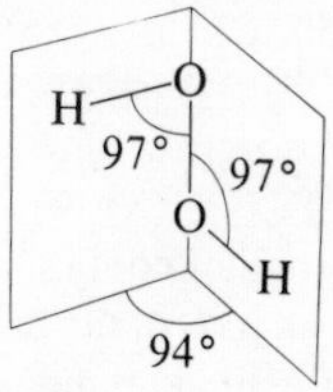

(9.25)

data on the solid show that the individual molecules take up the gauche structure (9.25). Like water, hydrogen peroxide has a high dipole moment (2·01 debyes), a high dielectric constant (89 at 0 °C) and also undergoes a slight self-ionization

$$2H_2O_2 \rightleftharpoons H_3O_2^+ + HO_2^-,$$

for which $[H_3O_2^+][HO_2^-] = 1{\cdot}5 \times 10^{-12}$. Several metal salts will crystallize from solution with 'hydrogen peroxide of crystallization'; for example, $Na_2CO_3.3H_2O_2$, which is used commercially in washing powders.

Hydrogen peroxide is synthesized on an industrial scale by passing air into a solution of 2-ethylanthraquinol in an organic solvent (9.26).

$$\text{2-ethylanthraquinol (OH, OH, } C_2H_5\text{)} \xrightarrow{O_2} \text{2-ethylanthraquinone (O, O, } C_2H_5\text{)} + H_2O_2$$

(9.26) (10 per cent solution)

The quinone is reduced back to the quinol by bubbling hydrogen through the solution in the presence of a catalyst of palladium metal. Periodically the hydrogen peroxide is extracted by shaking the organic solvent with water; the aqueous solution may be concentrated up to about 90 per cent by fractional distillation at low pressure.

On the laboratory scale hydrogen peroxide can be prepared by heating barium oxide in air or oxygen and treating the peroxide which results with sulphuric acid

$$BaO \xrightarrow{O_2} BaO_2 \xrightarrow{H_2SO_4} BaSO_4\downarrow + H_2O_2.$$

Although the heat of formation of liquid hydrogen peroxide is $-187{\cdot}8$ kJ mol^{-1}, indicating high stability with respect to decomposition into hydrogen and oxygen, it is thermodynamically unstable with respect to water and oxygen,

$$H_2O_2(l) \longrightarrow H_2O(l) + \tfrac{1}{2}O_2;\ \Delta H = -98{\cdot}3\ \text{kJ mol}^{-1},\ \Delta G = -121{\cdot}4\ \text{kJ mol}^{-1}.$$

At room temperature, oxygen evolution is very slow in the absence of catalysts (silver, platinum, several transition-metal oxides such as manganese dioxide) because of the reaction's high activation energy, but the heated liquid and vapour will explode well below the extrapolated boiling point (150 °C).

9.7 Peroxides and peroxy-salts

The alkali and alkaline-earth metals form solid, ionic peroxides containing the $[O{-}O]^{2-}$ ion, the electronic structure of which was given on p. 214; the oxygen–oxygen distance in the peroxide ion (1·49 Å) is typical of a single bond. A few alkali-metal derivatives containing the hydroperoxide ion $O{-}OH^-$ (akin to the hydroxide ion in the water system) are known; for example, sodium hydroperoxide NaOOH. Addition of cold, aqueous acid to both peroxides and hydroperoxides liberates hydrogen peroxide.

To a limited extent it is possible to substitute —OOH for —OH in ortho oxy-acids (9.27, 9.28) and —O—O— for the —O— bridge in pyro oxy-acids (9.29,

(9.27) sulphuric acid H_2SO_4

(9.28) peroxomonosulphuric acid H_2SO_5 (Caro's acid)

(9.29) pyrosulphate ion $S_2O_7^{2-}$

(9.30) peroxodisulphate ion $S_2O_8^{2-}$

9.30). Peroxophosphoric acid (H_3PO_5), peroxodicarbonates ($O_2CO—OCO_2^{2-}$) and peroxodiphosphates ($O_3PO—OPO_3^{4-}$) can be envisaged as arising via similar substitutions.

The preparation of peroxomonosulphuric acid is quite logical and involves the addition of hydrogen peroxide to chlorosulphonic acid

$$SO_2(OH)Cl + H_2O_2 \longrightarrow HCl + SO_2(OH)OOH$$

a reaction very similar to the ready hydrolysis of chlorosulphonic acid to sulphuric acid

$$SO_2(OH)Cl + H_2O \longrightarrow HCl + SO_2(OH)_2$$

However, salts of the peroxy-diacids have to be synthesized by means of electrolytic oxidation of the corresponding simple salts,

$$CO_3^{2-} \xrightarrow{-e} \left[\begin{matrix} O \\ O \end{matrix} \rangle CO\cdot \right]^- \xrightarrow{\text{dimerization}} C_2O_6^{2-} \quad \text{(also for } PO_4^{3-}, SO_4^{2-}\text{)}.$$

possible intermediate radical ion

9.8 Metallic oxides, sulphides, selenides and tellurides

All the group VI elements form alkali-metal derivatives which contain X^{2-} ions but, due to the increasing metallic character down the group, the stability of the X^{2-} ions decreases from oxygen to tellurium. For example, treatment of these compounds with water produces XH^- with oxygen and sulphur, but the hydrides XH_2 with selenium or tellurium. On warming, even solutions of hydrosulphides evolve hydrogen sulphide, demonstrating that the hydrosulphide ion HS^- is much less stable than the hydroxide.

The relative instability of the sulphide S^{2-}, selenide Se^{2-} and telluride Te^{2-} ions compared to the oxide ion O^{2-} is also illustrated by the structures adopted by metal derivatives of these elements. The alkali and alkaline-earth derivatives are ionic and contain X^{2-} ions, but few ionic sulphides, selenides or tellurides of the type $M_2^{III}X_3$ are known, whereas many $M_2^{III}O_3$ oxides are ionic. Only oxygen forms any ionic derivatives of the type $M^{IV}O_2$, the other X^{2-} ions being unstable in a crystal lattice containing M^{4+} ions.

The most common metallic oxide types are:

(a) M_2^IO. Alkali-metal oxides (also sulphides, selenides and tellurides) crystallize with the *anti*-fluorite structure (8:4 coordination), the O^{2-} ions taking the place of Ca^{2+} ions and M^+ replacing F^- in the normal fluorite lattice; see Figure 50.

(b) $M^{II}O$. The group II elements (with the exception of beryllium), cadmium, vanadium, manganese, cobalt and nickel have oxides (sulphides and selenides) which have this stoichiometry; they adopt the 6:6 coordination of the sodium chloride lattice; see Figure 45.

(c) $M_2^{III}O_3$. Oxides of this stoichiometry often have the corundum (Al_2O_3) structure, in which the metal is octahedrally coordinated by O^{2-} ions and each O^{2-} ion is tetrahedrally surrounded by M^{3+} ions.

(d) $M^{IV}O_2$. Typical structures adopted by these oxides are those of rutile (TiO_2, Figure 54) and fluorite (CaF_2, Figure 50).

(e) M_3O_4. These oxides are unusual in that the metal is present in two oxidation states:

either $M^{II}M_2^{III}O_4$, e.g. $Fe^{II}Fe_2^{III}O_4$,
or $M^{IV}M_2^{II}O_4$, e.g. $Pb^{IV}Pb_2^{II}O_4$.

9.9 Oxides of sulphur, selenium and tellurium

There are two main types of oxide within this group, the dioxides (SO_2, SeO_2, TeO_2, PoO_2) and the trioxides (SO_3, SeO_3, TeO_3). In addition, sulphur also forms disulphur monoxide S_2O, transient MO species are known in the gaseous phase for sulphur, selenium and tellurium, and polonium forms a black monoxide PoO.

9.9.1 *Disulphur monoxide* S_2O

This oxide is prepared by passing an electrical discharge through a mixture of sulphur vapour and sulphur dioxide. It is a bent molecule (9.31), being closely related, structurally and electronically, to ozone (9.32) and sulphur dioxide (9.33).

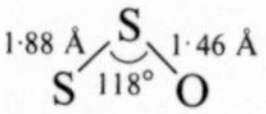

$$\underset{(9.31)}{S\overset{1.88\,Å}{—}S\overset{1.46\,Å}{—}O,\ \angle 118^\circ}\qquad \underset{(9.32)}{O—O—O,\ \angle 117^\circ}\qquad \underset{(9.33)}{O—S\overset{1.43\,Å}{—}O,\ \angle 119.5^\circ}$$

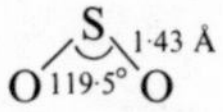

9.9.2 *Dioxides* MO_2

The dioxides may be prepared directly by heating the elements in air or oxygen. The decreasing M—O π-bond strength and the increasing metallic character from sulphur to polonium are reflected in the structures of the solid dioxides: solid sulphur dioxide contains discrete SO_2 molecules, selenium dioxide consists of infinite chains (9.34), whereas tellurium dioxide and polonium dioxide are

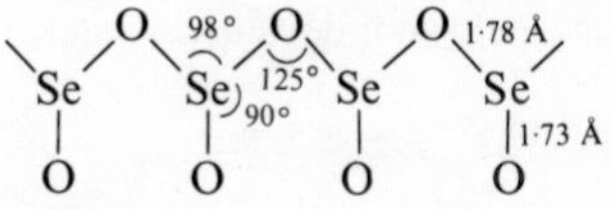

Se–O–Se chain with terminal Se–O; O–Se–O 98°, Se–O–Se 125°, 90°; Se–O (bridge) 1·78 Å; Se–O (terminal) 1·73 Å

(9.34)

essentially ionic solids having rutile and face-centred cubic structures (Figures 54 and 45) respectively. In the gaseous phase selenium dioxide is monomeric with an Se—O—Se bond angle of about 125°, the short selenium–oxygen distance of 1·61 Å being indicative of multiple bonding (compare the selenium–oxygen distances in solid selenium dioxide).

9.9.3 *Trioxides* MO_3

The trioxides are known for sulphur, selenium, tellurium and, possibly, polonium, although selenium trioxide is very difficult to prepare. The difficulties encountered in preparing oxygen derivatives of selenium(VI) and bromine(VII) (e.g. $HBrO_4$) have been interpreted in terms of relatively weak π-bonding between oxygen and the central atom due to the presence of the nodes in 4d orbitals of selenium and bromine; the nodes occur close to the regions where π-bonding to oxygen occurs, resulting in poor overlap and destabilization of the oxygen derivatives.

In the gaseous phase, the monomeric sulphur trioxide molecule has a zero dipole moment indicating a symmetrically planar molecule (9.35); this has been verified by electron diffraction studies, which show that the S—O bond lengths are 1·43 Å, much shorter than the single-bond value of about 1·60 Å. This is due to considerable p_π–d_π bonding between filled oxygen 2p orbitals and empty 3d orbitals on the sulphur atom. In the several modifications of sulphur trioxide which are known in the solid state, some of this π-bonding is sacrificed for further σ-bonding, which occurs via the formation of S—O—S bridges either in polymeric chains (9.36) or in cyclic trimers (9.37).

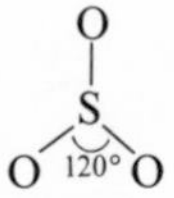

(9.35) (gas) 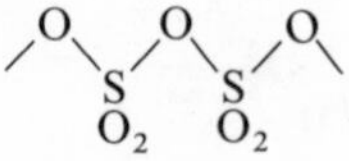 (9.36) (solid) 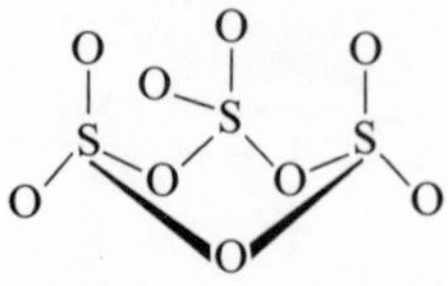 (9.37) (solid)

Little is known concerning the structures of the other trioxides.

9.10 Oxy-acids of sulphur, selenium and tellurium

There are many sulphur oxy-acids most of which may be predicted by carrying out two types of *theoretical* reactions on sulphuric acid: condensation reactions and substitution reactions.

9.10.1 *Condensation reactions*

To be quite correct, sulphuric acid should be called ortho-sulphuric acid since it contains the maximum known number of hydroxyl groups bound to one sulphur(VI) atom. The pyro-acid (9.38) may be thought of as derived from sulphuric acid by loss of water between two molecules of the ortho-acid

(9.38) dibasic pyro-sulphuric acid, $H_2S_2O_7$

In the pyro-acid the sulphur atoms share one oxygen atom; attempts to make meta-sulphuric acid, in which each sulphur shares two oxygen atoms, result only in the formation of sulphur trioxide (9.36). Although meta-sulphuric acid

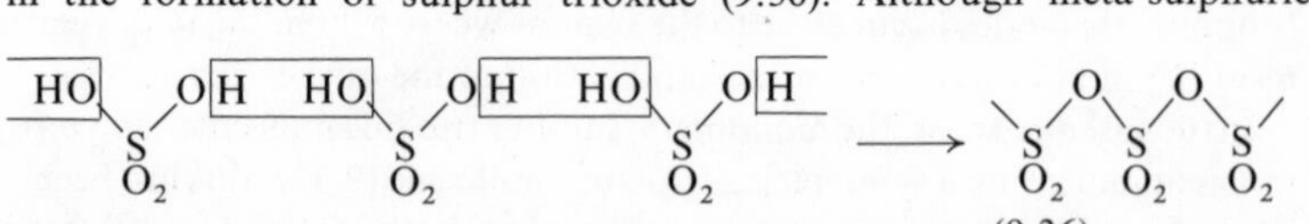

(9.36)

cannot exist, polymeric acids, in which some sulphur atoms share two oxygens and other sulphurs share only one oxygen atom, are possible (9.39). Salts are known for the acid which has $n = 1$; no other members of the series have yet been made.

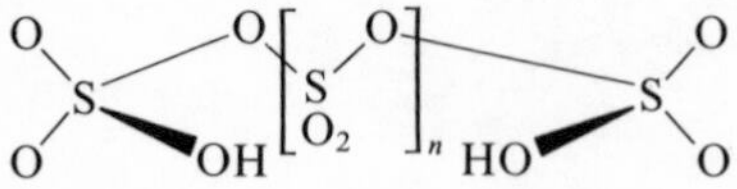

(9.39)

9.10.2 *Substitution reactions*

(a) *Substitution of a free oxygen atom.* Other elements of group VI should be capable of substitution for the two free oxygen atoms in the sulphuric acid molecule (9.40, 9.41). Examples of type (9.40) are thiosulphuric acid (9.42) and selenosulphate salts derived from selenosulphuric acid (9.43). No derivatives are yet known of type (9.41).

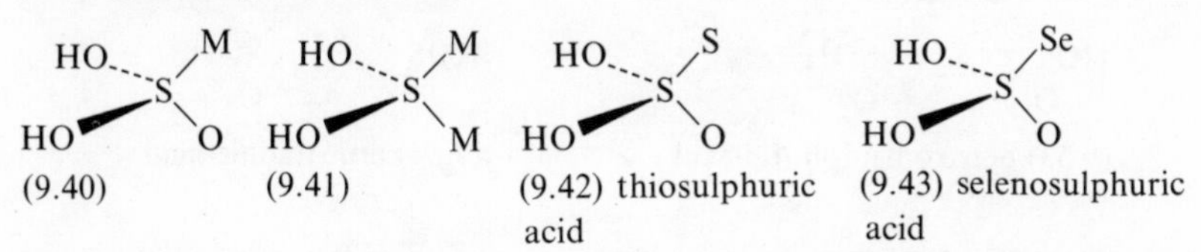

(9.40) (9.41) (9.42) thiosulphuric acid (9.43) selenosulphuric acid

(b) *Substitution of a hydroxyl group.* The hydroperoxide group —OOH is very similar to the hydroxyl group —OH and should be capable of at least partial substitution. Peroxomonosulphuric (9.44), also known as Caro's acid, does indeed exist, but of the doubly substituted acid (9.45) no derivatives are yet known.

HO O
S
HOO O
(9.44)

HOO O
S
HOO O
(9.45)

Groups iso-electronic to hydroxyl may also be substituted into sulphuric acid. For example, an amino group may be substituted for hydroxyl to give sulphamic acid (9.46) and substitution of fluorine and chlorine yield fluorosulphonic acid (9.47) and chlorosulphonic acid (9.48) respectively.

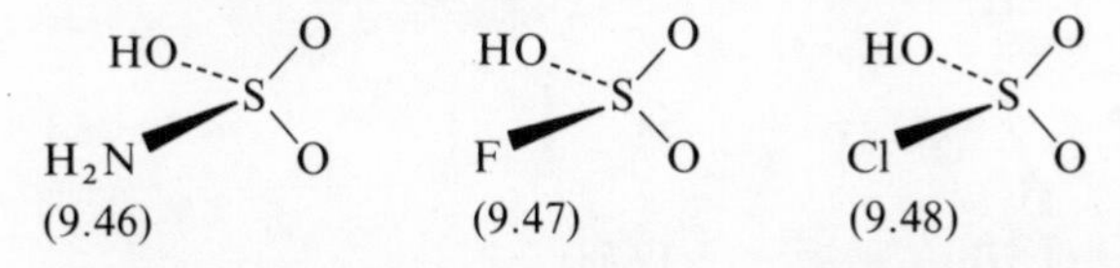

(9.46) (9.47) (9.48)

(c) *Replacement of the bridge oxygen atom in pyro-sulphuric acid.* The bridging oxygen atom in pyro-sulphuric acid can be replaced by other group VI elements (9.49–51) and also the number of group VI atoms in the bridge increased (9.52–56).

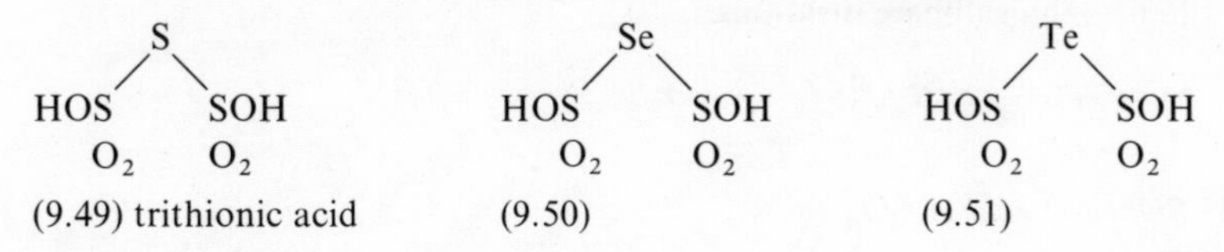

(9.49) trithionic acid (9.50) (9.51)

$(M)_n$
HOS SOH
O_2 O_2
(9.52)

O—O
HOS SOH
O_2 O_2
(9.53) peroxodisulphuric acid

S—S
HOS SOH
O_2 O_2
(9.54) tetrathionic acid

S—S—S
HOS SOH
O_2 O_2
(9.55) pentathionic acid

$S(Se)_nS$
HOS SOH
O_2 O_2
(9.56) ($n = 1-6$)

9.10.3 *Sulphurous acid* H_2SO_3

Although sulphur dioxide dissolves readily in water to give sulphurous acid, the free acid cannot be isolated from the solution. Two series of salts are known, for example, Na_2SO_3 and $NaHSO_3$. The bisulphite ion exists in two tautomeric forms in aqueous solution (9.57a and b). The sulphite ion is pyramidal in shape with one lone pair of electrons on the sulphur atom (9.58).

S^- O OH O ⇌ H S^- O O O S^{2-} O O O

(9.57a) (9.57b) (9.58)

Alkali-metal sulphites react slowly with sulphur on boiling in aqueous solution to give thiosulphates; iodine oxidizes thiosulphates to tetrathionates as in iodine–thiosulphate titrations.

$$SO_3^{2-} + S \longrightarrow S_2O_3^{2-} \xrightarrow{I_2} [O_3S{-}S{-}S{-}SO_3]^{2-}.$$

9.10.4 *Sulphuric acid* H_2SO_4

This is prepared on an industrial scale via the catalytic oxidation of sulphur dioxide to sulphur trioxide

$$SO_2 + O_2 \xrightarrow[\text{catalyst}]{\text{Pt}} SO_3 \xrightarrow[\text{in } H_2SO_4]{\text{dissolved}} H_2S_2O_7 \xrightarrow{H_2O} 2H_2SO_4.$$

The anhydrous acid, which is frequently used as a non-aqueous solvent, has a high electrical conductivity due to extensive self-ionization

$$\left.\begin{array}{l} 2H_2SO_4 \rightleftharpoons H_3SO_4^+ + HSO_4^- \\ 2H_2SO_4 \rightleftharpoons H_3O^+ + HS_2O_7^- \end{array}\right\} \text{both reactions occur simultaneously.}$$

It is widely used in organic chemistry both as a sulphonating agent and as a solvent for nitric acid in nitration reactions (when NO_2^+ is the electrophile),

$$HNO_3 + 2H_2SO_4 \longrightarrow NO_2^+ + H_3O^+ + 2HSO_4^-.$$

9.10.5 *Oxy-acids of selenium*

In general these are similar to the sulphur acids, although slight differences do occur. For example, selenium dioxide is (like sulphur dioxide) very soluble in water, but the selenious acid H_2SeO_3 which is formed is only negligibly dissociated (pK_1 for sulphurous acid is 1·8). The $HSeO_3^-$ and SeO_3^{2-} ions are readily formed when alkali is added to aqueous selenious acid.

The selenite and selenate ions are isostructural with the pyramidal sulphite and tetrahedral sulphate ions, respectively; as a result, many metal sulphates and selenates are found to be isomorphous; for example, the alums and seleno-alums $M_2SeO_4.Al_2(SeO_4)_3.24H_2O$, where M is an alkali metal.

9.10.6 *Oxy-acids of tellurium*

Tellurous acid H_2TeO_3 is similar to, but weaker than, selenious acid; both normal and acid salts are known.

Unlike sulphur and selenium, tellurium(VI) forms a hexahydroxy acid $Te(OH)_6$, which is prepared by the oxidation of tellurium dioxide in aqueous nitric acid using potassium permanganate. On titration with standard alkali using phenolphthalein as indicator, telluric acid behaves only as a monobasic acid giving salts $M^I TeO(OH)_5$, but other salts can be prepared in which two or six protons have been replaced; for example, $Li_2TeO_2(OH)_4$ and Ag_6TeO_6.

9.11 Halides of the group VI elements

Table 31 lists the known halides of this group. An unusual trend is that the iodides only become stable at tellurium and polonium, because normally M—X bonds decrease in strength as the atomic weight of M increases. Similarly, tellurium tetrachloride is the only tetrachloride which is stable in the gaseous state.

The monohalides M_2X_2 can, theoretically, adopt two possible structures (9.59, 9.60). Only the highly unstable sulphur monofluoride S_2F_2 has been found to exist in both forms; all the other monohalides adopt the configuration (9.59). The structures of the remaining halides are (9.61–63).

Table 31 Halides of the Group VI Elements

	Oxygen	*Sulphur*	*Selenium*	*Tellurium*	*Polonium*
M_2X	O_2Cl				
	O_2Br				
M_2X_2	O_2F_2	S_2F_2 (two isomers)			
		S_2Cl_2	Se_2Cl_2		
		S_2Br_2	Se_2Br_2		
M_nX_2	O_4F_2	S_nCl_2 (short chains of S atoms)			
MX_2	OF_2	SF_2			
	OCl_2	SCl_2		$TeCl_2$	$PoCl_2$
	OBr_2			$TeBr_2$	$PoBr_2$
MX_4		SF_4	SeF_4	TeF_4	
		SCl_4	$SeCl_4$	$TeCl_4$	$PoCl_4$
			$SeBr_4$	$TeBr_4$	$PoBr_4$
				TeI_4	PoI_4
MX_6		SF_6	SeF_6	TeF_6	
M_2X_{10}		S_2F_{10}			

Oxygen also forms Cl_2O_6, Cl_2O_7, I_2O_4, I_4O_9 and I_2O_5.

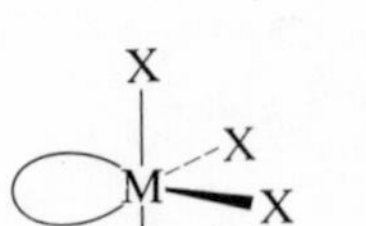

(9.59) (cf. H_2O_2)

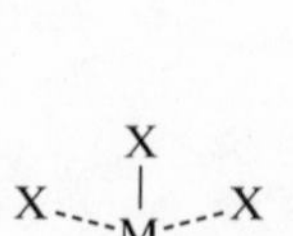

(9.60)

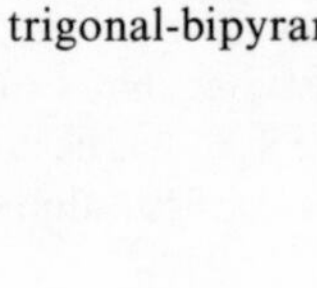

(9.61) MX_4 trigonal-bipyramidal

(9.62) MX_6 octahedral

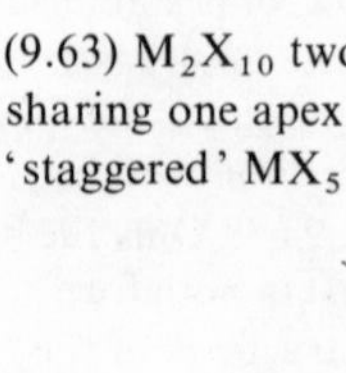

(9.63) M_2X_{10} two octahedra sharing one apex; 'staggered' MX_5 groups

In most cases the halides can be synthesized from the elements, using carefully controlled conditions when necessary,

$$S + F_2 \text{ (in excess)} \longrightarrow SF_6;$$

$$S + F_2 \text{ (diluted with } N_2) \xrightarrow[\text{temperature}]{\text{low}} SF_4 + SF_6;$$

$$O_2 + F_2 \xrightarrow[\text{at } -185\,°C]{\text{electrical discharge}} F_2O_2,\ F_2O_4.$$

Sulphur tetrafluoride, made more conveniently by treating sulphur dichloride with a suspension of sodium fluoride in acetonitrile

$$2SCl_2 + 4NaF \xrightarrow{70\,°C} SF_4 + 4NaCl\downarrow + S$$

is a very useful fluorinating agent in organic chemistry, replacing hydroxyl groups and the oxygen atoms of carbonyl groups by fluorine,

$$RC(=O)OH \xrightarrow{SF_4} RCF_3$$

$$R_2C{=}O \xrightarrow{SF_4} R_2CF_2$$

$$ROH \xrightarrow{SF_4} RF.$$

Sulphur hexafluoride is normally considered to be a highly stable molecule; apparently the stability must be purely kinetic in nature, because many reactions which are thermodynamically favoured do not occur. An example is the hydrolysis

$$SF_6(g) + 6H_2O(g) \longrightarrow SO_3(g) + 6HF(g); \quad \Delta G^\circ = -301{\cdot}2 \text{ kJ}.$$

The six fluorine atoms presumably sterically shield the sulphur atom from the close approach of possible attacking groups. The less sterically crowded sulphur tetrafluoride, in contrast, is exceedingly water sensitive

$$SF_4 + H_2O \longrightarrow SOF_2 + 2HF.$$

9.12 Organo-derivatives of the group VI elements

All the elements except polonium form MR_2 derivatives (those of oxygen, the ethers, belong to organic chemistry and will not be discussed here) which are easily prepared by treating a halide with the corresponding Grignard or lithium reagent in an ether solvent,

e.g. $MX_2 + 2LiR \longrightarrow MR_2 + 2LiX \quad (M = S, Se, Te;\ X = Cl).$

Organo-derivatives containing short chains of the group VI elements have also been isolated

$$RI + S \xrightarrow[\text{tube}]{\text{sealed}} SR_2 + RS{-}SR + RS{-}S{-}SR \quad (R = CF_3; C_6F_5).$$

The structure of the MR_2 compounds is similar to that of the water molecule, the two lone pairs of electrons making them useful donor molecules

$$BF_3 + :O(C_2H_5)_2 \longrightarrow F_3B \leftarrow O(C_2H_5)_2$$

(9.64) boron trifluoride etherate

Chapter 10
Group VII. The Halogens: Fluorine, Chlorine, Bromine, Iodine and Astatine

10.1 Introduction

Except for fluorine, the first ionization energies of the halogens are lower than that of hydrogen (1310 kJ) but, unlike H^+, the ions are large and hence cannot be stabilized by high lattice energies and high solvation energies. There are, therefore, no simple ionic salts of the halogens containing X^+ ions. Due to the relatively poor shielding effect of electrons in p-orbitals, the removal of one of the *n*p electrons from a halogen atom will not greatly affect the remaining four. This results in only a slight decrease in size of the halogen on cation formation; compare this to the substantial shrinking which occurs on removal of the well-shielded *n*s electron of the alkali metals.

A few solid compounds are known which contain I^+ stabilized as a linear complex ion $D—I^+—D$, where D is a Lewis base such as an aliphatic amine, pyridine or thiourea (10.1, 10.2). A salt, such as $I(pyridine)_2NO_3$, which contains

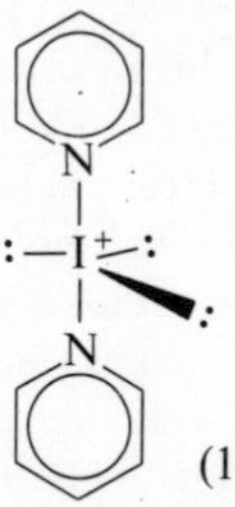

(10.1)

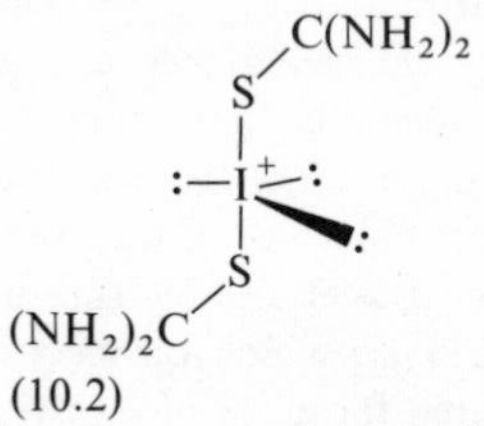

(10.2)

Table 32 The Halogens

Element		*1st IE* kJ mol^{-1}	*2nd IE* kJ mol^{-1}	*3rd IE* kJ mol^{-1}	*EA* kJ mol^{-1}
Fluorine F	$1s^2 2s^2 2p^5$	1682	3372	6039	332·7
Chlorine Cl	$[Ne]3s^2 3p^5$	1255	2297	3850	348·5
Bromine Br	$[Ar]3d^{10} 4s^2 4p^5$	1142	2083	3464	324·5
Iodine I	$[Kr]4d^{10} 5s^2 5p^5$	1009	1841	—	295·4
Astatine At	$[Xe]4f^{14} 5d^{10} 6s^2 6p^5$	912	—	—	255·3

I^+ yields iodine at the *cathode* on electrolysis in anhydrous chloroform. Typically, the positive iodine complexes are formed via the heterolytic cleavage of the I—I bond in I_2

$$I_2 \longrightarrow I^+ + I^-.$$

The reaction is forced to the right by adding the iodine to a silver salt in a non-aqueous solvent (e.g. ethanol or pyridine) when the iodide ions are removed as precipitated silver iodide, the I^+ cation being stabilized by solvation as $I(solvent)_2^+$

$$AgClO_4 + I_2 \xrightarrow[-80\,°C]{ethanol} AgI + IClO_4.$$

Addition of potassium iodide to a solution of solvated iodine perchlorate results in the quantitative formation of free iodine

$$I^+ + I^- \longrightarrow I_2.$$

In uncatalysed halogenations of organic molecules by chlorine, bromine and, probably, iodine, the substituting agents are the free molecules X_2; the halogens do *not* act as precursors of positive halogenating species such as X^+. The first ionization energies of the other halogen atoms are higher than that of iodine, but there is evidence to suggest that the $Br(pyridine)_2^+$ and $Cl(pyridine)_2^+$ complex ions are capable of existence. The fact that astatine follows iodine in tracer studies on the Ipy_2^+ ion demonstrates the existence of At^+, as would be expected from trends within the group.

The X_2^+ cations can be observed transiently in the gas phase for all the halogens, but I_2^+ and Br_2^+ are the only such cations which are stable in solution in highly acidic media. For example, iodine dissolves in fluorosulphonic acid to give the deep blue, paramagnetic ion I_2^+ as well as I_3^+ and I_5^+; a few solid salts of I_2^+ are known. There is little doubt that At_2^+ could be isolated were it not for the very short half-lives of the known astatine isotopes. Oxidation of iodine in the presence of acetic anhydride produces the rather unstable 'iodine triacetate' $I(OOCCH_3)_3$, which on electrolysis produces one mole of iodine atoms for every three faradays of electricity passed through the solution. This suggests that this compound (and several closely related species) is a derivative of iodine(III), but nothing is known about its structure or mode of bonding.

In many of their compounds the halogens achieve a rare-gas configuration on acquiring a further electron by the formation of either the X^- ion or a covalent bond M—X. The simplest covalent compound which the halogens can form is the X_2 diatomic molecule. The dissociation energies (D_{X_2}) of the halogen molecules given in Table 33 shows a distinct anomaly for D_{F_2}, which is much smaller than would be expected by direct extrapolation from the other halogens, but is in line with the N—N (88 kJ mol^{-1}) and O—O (146 kJ mol^{-1}) single-bond energies. Two possible contributing factors to the weakness of the F—F bond are (a) that chlorine, bromine, iodine and astatine can increase the strength

Table 33 Some Physical Properties of the Halogens

Element	D_{X_2} / kJ mol^{-1}	r_{ion} / Å	r_{cov} / Å	$D_{X_2^+}$ / kJ mol^{-1}
Fluorine	157·7	1·36	0·71	314
Chlorine	243·5	1·81	0·99	377
Bromine	192·9	1·96	1·14	314
Iodine	150·6	2·19	1·33	—
Astatine	115·9	—	—	230

of the X—X bond by using their d-orbitals in pd hybrids, and (b) that non-bonding and inner-shell electron–electron repulsions are calculated to be greater for the fluorine molecule than the other halogen molecules; it appears probable that the latter effect is the more important, see p. 27. The electron configurations of the X_2 molecules are essentially

$$[—]\sigma ns^2\sigma^* ns^2\sigma np^2\pi np^2\pi np^2\pi^* np^2\pi^* np^2.$$

Hence, on ionization to X_2^+, the electron is removed from an antibonding orbital, which results in an increase in the X—X bond order from 1·0 in X_2 to 1·5 in X_2^+; in agreement with this $D_{X_2^+}$ is larger than D_{X_2} (Table 33), but it will be noticed that $D_{F_2^+}$, like D_{F_2}, is still smaller than expected by comparison to $D_{X_2^+}$ of the other halogens. The increase in bond order in going from X_2 to X_2^+ results, as expected, in an increase in the X—X stretching frequency; for example, Br_2 316 cm^{-1} and Br_2^+ 360 cm^{-1}.

Another manner in which the halogens could attain a rare-gas electronic configuration is by forming the cation XR_2^+, in which two alkyl or aryl groups are covalently bound to X^+, such species being analogous to the '-onium' salts of groups V and VI (e.g. NR_4^+, PR_4^+ and SR_3^+). Only iodine has so far been proved capable of forming derivatives of this type when R is an organic group, although XY_2^+ interhalogen cations are known for chlorine, bromine and iodine (see Table 36). The halides are the best-known members of the series; the iodonium hydroxides behave as strong bases in aqueous solution, proving that they ionize to the (bent) IR_2^+ cation (10.5)

(10.3) group V: no lone pairs of electrons

(10.4) group VI: one lone pair

(10.5) group VII: two lone pairs resulting in a bent IR_2^+ ion

From the covalent radii given in Table 33 it can be seen that the volume occupied by a covalently bound fluorine atom is only about one-third of that of chlorine and less than one-sixth of that occupied by iodine. The small size of the fluorine atom and the low dissociation energy of the fluorine molecule are mainly responsible for the ability of fluorine to expand the covalence of many elements; for example, iodine heptafluoride IF_7, sulphur hexafluoride SF_6, uranium hexafluoride UF_6, tungsten hexafluoride WF_6 and many other transition-metal hexafluorides. The obvious influence of size is to allow a large number of fluorine atoms round the central element; the dissociation energy of the fluorine molecule is an important thermodynamic step in the production of a covalent fluoride:

$$\underset{\text{(standard state)}}{M} \xrightarrow{\text{atomization}} \underset{\text{(gaseous atoms)}}{M(g)} \xrightarrow{\text{excitation}} \underset{\text{(valence state)}}{M}$$

$$+ \longrightarrow MF_n$$

$$\tfrac{1}{2}n\,F_2 \xrightarrow{\frac{1}{2}nD_{F_2}} \underset{\text{(gaseous atoms)}}{nF}$$

For a stable fluoride to be formed

$B_{M-F} >$ [energy of atomization of M + excitation energy of M to valency state $+\frac{1}{2}nD_{F_2}$],

where B_{M-F} is the total M—F bond energy in MF_n. Clearly, formation of the fluoride MF_n is favoured by the small value of D_{F_2}.

The heats of formation of the halide ions from the gaseous X_2 molecules

$$\tfrac{1}{2}X_2(g) \longrightarrow X^-(g)$$

are listed in Table 34. They obviously show little variation within the group; therefore they play only a minor role in determining the differences in chemical

Table 34 Heats of Formation and Hydration Energies of the Halide Ions

Ion	*Heat of formation*/kJ mol^{-1} $\frac{1}{2}X_2(g) \rightarrow X^-(g)$	*Hydration energy*/kJ mol^{-1} $X^-(g) \rightarrow X^-(aq)$
Fluoride F^-	−259·8	506
Chloride Cl^-	−233·0	369
Bromide Br^-	−234·3	335
Iodide I^-	−226·0	293

behaviour between the halogens. Of more importance is the small size of the fluoride ion relative to the other halides (Table 33), because this results in the fluoride ion having the highest hydration energy (see Table 34) and in ionic fluorides having large lattice energies. The low dissociation energy of F_2 and the small size of the fluoride ion sometimes combine to allow the formation of an essentially ionic fluoride of a metal when the other halides are covalent; for example, cupric fluoride CuF_2. The substantial change in size accompanying the formation of X^- from X atoms is due to the increased mutual shielding of the electrons from the nucleus and to increased electron–electron repulsions which occur on addition of the extra electron to the system.

The radius of the fluoride ion is very similar to that of the oxide ion (1·40 Å) and it is often found that fluorides and oxides which have similar stoichiometries have similar structures. An example of this behaviour is shown by calcium oxide and potassium fluoride, both of which adopt the sodium chloride structure. However, the oxide always has the higher melting point and is very much less soluble than the fluoride, because the ions in a crystalline oxide necessarily carry a greater charge and this results in them having extremely high lattice energies.

10.2 Mixed halogen compounds

One of the most fascinating properties of the halogens is their ability to react directly with each other and form mixed halogen derivatives which can be cationic, neutral or anionic as shown in Table 35. It will be noticed that the neutral species (called the 'interhalogens') have the general formula YX_n, where n is any *odd* number from 1 to 7 and Y is heavier than X; the charged species

Table 35 Some Mixed Halogen Compounds

Neutral species			
ClF	ClF_3	ClF_5	
BrF	BrF_3	BrF_5	
IF	IF_3	IF_5	IF_7
$BrCl$			
ICl	$(ICl_3)_2$		
IBr			
Cationic species (central atom on left)			
$ClFCl^+$	IF_4^+	IF_6^+	
ClF_2^+			
ICl_2^+			
Anionic species (central atom on extreme left)			
$BrCl_2^-$	ClF_4^-	IF_6^-	
ICl_2^-	BrF_4^-		
IBr_2^-	ICl_4^-		
$IBrCl^-$	ICl_3F^-		
$IICl^-$			
$BrBr_2^-$			
II_2^-			

have the general formulae YX_m^+ or YX_m^-, where *m* is any *even* number from 2 to 6 and Y is usually, but not always, heavier than X. The unique ability of fluorine to force an element into unusually high oxidation states (see above) is clearly demonstrated by several of the mixed halogen compounds which are known only for fluorine as the outer halogen. It is possible to make mixed anions which contain more than two halogen atoms, as in $IBrCl^-$ and ICl_3F^-. There are also a number of polyhalide anions, especially of iodine, which contain only the one halogen, perhaps the best-known example being the I_3^- ion encountered in iodine–thiosulphate titrations. Polyiodides containing the ions I_3^-, I_5^-, I_7^- and I_9^- have also been isolated; I_3^- has a linear structure like all YX_2^- ions but the other polyiodides have complex structures which will not be discussed here.

Molecular structures of the various types of interhalogen compounds are shown in Table 36. Unfortunately, experimental data (and several theoretical calculations) are conflicting as to the significance of the d-orbitals of the central atoms in the bonding of these compounds. Because of this, two rationalizations of the interhalogen structures are given; one of these involves the use of 3c, 4e bonds (p. 33) and the other suggests participation of d-orbitals on the central atom.

Table 36 Structures of the Mixed Halogen Compounds

Neutral species

YX_3

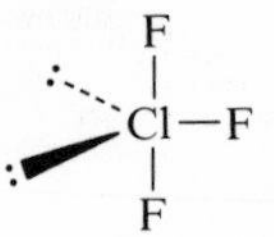

F
Cl—F
F

F
Cl—F
F

Trigonal-bipyramidal arrangement of orbitals (two lone pairs). Axial fluorine atoms held by 3c, 4e bonds; chlorine atom uses sp^2 hybrid orbitals to hold lone pairs and to bond to equatorial fluorine atom (or sp^3d hybridization of chlorine orbitals before their interaction with the fluorine 2p orbitals).

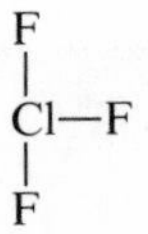

Molecule is planar and T-shaped.

YX_5

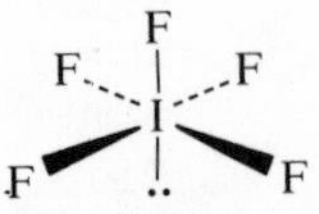

F
F F
I
F F

F
F F
I
F F

Octahedral arrangement of orbitals (one lone pair). Iodine atom uses sp hybrid orbitals to bond to the axial fluorine atom and to hold the lone pair. The equatorial fluorine atoms are held by two 3c, 4e bonds at right angles to each other (or sp^3d^2 hybridization of the orbitals of the iodine atoms).

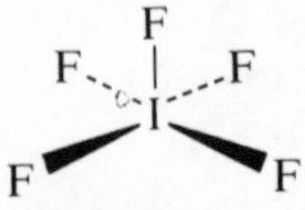

Molecule is square pyramidal.

YX_7

The molecular structure of iodine heptafluoride is not known with any certainty. One possibility is that the fluorine atoms are arranged in the shape of a pentagonal bipyramid round the iodine (as shown), although a distorted version of this structure would appear to be more likely.

Molecule is possibly pentagonal bipyramidal.

Cationic species

(N.B. I^+ is iso-electronic with tellurium.)

YX_2^+

Essentially pure p-orbitals used in bonding, giving an X—Y—X bond angle of about 90°.

Ion is non-linear.

YX_4^+

Trigonal-bipyramidal arrangement of orbitals (one lone pair). Axial fluorine atoms held by 3c, 4e bond; iodine then uses sp^2 orbitals, one to hold the lone pair and the other two to bond to the equatorial fluorines (or sp^3d hybridization of the iodine orbitals).

Ion has 'distorted tetrahedral' shape.

Table 36 – *continued*

YX_6^+	
F / F..+..F / I / F◄ ►F / F	F / F..+..F / I / F◄ ►F / F
Octahedral arrangement of orbitals (no sterically important lone pairs). Three 3c, 4e bonds hold the six fluorine atoms to the iodine atom. One lone pair on the iodine atom is in the spherically symmetrical 5s orbital (or sp^3d^2 hybridization of the iodine orbitals).	Ion is octahedral.
Anionic species	
YX_2^-	
Cl / :..I—: / :◄ / Cl	Cl—I^-—Cl
Trigonal-bipyramidal arrangement of orbitals (three lone pairs).	Linear ion.
YX_4^-	
Cl..I..Cl / Cl◄ ►Cl (two lone pairs)	Cl..I⁻..Cl / Cl◄ ►Cl
Octahedral arrangement of orbitals (two lone pairs)	Ion is square planar.
YX_6^- The structures of BrF_6^- and IF_6^- have not yet been determined, but spectroscopic studies suggest that they are not octahedral and, therefore, there is the possibility that the lone-pair electrons are sterically important.	

The structures given in Table 36 are idealized, and slight variations in some of the molecules do occur due to the steric influence of the lone-pair electrons. For example, although chlorine trifluoride ClF_3 is indeed T-shaped, the F—Cl—F bond angles are not exactly 90° (10.6). In a similar manner the lone-pair elec-

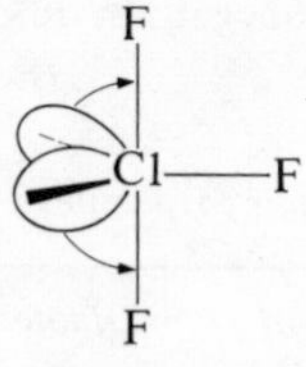

F
Cl—F
F

(10.6a)

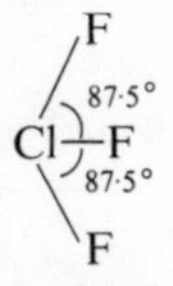

F
87·5°
Cl—F
87·5°
F

(10.6b)

trons in bromine pentafluoride BrF_5 force the fluorine atoms slightly out of the plane containing the bromine atom (10.7).

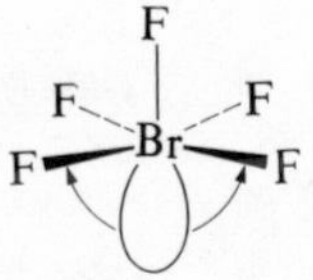

F
F F
Br
F F

(10.7a)

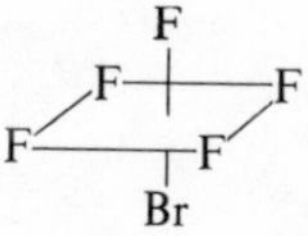

F
F F
F F
Br

(10.7b)

Iodine trichloride is too unstable for its structure to be determined in the gaseous state (it readily disproportionates into ICl and Cl_2); in the solid state two T-shaped ICl_3 molecules share two chlorine atoms to give the planar I_2Cl_6 (10.8).

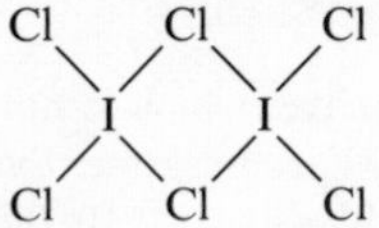

Cl Cl Cl
I I
Cl Cl Cl

(10.8)

The small entropy increase (between 4·6 and 5·4 $J K^{-1} mol^{-1}$ for all the diatomic interhalogens) associated with the reaction

$$\tfrac{1}{2}X_2\ (\text{gas}) + \tfrac{1}{2}Y_2(\text{gas}) \xrightarrow[1\ \text{atm}]{25\ ^\circ\text{C}} XY\ (\text{gas})$$

is close to the value calculated for homonuclear diatomic molecules changing into heteronuclear diatomic molecules

$$\Delta S = R \log_e 2 = 5{\cdot}9\ J\,K^{-1}\,mol^{-1}.$$

Since ΔS is constant for all the diatomic interhalogens, the stability of XY is determined essentially by ΔH measured under the same conditions (gaseous reactants at 25 °C and 1 atm). The values of ΔH are given in Table 37, and show that the stability sequence is IF > BrF > ClF > ICl > IBr > BrCl, which parallels the magnitude of the electronegativity difference between the two halogens in XY. This is due to an electrostatic contribution to the bonding arising from the charge separation $Y^{\delta+}$—$X^{\delta-}$, which occurs when X is more electronegative than Y. For the same reason the bond length X—Y is close to the sum of the covalent radii of X and Y only when the electronegativity difference between the two halogens is small (Table 37).

Table 37 Some Physical Data for the Diatomic Interhalogen Molecules

Molecule	ΔH/kJ mol^{-1} $\frac{1}{2}X_2(g)+\frac{1}{2}Y_2(g) \xrightarrow[1\ atm]{25\ ^\circ C} XY(g)$	*Bond length* d_{X-Y}/Å	*Sum of covalent radii* r_X+r_Y/Å	*Dipole moment (gas phase)* μ/*debye*
IF	−95·0	—	—	—
BrF	−77·0	1·96	1·85	1·29
ClF	−56·1	1·63	1·70	0·88
ICl	−14·2	2·32	2·33	0·63
IBr	−5·9	2·47	2·47	—
BrCl	−0·8	2·14	2·14	—

Although the heat of formation of IF shows that it is the most stable diatomic interhalogen relative to the free parent halogen molecules it is highly unstable towards disproportionation into IF_5 (and a little IF_7)

$$5IF \longrightarrow 2I_2 + IF_5,$$

for which ΔG° is $-166{\cdot}5$ kJ (mol IF_5)$^{-1}$.

All twenty reported isotopes of astatine are radioactive, which is not completely unexpected since all nuclides above bismuth-209 are unstable. Even the most stable isotopes, astatine-211 (half-life 7·5 h) and astatine-210 (half-life 8·3 h) are very short-lived, so that the chemistry of astatine has had to be mapped out using tracer techniques, often with amounts of the order of 10^{-14} mole or less. Chemically, astatine behaves like iodine, even to the extent of being selectively concentrated in thyroid tissue. In the -1 oxidation state it coprecipitates with silver iodide; in the zero oxidation state it is extracted (presumably as At_2) into organic solvents and forms interhalogens such as astatine bromide AtBr and astatine iodide AtI; in the $+5$ state it precipitates with insoluble iodates. Mixed halogen anions such as $AtCl_2^-$, $AtBr_2^-$, $AtICl^-$ and $AtIBr^-$ appear to be formed, but no compounds containing fluorine have yet been detected – probably only because of difficulties encountered in studying the reactions.

10.3 Pseudohalides and pseudohalogens

There are a number of uni-negative ions known which have many of the properties expected of halide ions; because of these similarities to the group VII elements they are often called pseudohalides. Of the pseudohalides listed in Table 36, probably the best-known example is the cyanide ion, which has the following properties, among others, in common with chloride, bromide and iodide ions.

(a) It can be oxidized to cyanogen N≡C—C≡N, which is volatile and has two cyanide radicals joined in a symmetrical fashion. Cyanogen is, therefore, a pseudohalogen.

(b) The cyanide group is able to form 'interhalogen' derivatives such as FCN, ClCN, BrCN, ICN and $CN.N_3$, the latter compound being an 'interpseudohalogen' formed between cyanide and azide groups.

(c) Silver cyanide is insoluble in water but dissolves on addition of ammonia (cf. silver chloride and silver bromide).

(d) A wide variety of complex anions containing CN^- have been prepared and which have their analogous halo-complexes (e.g. $Cu(CN)_4^{2-}$, $CuCl_4^{2-}$ and $CuBr_4^{2-}$).

(e) Cyanogen reacts with alkali to form cyanide and cyanate

$$(CN)_2 + 2KOH \longrightarrow KCN + KCNO + H_2O,$$

$$\text{cf.}\quad Cl_2 + 2KOH \longrightarrow KCl + KClO + H_2O.$$

Note, however, that the cyanate ion CNO^- is another pseudohalide and has properties quite unlike the hypochlorite ion ClO^-.

(f) The acid HCN is known, but it is very weak relative to the halogen acids ($pK = 8{\cdot}9$).

(g) Copper(II) cyanide is unstable and readily loses cyanogen

$$Cu^{2+} + 2KCN \longrightarrow Cu(CN)_2 \longrightarrow CuCN + \tfrac{1}{2}(CN)_2.$$

Compare this with the rapid decomposition of copper(II) iodide into copper(I) iodide and iodine.

Table 38 Some Pseudohalogens and Pseudohalides

*Pseudohalogen**	*Pseudohalide*
Cyanogen $(CN)_2$	Cyanide CN^-
—	Cyanate OCN^-
Thiocyanogen $(SCN)_2$	Thiocyanate SCN^-
Selenocyanogen $(SeCN)_2$	Selenocyanate $SeCN^-$
—	Tellurocyanate $TeCN^-$
—	Azide N_3^-
—	Isocyanate ONC^-

*In addition to the X_2 molecule the pseudohalogens can also be obtained as highly polymeric materials, e.g. $(CN)_x$, a black solid called paracyanogen and $(SCN)_x$, an orange solid called parathiocyanogen or polythiocyanogen.

10.4 Occurrence of the halogens

10.4.1 *Fluorine*

270 p.p.m. of the Earth's crust; occurs fairly commonly as fluorspar CaF_2 and fluorapatite $CaF_2.3Ca_3(PO_4)_2$.

10.4.2 *Chlorine*

480 p.p.m. overall abundance in the Earth's crust, but 15 000 p.p.m. in sea-water; main source is sodium chloride, which is either mined (rock salt) or obtained from sea-water.

10.4.3 *Bromine*

30 p.p.m.; principal source is the bromides present in sea-water.

10.4.4 *Iodine*

0·3 p.p.m.; occurs as sodium iodate to the extent of about 1 per cent in Chilean deposits of sodium nitrate.

10.4.5 *Astatine*

Does not occur in nature. All its isotopes are made artificially, for example, by the reaction $^{209}Bi\ (\alpha, 2n)^{211}At$; after irradiation by the α-particles, the bismuth target is heated to 430 °C in a stream of gaseous nitrogen and the astatine collected (as At_2) on a cold finger held at liquid-nitrogen temperatures.

10.5 **Extraction**

10.5.1 *Fluorine*

Produced by the electrolysis of dry, fused potassium hydrogen fluoride KHF_2 (m.p. 217 °C) carried out in a copper or steel cell. The cell must be provided with a diaphragm (or similar device) to prevent explosive recombination between fluorine and the hydrogen which is liberated simultaneously at the cathode. The main impurity is about 10 per cent of hydrogen fluoride which is removed using sodium fluoride

$$NaF + HF \longrightarrow NaHF_2 \quad (\text{i.e. } NaF.HF).$$

Essentially, this preparation is the electrolysis of hydrogen fluoride which has been made conducting by the presence of potassium fluoride, pure anhydrous hydrogen fluoride being a very poor conductor of electricity.

10.5.2 *Chlorine*

Since chlorine is not so reactive as fluorine and does not attack water to any meaningful extent, it may be prepared by the electrolysis of salt brine and practically all commercially required chlorine is made in this way; the secondary product of the electrolysis is a solution of sodium hydroxide.

10.5.3 *Bromine*

Bromine is released from sea-water by chlorinating it at pH 3·5 (mainly to prevent the extensive formation of oxy-salts which would occur at higher pH). Approximately 8000 l of sea-water yield 0·5 kg of bromine.

10.5.4 *Iodine*

Sodium iodate is reduced in aqueous solution (e.g. with sodium hydrogen sulphite) to iodine which is purified by sublimation.

10.6 The elements

The normal state of the halogens is that of a diatomic molecule X_2 although there are indications that iodine vapour does contain a small percentage of I_4 molecules. There may also be some interaction between the I_2 molecules in solid iodine because the distance between non-bonded iodine atoms (3·56 Å) is considerably shorter than twice the accepted van der Waals radius of iodine (2 × 2·15 Å).

The shift to longer wavelengths of charge-transfer bands in the absorption spectrum gives rise to the progressive change in colour of the free halogens: fluorine, yellow; chlorine, green; bromine, brown; iodine, violet. It has been known for many years that iodine forms brown solutions in some solvents and violet solutions in others, cryoscopic data showing that the iodine dissolves as 'I_2' in both cases. In essentially non-polar solvents the iodine–solvent interaction is weak and such solutions are violet. Those solvents forming brown solutions are often, though not always, polar and form very weak donor–acceptor complexes with I_2 molecules. (The colour of these complexes is associated with a strong absorption band arising from transfer of electronic charge from solvent to iodine, $S \rightarrow I_2$, and for this reason the complexes are often referred to as charge-transfer complexes.)

One-to-one complexes formed between amines and iodine have a linear N—I—I skeleton in which the nitrogen–iodine separation is about 2·30 Å, only 0·2–0·25 Å longer than expected for a normal covalent bond. A corresponding weakening of the I—I bond is shown up by the increased iodine–iodine distance from 2·67 Å in I_2 to about 2·83 Å in the complexes; for example, with trimethylamine (10.9). With a solvent molecule containing two donor atoms (e.g. 1,4-dioxan) a polymeric chain is built up in which both of the iodine atoms in the I_2 solute act as acceptor sites (10.10). For this type of interaction the donor mole-

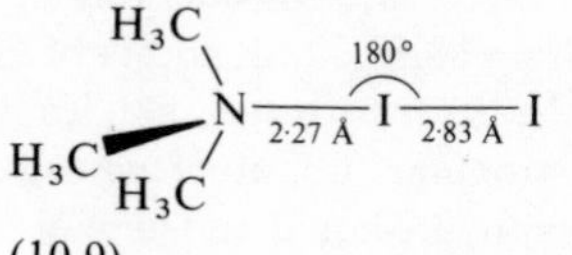

H_3C 180° $(H_3C)_3N$—I—I, N—I 2·27 Å, I—I 2·83 Å

(10.9)

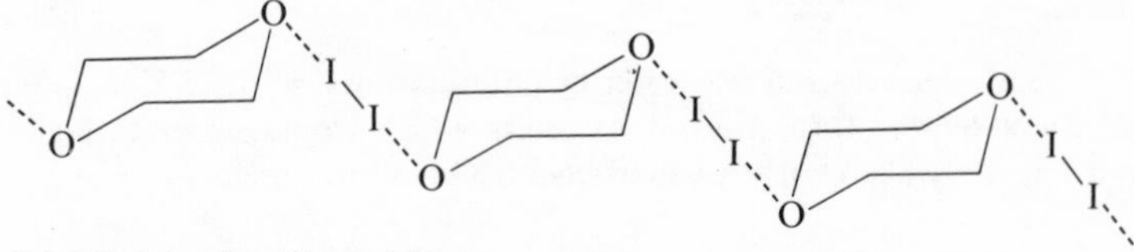

(10.10) (also for Cl_2 and Br_2)

cule must have weakly bound electrons (e.g. an sp^3 lone pair on nitrogen or oxygen atom), which it can donate to low-lying empty orbitals on the acceptor (iodine, bromine and chlorine molecules). Aromatic π-electrons are also capable of interacting with the halogen molecules, and benzene, for example, gives 1 : 1 addition compounds with chlorine and bromine which have the structure (10.11),

---- ⬡ ----X—X-- ⬡ ----X—X----

(10.11)

the X_2 molecules being symmetrically placed perpendicular to the benzene rings. The X—X distance is apparently unchanged relative to the free halogen, whilst the distance from each halogen atom to the plane of the nearest benzene ring is considerably shorter than the van der Waals separation (e.g. C_6H_6---Br— = 3·36 Å; van der Waals distance = 3·65 Å).

In many solvents it is probable that the halogens are in equilibrium between weak solute–solvent interaction and strong solute–solvent interaction states. The position of the equilibrium can be changed by thermal energy. Thus some brown iodine solutions (strong interaction) revert to the violet type (weak interaction) on heating, the reverse occurring when the solution is cooled.

It seems probable that the weakly bound I_4 molecules detected in iodine vapour are of the charge-transfer type in which one I_2 molecule acts as the donor and the other as the acceptor. Similarly, the structures of the polyiodides (I_3^-, I_5^-, I_7^- and I_9^-) contain I^- ions weakly interacting with one or more I_2 molecules.

10.7 Halogen hydrides

All the halogens form a hydride, HX, the stabilities of which fall from HF to HI due to the increasing incompatibility between the hydrogen 1s and the halogen *n*p orbitals. Fluorine has the highest electronegativity of all the elements and consequently many of the physical properties of HF are considerably modified by hydrogen bonding. For example, the melting and boiling points of HF are higher than expected by extrapolation from the other hydrogen halides (Figure 39). Solid hydrogen fluoride has a polymeric structure built up of zig-zag chains (2.9) (see section 2.5.7) and similar chains occur in the liquid and gaseous states, as well as in the hydrogen fluoride solvates of metallic fluorides such as KF . HF (10.12), KF . 2HF (10.13), KF . 3HF and KF . 4HF. In the vapour state below

$K^+ [F\text{---}H\text{---}F]^-$

(10.12) symmetrical H --- F bonds

$$K^+ \left[\begin{array}{c} F \\ H \quad\quad H \\ F \quad\quad\quad F \end{array} \right]^-$$

(10.13) not known whether the H --- F and H—F bonds are equal in length

about 60 °C the main polymers present are thought to be $(HF)_4$ and $(HF)_6$. At higher temperatures, hydrogen fluoride, like the other hydrogen halides, is monomeric.

In aqueous solution hydrogen chloride, bromide and iodide are completely dissociated and behave as strong acids, whereas hydrogen fluoride is a relatively weak acid in dilute solution,

$$HF \rightleftharpoons H^+(aq) + F^-; \; K_1 = \frac{[H^+][F^-]}{[HF]} = 6{\cdot}7 \times 10^{-4} \text{ at } 25 \text{ °C}.$$

The situation is not as simple as shown by the above equilibrium, because even dilute solutions of hydrogen fluoride contain the HF_2^- ion

$$HF + F^- \rightleftharpoons HF_2^-, \; K_2 = \frac{[HF_2^-]}{[F^-][HF]} = 3{\cdot}86 \text{ at } 25 \text{ °C},$$

and as the concentration of hydrogen fluoride rises the formation of other ions such as $H_2F_3^-$ and $H_3F_4^-$ becomes increasingly important. The difference between hydrogen fluoride and the other hydrogen halides in dilute aqueous solution is largely due to the greater bond strength of hydrogen fluoride. However, several other factors contribute to the acidity and the outcome is so finely balanced that if the dissociation energy of hydrogen fluoride were 35–40 kJ mol^{-1} lower, then aqueous hydrofluoric acid would also be a strong acid in line with the other hydrogen halides.

It has been stated by some chemists that hydrogen fluoride is a strong acid in concentrated (5–15 molar) solutions, but this is not strictly true because a high proportion of the hydrogen fluoride is present as undissociated molecules – hydrogen-bonded to fluoride ions as $F(HF)_n^-$ anions. However, it can be argued that, since H_2F_2, H_3F_3 and H_4F_4 molecules cannot be detected in such solutions, these polymers of hydrogen fluoride must themselves be strong acids because they undergo complete dissociation

$$H_2F_2 \longrightarrow HF_2^- + H^+(aq),$$
$$H_3F_3 \longrightarrow H_2F_3^- + H^+(aq), \text{ etc.}$$

When the hydrogen halides are dissolved in a solvent which is more acidic than water (e.g. glacial acetic acid) their proton-donating ability (acidity) can be more usefully studied and shown to be in the order HI > HBr > HCl ≫ HF.

10.8 Oxides and oxy-acids of the halogens

Most of the oxides formed by the halogens are unstable and tend to detonate on shock or exposure to light.

10.8.1 *Dihalogen monoxides* X_2O (X = F, Cl, Br)

These are bent triatomic molecules similar to water, but with the larger halogens the X—O—X bond angle is greater than 109° 28′, no doubt due to steric crowding of the two halogen atoms. In difluorine monoxide the bond angle is 103·2°, the steric requirements of the two sp^3 lone pairs of electrons on the oxygen being the major contributing factor to the shape, as in the water molecule.

Difluorine monoxide. This, the most stable monoxide of the group, may be prepared by bubbling fluorine through 2 per cent aqueous sodium hydroxide (cf. the action of the other halogens with a base) but on prolonged contact with hydroxide ions fluorine produces oxygen

$$F_2 + 2OH^- \longrightarrow 2F^- + H_2O + \tfrac{1}{2}O_2.$$

Difluorine monoxide behaves as an oxidizing agent and will oxidize bromide and iodide to the free halogen.

Dichlorine monoxide. When chlorine and dry air are passed over mercuric oxide and the products condensed at liquid nitrogen temperatures, dichlorine monoxide is obtained; it is extraordinarily unstable and explodes on shock or impact. A kinetic study has shown the photo-decomposition of dichlorine monoxide to occur via the following steps:

$$\begin{aligned} Cl_2O + h\nu &\longrightarrow ClO + Cl, \\ Cl_2O + Cl &\longrightarrow ClO + Cl_2, \\ ClO + ClO &\longrightarrow Cl_2 + O_2. \end{aligned}$$

Both dichlorine and dibromine monoxides react with water giving the corresponding hypohalous acids (thus acting as acid anhydrides) and with alkalis to produce hypohalites – compare this with difluorine monoxide, which is made in the presence of base.

10.8.2 *Other fluorine oxides*

A short series of oxides is known which have the general formula FO_nF, with $n = 2, 4, 5$ and 6. All members of the series are thermally very, very unstable, the most robust being F_2O_2, which decomposes at −57 °C into oxygen and fluorine. Their synthesis is accomplished by subjecting mixtures of oxygen and fluorine to an electrical discharge, the discharge cell being immersed in liquid oxygen (~ −185 °C), so that the reaction products may be frozen out of the gas phase before they have time to decompose (Figure 100). The structure of F_2O_2, determined by continuously pumping the vapour through a Teflon-lined microwave cavity and observing the microwave spectrum, is similar to that of hydrogen peroxide but with a much shorter O—O bond length (1·22 Å); the reason for this short bond is not clearly understood. The compound O_4F_2 appears to be a loosely bound dimer of the radical OOF; i.e. $(OOF)_2$, and its solutions exhibit enormous electron spin resonance signals due to extensive dissociation into OOF.

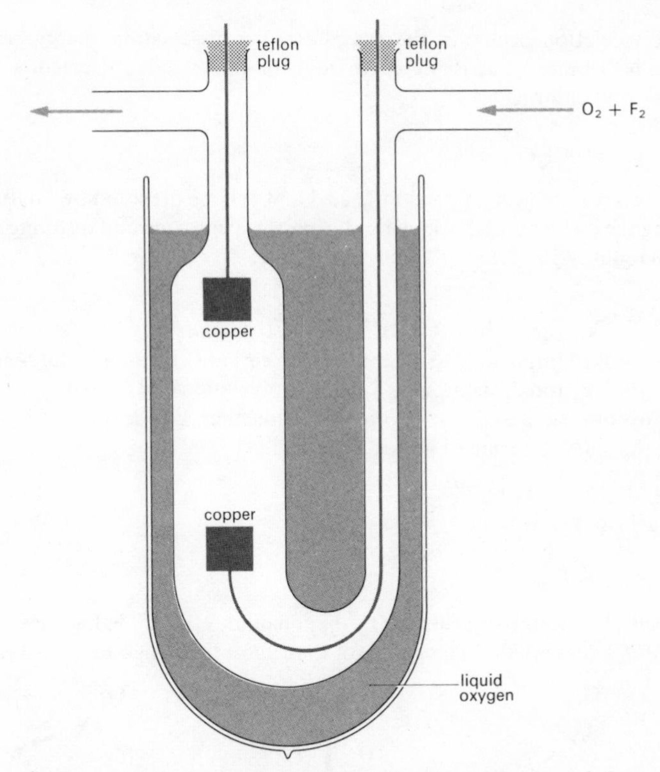

Figure 100 Discharge apparatus used for the preparation of dioxygen difluoride O_2F_2

Earlier reports of O_3F_2 have now been disproved. Evidence for O_5F_2 and O_6F_2 has been published, but little is known about them; they are of interest in that they presumably contain chains of five and six oxygen atoms.

10.8.3 *Chlorine dioxide* ClO_2 *and bromine dioxide* BrO_2

Bromine dioxide is highly unstable and above −40 °C decomposes into bromine, oxygen and dibromine monoxide. Chlorine dioxide is a paramagnetic, yellowish gas which shows little tendency to dimerize due to the unpaired electron being extensively delocalized over the whole molecule. It is prepared by the action of sulphuric acid on potassium chlorate in the presence of a reducing agent; usually oxalic acid is used so that the dangerously explosive chlorine dioxide is diluted with carbon dioxide, making manipulations somewhat less hazardous. Hydrolysis yields chlorite and chlorates as the final products.

10.8.4 *Chlorine trioxide* ClO_3

Liquid chlorine trioxide consists mainly of the dimer Cl_2O_6 (though the paramagnetic monomer can be detected by magnetic measurements), but virtually complete dissociation occurs in the gas phase, the dissociation energy being only about 6·3 kJ mol^{-1}. It is prepared from the photo-induced reaction between ozone and chlorine.

10.8.5 *Chlorine heptoxide* Cl_2O_7

This oxide is the anhydride of perchloric acid and may be prepared by carefully dehydrating the concentrated acid with phosphorus pentoxide and distilling the product at about −35 °C under 1 mmHg pressure.

10.8.6 *Oxides of iodine*

So far it has proved impossible to determine with certainty the number of oxides formed by iodine. Iodine pentoxide I_2O_5, a polymeric solid, behaves as the anhydride of iodic acid and can be observed (together with a trace of iodine dioxide) in the mass spectrum of heated iodic acid.

$$2HIO_3 \underset{}{\overset{240\ °C}{\rightleftharpoons}} I_2O_5 + H_2O$$

I_2O_4 and I_4O_9 may also exist.

10.8.7 *Oxy-acids of the halogens*

The halogens form four oxy-acids: HOX^{I} hypohalous, HO_2X^{III} halous, HO_3X^{V} halic and HO_4X^{VII} perhalic. The anions of these have the structures (10.14–17).

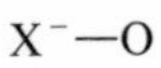

X^-—O

~110°

~107°

(10.14) (p) 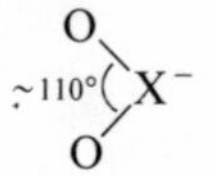 (10.15) (sp^3; two lone pairs)  (10.16) (sp^3; one lone pair) 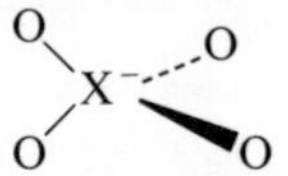 (10.17) (sp^3; no lone pairs)

The energy separation of the *n*s and *n*p atomic orbitals of the halogens is large, so that the σ-bonds formed by sp^3-hybridized halogen atoms are quite weak, and theoretical calculations show that halite XO_2^-, halate XO_3^- and perhalate XO_4^- anions owe their stability almost completely to strong p_π–d_π bonding between filled oxygen 2p orbitals and vacant halogen *n*d orbitals. For this reason fluorine, which has no d-orbitals, would not be expected to form fluorous, fluoric or perfluoric acids (or their salts) and none are known. There seems to be no valid reason why hypofluorous acid and hypofluorites should not exist, and indeed the heat of formation of hydrofluorous acid HOF has been calculated to be 109±20 kJ mol^{-1}. However, all attempts to make the acid under normal conditions have failed.† A few compounds are known which contain the co-

†Photolysis of fluorine–water mixtures suspended in a matrix of solid nitrogen held at 14–20 K yields a product which has vibrational spectra consistent with its being hypofluorous acid; the compound is stable only in matrix isolation. It is possible that hypofluorous acid decomposes into hydrogen fluoride and oxygen; if so, the free-energy change of the reaction $2HOF \rightarrow 2HF + O_2$ is so highly negative (~−335 kJ mol^{-1}) that such a decomposition route would easily explain the high instability of hypofluorous acid.

valently bound OF group and can thus be considered to be derivatives of hypofluorous acid; perhaps the most studied is trifluoromethylhypofluorite CF_3OF

$$F_2 + CO + AgF_2 \text{ (as catalyst)} \longrightarrow CF_3OF \xrightleftharpoons{275\,°C} COF_2 + F_2.$$

The complex interrelations between the oxy-acids of a particular halogen are best summed up by a list of the various standard oxidation potentials which operate in aqueous solution (Table 39). Few other non-transition elements exist

Table 39 Some Oxidation–Reduction Potentials Pertinent to the Chemistry of the Halogens in Aqueous Solution

Acid solutions

$O_2 + 4H^+ + 4e \longrightarrow 2H_2O$; 1·23 V

	F	Cl	Br	I	At
$\frac{1}{2}X_2 + e \rightarrow X^-$	2·85	1·36	1·06	0·53	−0·2
$\frac{1}{2}F_2 + H^+ + e \rightarrow HF$	3·03	—	—	—	—
$HXO + H^+ + e \rightarrow \frac{1}{2}X_2 + H_2O$	—	1·63	1·59	1·45	0·7
$HXO_2 + 3H^+ + 4e \rightarrow X^- + 2H_2O$	—	1·56	—	—	—
$HXO_2 + 3H^+ + 3e \rightarrow \frac{1}{2}X_2 + 2H_2O$	—	1·63	—	—	—
$XO_3^- + 6H^+ + 5e \rightarrow \frac{1}{2}X_2 + 3H_2O$	—	1·47	1·52	1·20	—
$XO_4^- + 8H^+ + 7e \rightarrow \frac{1}{2}X_2 + 4H_2O$	—	1·34	—	—	—

Basic solutions

$O_2 + 2H_2O + 4e \longrightarrow 4OH^-$; 0·40 V

	Cl	Br	I	At
$XO^- + H_2O + 2e \rightarrow X^- + 2OH^-$	0·94	0·76	0·49	—
$XO_2^- + 2H_2O + 4e \rightarrow X^- + 4OH^-$	0·76	—	—	—
$XO_3^- + 3H_2O + 6e \rightarrow X^- + 6OH^-$	0·62	0·61	0·26	—
$XO_4^- + 4H_2O + 8e \rightarrow X^- + 8OH^-$	0·56	—	0·39	—
$\frac{1}{2}X_2 + e \rightarrow X^-$	1·36	1·06	0·53	−0·2
$XO_2^- + H_2O + 2e \rightarrow XO^- + 2OH^-$	0·59	—	—	—
$XO_3^- + 2H_2O + 4e \rightarrow XO^- + 4OH^-$	—	0·46	0·56	0·50
$XO_3^- + H_2O + 2e \rightarrow XO_2^- + 2OH^-$	0·35	—	—	—

in a variety of oxidation states as do the halogens in their oxy-acids, so that this is the first time in this book we have been able to use standard potentials as an aid to understanding chemical reactions.

From the data given in Table 39:

$$X_2 + 2e \longrightarrow 2X^- \qquad E_1$$
$$H_2O \longrightarrow \tfrac{1}{2}O_2 + 2H^+ + 2e \qquad -1{\cdot}23\ V$$

$$X_2 + H_2O \longrightarrow \tfrac{1}{2}O_2 + 2H^+ + 2X^- \quad E_2 = (E_1 - 1{\cdot}23\ V)$$

E_2 has the values 1·62 V (F_2), 0·13 V (Cl_2), −0·17 V (Br_2), −0·70 V (I_2) and −1·43 V (At_2). Since the potential for a reaction is directly proportional to the standard free-energy change† ΔG° for that reaction, we observe that the reaction of fluorine and chlorine with water under acid conditions to give oxygen is thermodynamically favoured. In practice, however, it is usually found that a reaction proceeds very slowly if the potential is less than about half a volt, and consequently only fluorine is found to release oxygen from water (even with fluorine the reaction is fairly sluggish as evidenced by the preparation of difluorine monoxide by passing fluorine through dilute aqueous alkali). Conversely, we note from the high negative potentials of iodine and astatine for the equation as written, that oxygen is capable of releasing halogen from acidic solutions of iodide or astatide.

Chlorine, bromine and iodine are slightly soluble in water (saturated solutions at 25 °C are about 0·06, 0·21 and 0·0013 molar respectively). The equilibrium set up in solution is

$$X_2 + H_2O \rightleftharpoons HOX + HX.$$

With chlorine, about half is present as hypochlorous and hydrochloric acids, the rest as free chlorine. This reaction can be made into a preparation of hypochlorous acid by having present in the system an insoluble compound which will remove the chloride ions by forming an insoluble or sparingly ionized chloride (usually mercuric oxide is used). Hypochlorous acid is a far weaker acid than is acetic acid (K_a is about $3{\cdot}7 \times 10^{-8}$ at 20 °C); hypobromous and hypoiodous acids are even weaker, their dissociation constants being 2×10^{-9} and $4{\cdot}5 \times 10^{-13}$ respectively.

The potentials for basic solutions show that bromine is a stronger oxidizing agent than hypobromite:

$$\begin{array}{lll} Br_2 + 2e & \longrightarrow 2Br^- & 1{\cdot}06\ V, \\ Br^- + 2OH^- & \longrightarrow BrO^- + H_2O + 2e & -0{\cdot}76\ V, \\ \hline Br_2 + 2OH^- & \longrightarrow BrO^- + H_2O + Br^- & +0{\cdot}30\ V. \end{array}$$

Hence ΔG° is negative so that the reaction as written is favoured and the equilibrium constant, K, may be calculated from the expression

$$\log K = \frac{zFE}{2{\cdot}303RT},$$

$$K = \frac{[BrO^-][Br^-]}{[Br_2][OH^-]^2} \simeq 2 \times 10^5.$$

Therefore the reaction of bromine with a base should give a mixture of hypo-

† $-\Delta G^\circ = zFE$, where F is the faraday (96 500 coulombs), E is the potential in volts and z is the number of electrons associated with the reaction. If E is positive, then the free-energy change for the reaction is favourable.

bromite and bromide, the extent of the conversion being dependent on the *square* of the hydroxyl ion concentration.

However, the situation is not quite so simple because hypobromite is thermodynamically unstable towards disproportionation

$$\begin{array}{rcll} 3BrO^- + 3H_2O + 6e & \longrightarrow & 3Br^- + 6OH^- & 0{\cdot}76\text{ V}, \\ Br^- + 6OH^- & \longrightarrow & BrO_3^- + 3H_2O + 6e & -0{\cdot}61\text{ V}, \\ \hline 3BrO^- & \longrightarrow & 2Br^- + BrO_3^- & +0{\cdot}15\text{ V}. \end{array}$$

This disproportionation is slow at 0 °C so that the reaction of bromine with ice-cold base can be used to prepare hypobromite, but at room temperature or above the products are bromide and bromate.

It is obvious from Table 39 that in basic solution all the halogens except astatine are capable of releasing oxygen from water, but this reaction occurs only relatively slowly even with fluorine (which initially forms difluorine monoxide). Hence, in the case of bromine reacting with a base, three reactions are thermodynamically favoured, one of which (oxygen evolution) is not kinetically important, but the balance between the other two is temperature sensitive, due to the rate of the hypobromite disproportionation being so markedly temperature dependent. Similar products are formed at similar temperatures in the reaction of chlorine with a base, but iodine reacts to give iodate and iodide at all temperatures, because hypoiodite ions disproportionate too rapidly to be isolated.

The disproportionation of aqueous chlorate to perchlorate and chloride

$$4ClO_3^- \longrightarrow 3ClO_4^- + Cl^-$$

is energetically favoured, but it occurs so very slowly that perchlorate is not formed even at the boiling point of water; however, careful heating of solid potassium chlorate does give the perchlorate

$$4KClO_3 \longrightarrow 3KClO_4 + KCl.$$

Only recently has it been proved possible to prepare perbromates by the action of the very powerful oxidizing agent xenon difluoride XeF_2 on bromates; acidification of perbromate solutions gives perbromic acid $HBrO_4$, which was detected as the parent peak $HBrO_4^+$ in the mass spectrum. No mass-spectral evidence for bromine heptoxide Br_2O_7 was obtained on heating perbromic acid, but the presence of bromine dioxide in the decomposition products was noted.

10.8.8 *Periodates*

Although periodic acid HIO_4, and several salts containing the tetrahedral IO_4^- periodate ion are known, a variety of other periodic acids and their derivatives exist. Iodine resembles tellurium to some extent by forming hydroxy-derivatives in which the iodine atoms are octahedrally coordinated, and aqueous solutions of periodate at low pH contain at least four species, viz. H_5IO_6, $H_4IO_6^-$, $H_3IO_6^{2-}$ and IO_4^-

$$IO_4^- \rightleftharpoons HIO_4 + 2H_2O \rightleftharpoons H_5IO_6 \rightleftharpoons H_4IO_6^- \rightleftharpoons H_3IO_6^{2-}.$$

Under the correct conditions salts of these periodate species can be isolated (e.g. $NaIO_4$, $AgIO_4$, NaH_4IO_6, $Na_2H_3IO_6$ and $(NH_4)_2H_3IO_6$). Evaporation of aqueous solutions of periodic acid gives crystals of H_5IO_6 which can be dehydrated by careful heating,

$$H_5IO_6 \text{ (i.e. 'HIO}_4.2H_2O\text{')} \xrightarrow[12\ \text{mmHg}]{100\ °C} HIO_4 \xrightarrow{138\ °C} HIO_3 + O_2.$$

The acid H_5IO_6 has the structure (10.18), in which the iodine atom is octahedrally coordinated to six oxygen atoms. Since the iodine is surrounded by the maximum number of hydroxyl groups for this oxidation state (VII) we may term it the *ortho* periodic acid. It is an essentially dibasic acid with $K_1 = 0{\cdot}02$ and $K_2 \simeq 10^{-8}$, giving two main series of salts $M^IH_4IO_6$ and $M^I_2H_3IO_6$, although other salts containing $H_2IO_6^{3-}$ and IO_6^{5-} ions can be prepared,

$$\text{e.g.} \quad 5Ba(IO_3)_2 \xrightarrow{\text{heat}} Ba_5(IO_6)_2 + 4I_2 + 9O_2.$$

At high pH (~ 10–11) dimerization of the periodate anions occurs to give $H_2I_2O_{10}^{4-}$, in which each iodine is octahedrally surrounded by six oxygen atoms, the two octahedra having one edge in common. On crystallization from potassium hydroxide solutions the potassium salt $K_4H_2I_2O_{10}.8H_2O$ is obtained, the anion of which has been shown to have the structure (10.19).

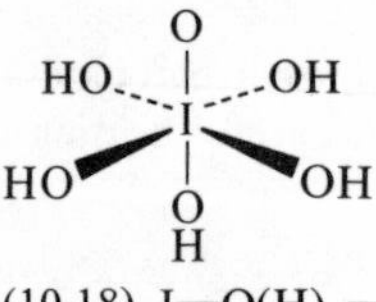

(10.18) I—O(H) = 1·89 Å; I—O = 1·78 Å

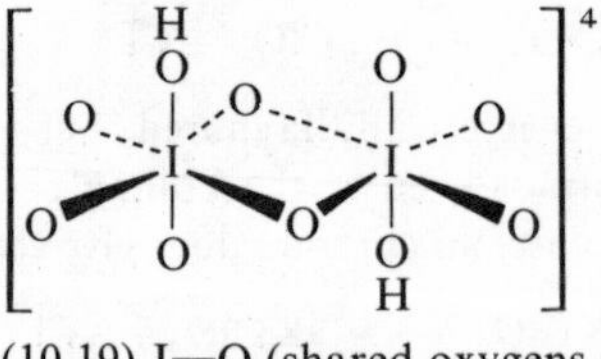

(10.19) I—O (shared oxygens and hydroxyls) = 2·00 Å; I—O = 1·81 Å

It has been suggested that perbromic acid and the perbromates owe their instability to the relatively weak π-bonding which occurs between the filled oxygen 2p orbitals and the empty bromine 4d orbitals; there is a node of zero electron density in the 4d orbitals near to the region where overlap of the two orbitals should take place. A similar weak π-bonding situation would have been expected in the periodates; the fact that periodates exist and are stable may be due to the participation of iodine 4f orbitals in the π-bonding. It is interesting to note that the partial polymerization of oxy-anions in groups IV, V and VI of the periodic table is thought to be due, at least partially, to weak π-bonding.

Spectroscopic evidence suggests that orthoperiodic acid H_5IO_6, can be protonated under the very acidic conditions existing in 10 molar perchloric acid to give the cationic species $I(OH)_6^+$.

10.9 Fluorocarbons

Strictly, organo-derivatives of the halogens belong to the realm of organic chemistry and a discussion of them would be out of place in this book. However, perhaps some mention ought to be made of highly fluorinated carbon derivatives which are now becoming widely used in polymers (e.g. 'Teflon' and 'Fluon'), refrigerant fluids, aerosol propellents and non-inflammable anaesthetics.

Of the groups attached to carbon in organic compounds, fluorine is rapidly becoming established as being second only to hydrogen in importance, and a host of highly fluorinated analogues of the hydrocarbons and their derivatives, especially in the aromatic field, have been synthesized in recent years. Fluorine, like hydrogen, is always monovalent and, due to shielding effects operative from lithium to fluorine as the $n = 2$ quantum shell is filled, it has an exceptionally small van der Waals radius of only 1·35 Å (that of hydrogen is 1·2 Å). This means that steric crowding round a carbon atom is minimized relative to the other halogen atoms. The C—F bond energy varies between 460 and 500 kJ mol^{-1} (C—H bond energy $\sim$414 kJ mol^{-1}), so that polymers containing a $—CF_2—$ chain are very stable from both energetic and kinetic viewpoints – the fluorine atoms, being slightly larger than hydrogen atoms, form a more effective shield round the —C—C— backbone of the polymer and inhibit the approach of attacking reagents. Fluorine atoms are not very polarizable, so that intermolecular forces are weaker than between hydrocarbons, resulting in higher volatilities and solubilities of the fluorocarbons for a given number of carbon atoms.

Three main methods of forming fluorocarbons are available:

10.9.1 *Using fluorides of metals in high oxidation states, for example, cobalt(III) fluoride*

$$2CoF_2 + F_2 \longrightarrow 2CoF_3,$$
$$R—H + 2CoF_3 \longrightarrow R—F + 2CoF_2 + HF \quad (R = \text{alkyl, aryl}).$$

When aromatic hydrocarbons are used, the main products of fluorination by cobalt(III) fluoride are the *saturated* alicyclic fluorides (e.g. 10.20)

$$C_6H_6 + CoF_3 \longrightarrow C_6F_{12}$$

(10.20)

The aromatic ring can be regenerated by passing the alicyclic fluorocarbons over heated iron or nickel

$$C_6F_{12} \xrightarrow{Ni} C_6F_6 + 3NiF_2$$

(hexafluorobenzene)

10.9.2 *Halogen-exchange reactions*

A variety of halogen-exchange reagents can be used, the two finding the widest application are potassium fluoride and antimony trifluoride

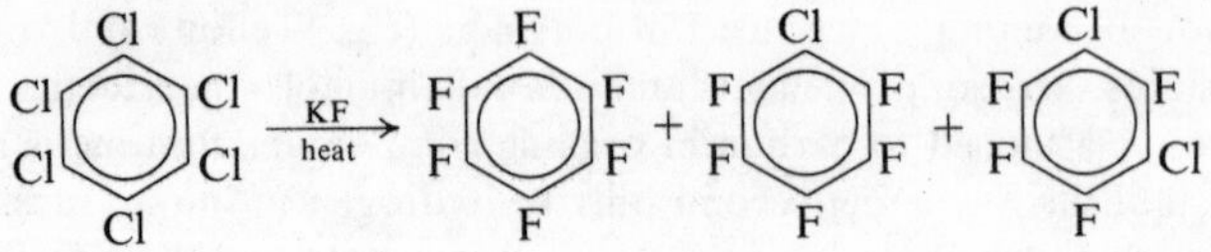

$$C_6Cl_6 \xrightarrow[\text{heat}]{KF} C_6F_6 + C_6F_5Cl + C_6F_4Cl_2$$

10.9.3 *Electrochemical fluorination*

The organic compound is dissolved in liquefied anhydrous hydrogen fluoride and the solution electrolysed at low voltage

$$\left.\begin{array}{l} R{-}H \longrightarrow R_F{-}F \\ OR_2 \longrightarrow O(R_F)_2 \\ NR_3 \longrightarrow N(R_F)_3 \\ NHR_2 \longrightarrow FN(R_F)_2 \end{array}\right\} \quad (R_F = C_nF_{2n+1}).$$

The highly fluorinated aromatic derivatives are the most amenable to further study. For example, C_6F_5Cl synthesized by halogen-exchange methods can be used to form a wide variety of new products including organometallic derivatives.

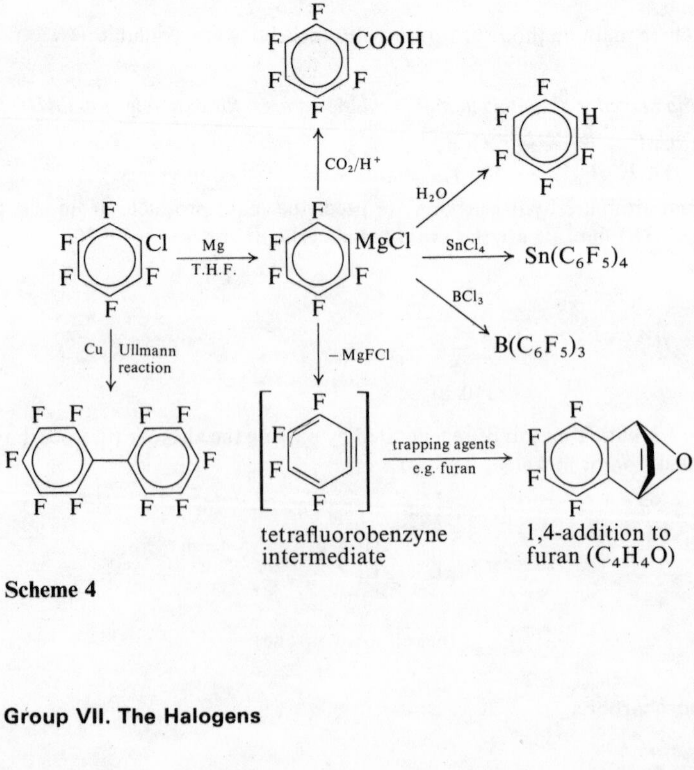

$$C_6F_5Cl \xrightarrow[\text{T.H.F.}]{Mg} C_6F_5MgCl$$

$$C_6F_5MgCl \xrightarrow{CO_2/H^+} C_6F_5COOH$$

$$C_6F_5MgCl \xrightarrow{H_2O} C_6F_5H$$

$$C_6F_5MgCl \xrightarrow{SnCl_4} Sn(C_6F_5)_4$$

$$C_6F_5MgCl \xrightarrow{BCl_3} B(C_6F_5)_3$$

$$C_6F_5MgCl \xrightarrow{-MgFCl} [C_6F_4] \text{ (tetrafluorobenzyne intermediate)} \xrightarrow[\text{e.g. furan}]{\text{trapping agents}} \text{1,4-addition to furan } (C_4H_4O)$$

$$C_6F_5Cl \xrightarrow[\text{Ullmann reaction}]{Cu} C_6F_5{-}C_6F_5$$

Scheme 4

Chapter 11
Group 0. The Rare Gases: Helium, Neon, Argon, Krypton, Xenon and Radon

11.1 Introduction

The rare gases are characterized by both their high ionization energies and the fact that all the available s and p energy levels in the outer quantum shell contain two electrons, the maximum number allowed by the Pauli exclusion principle. Furthermore, the energy required to promote electrons either into orbitals of the next (i.e. $n+1$) quantum shell or into d- and f-orbitals of the n shell is high, being of the order of 960 kJ mol^{-1} for the promotions

$$5s^2 5p^6 \longrightarrow 5s^2 5p^5 6s^1 \quad \text{and} \quad 5s^2 5p^6 \longrightarrow 5s^2 5p^5 5d^1$$

in the case of xenon. Thus covalent bond formation involving promotion and hybridization to, for example, the sp^{3}d or sp^{3}d^2 states will be highly unfavourable energetically, since the compounds formed would be destabilized by an amount similar to the (very high) promotion energy.

For many years the rare gases were considered to be completely unreactive and indeed their inertness led chemists to the concept of the octet theory of chemical bonding; even van der Waals forces between the rare-gas atoms are very weak and this leads to the observed low melting and boiling points. The forces of attraction in solid helium are so slight that the energy of vibration of the helium atoms in the

Table 40 The Rare Gases

Element		*1st IE* / kJ mol^{-1}	*Melting point*/K	*Boiling point*/K
Helium He	$1s^2$	2372	1 (25 atm)	4·2
Neon Ne	$1s^2 2s^2 2p^6$	2080	24·5	27·2
Argon Ar	$[Ne]3s^2 3p^6$	1515	83·8	87·9
Krypton Kr	$[Ar]3d^{10} 4s^2 4p^6$	1347	105·9	120·9
Xenon Xe	$[Kr]4d^{10} 5s^2 5p^6$	1162	161·3	165·1
Radon Rn	$[Xe]4f^{14} 5d^{10} 6s^2 6p^6$	—	202	211

liquid even at the lowest attainable temperature (i.e. the zero-point energy) is sufficient to stop the solid forming; only by restraining the atomic motion by application of external pressure is it possible to cause liquid helium to freeze. The formation of crystalline derivatives of these gases (except helium) with such compounds as water and phenols caused some interest in the 1920s and 1930s, but these are not real compounds in the sense that no covalent bonds are formed by the rare gases; the latter are actually trapped mechanically in holes which occur in the rather open, hydrogen-bonded structures of the host solids. A typical 'clathrate' of the above type is $Xe.xH_2O$ ($x \simeq 5$ or 6) which has a melting point of 24 °C and a xenon dissociation pressure of 1·3 atmospheres at 0 °C.

From the ionization energies given above xenon would appear to be the most favourable element for attempting the formation of a unipositive rare-gas ion R^+. Using Born–Haber calculations one can calculate the heat of formation of solid xenon monofluoride Xe^+F^-.

$$\begin{array}{ccc} Xe + \frac{1}{2}F_2 & \xrightarrow[1\ atm]{25\ ^\circ C} & Xe^+F^-(s); \qquad \Delta H_f \\ \big\downarrow {\scriptstyle +\frac{1}{2}D} & & \big\uparrow {\scriptstyle -L_{XeF}} \\ Xe + F(g) & \xrightarrow[-E_F]{+I_{Xe}} & Xe^+(g) + F^-(g) \end{array}$$

D is the dissociation energy of the F_2 molecule, I_{Xe} is the first ionization energy of xenon, E_F is the electron affinity of the fluorine atom and L_{XeF} is the lattice energy of ionic xenon monofluoride. (The sublimation energy of xenon may be taken as zero since the element is a gas at 25 °C.)

All the quantities in this cycle are known with the exception of L_{XeF} which, to a close approximation, can be equated to the lattice energy of caesium fluoride. In fact the radius of Xe^+ will be slightly smaller than that of the caesium ion, so that L_{XeF} will probably be about 40 kJ mol^{-1} or so greater in energy.

$$\begin{aligned} \Delta H_f &= \tfrac{1}{2}D + I_{Xe} - E_F - L_{XeF} \\ &= \tfrac{1}{2} \times 157{\cdot}7 + 1162 - 332{\cdot}7 - 719{\cdot}7 \\ &= +198{\cdot}4 \text{ kJ mol}^{-1}. \end{aligned}$$

Thus, unless the reaction has a large positive entropy term, ΔG_f for the formation of Xe^+F^- will be positive and the compound is most unlikely to be stable at 25 °C. The chloride, bromide and iodide are even less thermodynamically stable, due to the high dissociation energies of the halogen molecules and the comparatively low lattice energies of the products. If the electron affinity (i.e. the oxidizing power) of the species reacting with the xenon could be vastly increased, then it is possible that stable ionic compounds of univalent xenon might be formed, since this could possibly make ΔH_f (and ΔG_f) negative. This has recently been achieved with the very powerful oxidizing agent platinum hexafluoride, which reacts with xenon gas at room temperature forming $Xe^+Pt^VF_6^-$ (see below).

Bombardment of a rare gas with electrons in a mass spectrometer causes an electron to be knocked off the gas atom to give a gaseous ion, e.g. $He + e^- \longrightarrow He^+ + 2e^-$.

This ion is stable in the isolation of high vacuum, but on contact with the containing walls of the vacuum system it will readily pick up an electron to revert to the neutral atom; in a mass spectrometer the ion is kept from contact with the wall of the spectrometer by the use of magnetic and electric fields until it reaches the collector plates of the detector. The reactivity of these ions is so high that if the

Table 41 Bond Energies of X_2^+ Ions for the Rare Gases

Ion	*Bond energy* / kJ mol^{-1}
He_2^+	126
Ne_2^+	67
Ar_2^+	104
Kr_2^+	96
Xe_2^+	88

pressure in the mass spectrometer is allowed to rise slightly (to about 10^{-5} mmHg), they will combine with neutral gas atoms to give ionic species such as $[He{-}He]^+$ which are, again, only stable in a vacuum.

The bonding in the He_2^+ ion is shown in Figure 101. Although there is one bonding and one antibonding molecular orbital available for the three electrons in the system, the *Aufbau* principle states that the electrons must be used to fill up the lowest energy orbital first, in this case the bonding σ1s. Thus there is only one electron left to go into the antibonding σ^*1s orbital and the resultant state is a bonding one: $(\sigma 1s)^2(\sigma^* 1s)^1$. One might call this a half-bond as there is only one electron more in the bonding than in the antibonding molecular orbitals; however, the heat of dissociation of He_2^+ is still quite large, being about 126 kJ mol^{-1}. If a mass spectrometer is filled with a mixture of two rare gases it is possible to detect, besides the homonuclear ions R_2^+, 'mixed' ions such as $[Ne{-}Ar]^+$ and $[Xe{-}Ar]^+$.

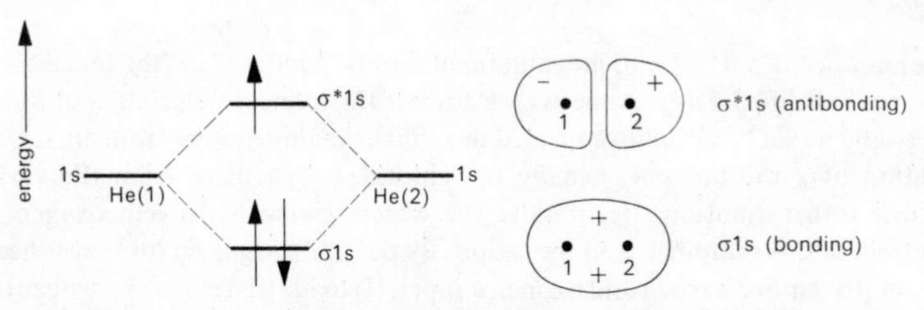

Figure 101 Energy-level diagram and molecular orbitals of the He_2^+ ion

The spectra of electric discharges operating in an atmosphere of any one of the rare gases consist of sharp lines characteristic of a monatomic gas; if traces of oxygen are admitted to a discharge tube in which either xenon or krypton is being electrically excited, band spectra are also observed due to vibrational and rotational energy transitions within the rare-gas oxides RO. These diatomic molecules are highly unstable and decay very rapidly when the discharge is switched off.

Another approach to the problem of making covalent bonds to the rare gases has been to study the decay products, using a special mass spectrometer, of radioactive halogen compounds. When $CH_3{}^{131}I$ was allowed to decay by β^--emission it was possible to detect a 70 per cent yield of CH_3Xe^+:

$$CH_3{}^{131}I - \beta^- \text{ (i.e. an increase of } +1 \text{ in nuclear charge)} \longrightarrow CH_3Xe^+.$$

The product was again only stable in the isolation of high vacua but was apparently much more robust than CH_3He^+ because virtually none of the helium methyl ion remained intact when $C^1H_3{}^3T$ decayed under the same conditions.

11.2 Occurrence of the rare gases

11.2.1 *Helium*

5 p.p.m. by volume of the atmosphere. Due to its extreme lightness, helium is able to exceed the escape velocity of 11·2 km s^{-1}, and hence all the helium now present on Earth has come from the radioactive decay of other elements (α-particles are doubly charged helium nuclei). The principal source of helium is natural gas, which can contain 0·1–2 per cent of helium; the hydrocarbons are removed by liquefying them and pumping the helium gas away.

11.2.2 *Neon, argon, krypton and xenon*

These four gases are present in the atmosphere to the extents of 0·002, 0·93, 0·0001 and 0·000 01 per cent by volume, respectively. They may be separated from liquid air by fractional distillation.

11.2.3 *Radon*

Does not occur naturally in meaningful amounts; radon-222 is the longest-lived isotope (half-life 3·8 days; α-active). Radon is the immediate product of radium decay and as such is usually collected in a glass vacuum system from an aqueous solution of a radium salt, usually the chloride; α-particles from the various nuclear transformations decompose the water, giving hydrogen, oxygen and ozone as gaseous impurities in the radon. By passing the gas mixture over heated copper (to remove oxygen and ozone), copper(II) oxide (to remove hydrogen) and phosphorus pentoxide (to remove water), the radon can be obtained pure.

11.3 The elements

These elements are monatomic in all phases and, as gases, have ratios of their specific heats C_p/C_v which are close to the theoretical value of 1·67 required for an ideal monatomic gas. In the solid state the rare-gas atoms are held together by non-directional van der Waals (or London) forces, which vary according to the inverse sixth power of the distance between atoms. For this reason the atoms tend to pack in the most space-economizing manner; this is the face-centred cubic (i.e. cubic close packed) arrangement which is adopted by neon, argon, krypton and xenon. When the mass of an atom is very small, its zero-point energy also exerts an influence on the crystal structure: thus at 1 K and 25 atm helium-4 has a hexagonal close-packed structure, whereas the lighter isotope, helium-3 adopts the body-centred cubic arrangement. Above 1100 atm helium-4 can be forced to assume the face-centred cubic structure like other group 0 atoms.

11.4 Compounds of the rare gases

11.4.1 *Fluorine compounds of krypton and xenon*

Before 1962 the highly reactive ions described above were the only known cases of bonding exhibited by the rare gases, but in the middle of that year it was argued that, as the highly oxidizing compound platinum hexafluoride would react with oxygen to give $O_2^+Pt^VF_6^-$, then it should be possible to oxidize xenon to Xe^+ since the first ionization potential of O_2 (to O_2^+) is slightly higher than that of xenon. Experiments showed that xenon and platinum hexafluoride do in fact react very readily at room temperature to give a yellow-to-deep-red solid but, although the first report of reaction suggested the formulation of this solid as $Xe^+Pt^VF_6^-$, xenon hexafluoroplatinate(V), further work has shown that the reaction is much more complex, the structure of the products (11.1) and (11.2) being in some doubt at the present time. A compound of the same formula as (11.2) is obtained by burning a platinum wire in a xenon–fluorine atmosphere.

$$Xe + PtF_6 \longrightarrow \underset{(11.1)}{Xe^+Pt^VF_6^-} + \underset{(11.2)}{Xe(Pt^VF_6)_2}$$

$$\downarrow \text{heat}$$

$$\underset{(11.3)}{XePt_2F_{10}} \text{ (diamagnetic)}$$

(11.1 and 11.2 are paramagnetic)

By reacting xenon and platinum hexafluoride in an atmosphere of fluorine at 180–220 °C two compounds are obtained, the principal product being yellow crystals of composition $XePtF_{11}$ in which magnetic measurements indicated the presence of platinum(V). An X-ray crystal-structure determination showed that this is an ionic hexafluoroplatinate(V) with XeF_5^+ as the cation

$$Xe + PtF_6 + F_2 \xrightarrow{1:1:1 \text{ ratio}} XeF_5^+Pt^VF_6^- + Xe_2F_{11}^+Pt^VF_6^-.$$

The pentafluoroxenonium ion is in the shape of a square pyramid with the xenon about 0·2 Å below the base, as expected if a sterically important lone pair of electrons makes up the sixth coordination position around the xenon (11.4).

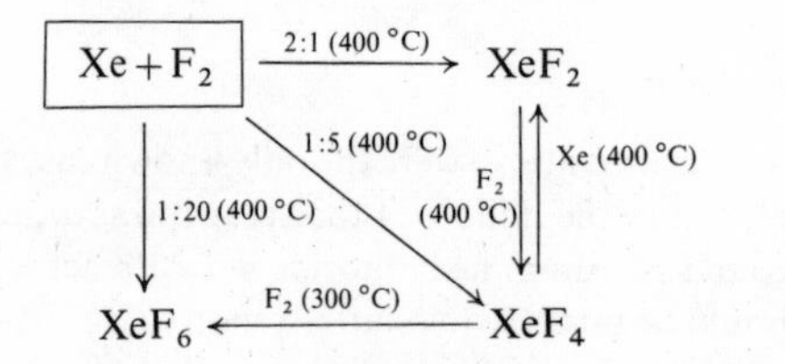

(11.4) Iso-electronic with IF_5

Since 1962 it has become apparent that xenon will react with fluorine on heating to about 400 °C in a sealed nickel vessel.

$Xe + F_2$ —2:1 (400 °C)→ XeF_2

$Xe + F_2$ —1:5 (400 °C)→ XeF_4

$Xe + F_2$ —1:20 (400 °C)→ XeF_6

XeF_2 —F_2 (400 °C)→ XeF_4

XeF_4 —Xe (400 °C)→ XeF_2

XeF_4 —F_2 (300 °C)→ XeF_6

The method is not very good for xenon difluoride because it readily reacts further with fluorine under the experimental conditions to give xenon tetrafluoride. To arrest the reaction at the difluoride stage a closed circulatory system of nickel can be used (Figure 102). The xenon and fluorine mixture, after being heated at

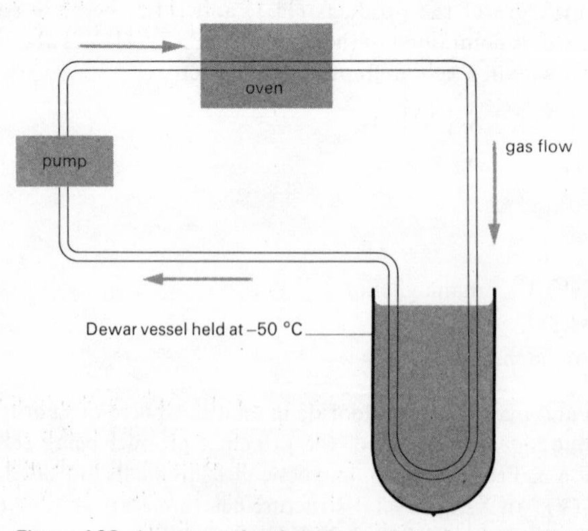

Figure 102 Apparatus for the preparation of xenon difluoride

400 °C in an oven, is passed through a U-tube cooled to – 50 °C, before recycling through the oven. With a suitable flow rate only xenon difluoride will be formed during passage through the oven and it will then be condensed out of the gas stream at – 50 °C on passing through the cold trap. A more convenient preparation for xenon tetrafluoride is to pass a xenon–fluorine mixture through a heated nickel tube

$$\underset{1:4}{Xe+F_2} \xrightarrow{300\,^{\circ}C} XeF_4 \quad \text{(100 per cent yield).}$$

A mixture of either xenon and fluorine or xenon and oxygen difluoride, when allowed to stand in daylight at room temperature in a sealed pyrex vessel, will slowly form xenon difluoride, the reaction being thought to proceed via the photochemical formation of fluorine atoms (minute amounts of krypton difluoride KrF_2 have been synthesized under similar conditions)

$$F_2 \xrightarrow{h\nu} 2F + Xe \longrightarrow XeF_2,$$

$$OF_2 \xrightarrow{h\nu} F + Xe \longrightarrow XeF_2.$$

The simplicity of the latter syntheses is remarkable when it is remembered that xenon was long considered to be an inert gas incapable of forming any compounds. Possibly the scarcity of xenon before the Second World War had much to do with this situation, as it made a careful study of the chemistry of xenon very difficult. In 1933 Yost and Kaye must have been very close to synthesizing the xenon fluorides when they passed xenon and fluorine through an electrical discharge but, no doubt somewhat polarized in their chemical thinking, they came to the conclusion that no compounds were formed. Modifications of their discharge method have since yielded xenon difluoride and krypton difluoride.

Attempts to prepare platinum hexafluoride derivatives of helium, neon, argon and krypton have so far failed, as have syntheses of their covalent fluorides using the action of heat, ultraviolet light and electrical discharges on mixtures of these gases with fluorine, except for the single instance of krypton difluoride mentioned above. Even so, krypton difluoride is so unstable that a special discharge cell operating in a Dewar flask of liquid oxygen has to be used to synthesize useful quantities (see Figure 100 for a similar discharge cell used in the preparation of O_2F_2). As the difluoride is formed in the discharge it immediately condenses on the cold surfaces of the discharge cell while the reactants are still gaseous at this temperature. Krypton difluoride decomposes into its elements slowly at room temperature and rapidly at 60 °C.

Radon has the lowest first ionization energy of the rare gases, which suggests that it may be more reactive and have a more extensive chemistry than even xenon. However, all the known isotopes of radon are radioactive, by far the most stable isotope being the α-emitter, radon-222, which has a half-life of only 3·8 days. This means that not only is work with radon-222 hazardous due to the intense radioactivity, but the half-life is such that only minute amounts of radon are available. Initial experiments have shown that when radon and fluorine are heated in a sealed nickel vessel a radon fluoride (or mixture of radon fluorides) is formed. Strong oxidizing agents in aqueous solution, such as hydrogen peroxide, persulphates and permanganates, oxidize radon to a species which is uncharged in neutral or acid solution and which is negatively charged in alkaline solution; it is thought that the radon is present as RnO_3 and $HRnO_4^-$ respectively.

11.4.2 *Chlorine compounds of xenon*

There is evidence that chlorine and xenon, when passed through an electrical discharge, or when sealed under pressure in a tube at room temperature, react partially to form xenon dichloride, but no chemistry of the compound has been reported. The β^- decay of radioactive iodine-129 has also been used to prepare minute amounts of xenon dichloride and tetrachloride

$$^{129}ICl_2^- \longrightarrow {}^{129}XeCl_2,$$
$$^{129}ICl_4^- \longrightarrow {}^{129}XeCl_4.$$

11.4.3 *Oxygen compounds of xenon*

During manipulation of xenon hexafluoride, several explosions have occurred from time to time, due to accidental hydrolysis of the hexafluoride to xenon trioxide, a highly explosive white solid

$$XeF_6 + 3H_2O \longrightarrow XeO_3 + 6HF.$$

With small quantities of water, or when reacted with silica, xenon hexafluoride gives the colourless liquid, xenon oxide tetrafluoride:

$$XeF_6 + H_2O \longrightarrow XeOF_4 + 2HF,$$

$$2XeF_6 + SiO_2 \xrightarrow{50\ ^\circ C} 2XeOF_4 + SiF_4.$$

The reaction of xenon hexafluoride with water is further complicated by two side reactions involving xenon trioxide

$$XeO_3 + 2XeF_6 \longrightarrow XeOF_4$$
$$\text{and}\quad XeO_3 + XeOF_4 \longrightarrow 2XeO_2F_2.$$

The latter compound is a colourless, very slightly volatile solid.

Xenon difluoride can be dissolved unchanged in water, but a slow reaction does occur on standing with the liberation of xenon gas

$$XeF_2 + H_2O \longrightarrow Xe + \tfrac{1}{2}O_2 + 2HF \quad (t_{1/2} \simeq 7\ \text{h at}\ 0\ ^\circ\text{C}).$$

On the other hand, only a partial evolution of xenon is observed when the tetrafluoride hydrolyses, as part of the xenon remains in solution as xenon trioxide,

$$XeF_4 + 6H_2O \longrightarrow 2Xe + 1\tfrac{1}{2}O_2 + XeO_3 + 12HF.$$

Xenon trioxide is not ionized in aqueous solution and concentrations as high as

4 molar in XeO_3 can be achieved, but in strongly alkaline solution above pH 10·5 the oxide behaves as a weak acid and exists as the xenate ion $HXe^{VI}O_4^-$,

$$HXeO_4^- \rightleftharpoons XeO_3 + OH^-; \quad K \simeq 7 \times 10^{-4}.$$

Alkali-metal salts such as sodium xenate $NaHXeO_4 . 1{\cdot}5H_2O$ can be isolated. In solution the xenate ion $HXeO_4^-$ slowly disproportionates to perxenate and xenon gas

$$2HXeO_4^- + 2OH^- \longrightarrow Xe^{VIII}O_6^{4-} + Xe + O_2 + 2H_2O.$$

Alkali-metal and alkaline-earth perxenates have been isolated, some of which, for example, $Na_4XeO_6 . 6H_2O$ and $Na_4XeO_6 . 8H_2O$, have been studied by X-ray crystallography to determine the geometry of the perxenate ion, see Table 42. In solution xenon(VIII) is an extremely powerful oxidizing agent and will oxidize iodate to periodate, manganese(II) to permanganate, water to oxygen (although only slowly in alkaline media) and concentrated hydrochloric acid to chlorine.

With concentrated sulphuric acid, sodium and barium perxenates give xenon tetroxide XeO_4, a volatile, yellow, highly unstable solid which decomposes explosively in the solid state even at temperatures as low as −40 °C

$$XeO_4 \longrightarrow Xe + 2O_2.$$

Halogen-substituted xenates are produced when an aqueous solution of xenon trioxide is mixed with either caesium fluoride or caesium chloride

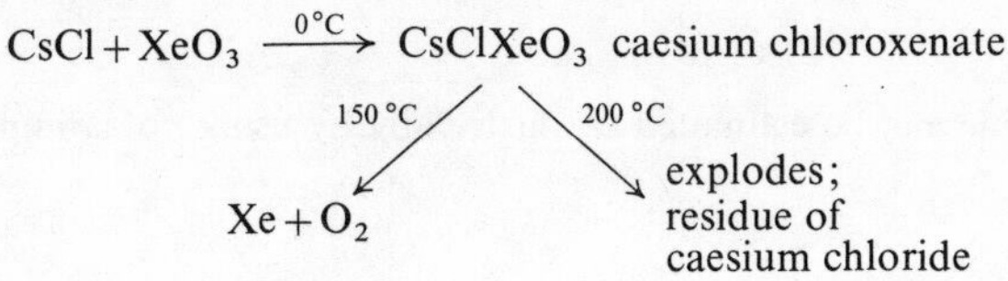

$$CsCl + XeO_3 \xrightarrow{0\,°C} CsClXeO_3 \text{ caesium chloroxenate}$$

$CsClXeO_3 \xrightarrow{150\,°C} Xe + O_2$

$CsClXeO_3 \xrightarrow{200\,°C}$ explodes; residue of caesium chloride

11.4.4 *Oxidation potentials for xenon in aqueous solution*

Acid solution.

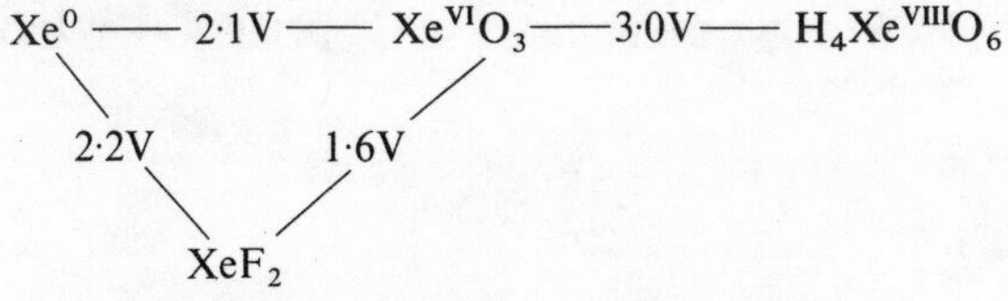

Xe^0 —— 2·1V —— $Xe^{VI}O_3$ —— 3·0V —— $H_4Xe^{VIII}O_6$

Xe^0 —— 2·2V —— XeF_2 —— 1·6V —— $Xe^{VI}O_3$

Xenon trioxide is one of the strongest oxidants known in aqueous media; its solutions should (thermodynamically) oxidize water; the non-occurrence of this reaction indicates a high activation barrier which probably arises from the absence of any stable intermediate oxidation states of xenon between free xenon and xenon trioxide.

Alkaline solution.

$$Xe^0 \xrightarrow{1{\cdot}3V} XeO(?) \xrightarrow{0{\cdot}7V} HXeO_4^- \xrightarrow{0{\cdot}9V} HXeO_6^{3-}$$

$$Xe^0 \xrightarrow{1{\cdot}2V} HXeO_4^-$$

11.4.5 *Reactions of xenon fluorides*

The three xenon fluorides are colourless, volatile solids which readily grow large crystals when allowed to stand in closed vessels; only the hexafluoride attacks silica in the absence of moisture so that xenon difluoride and tetrafluoride can be handled in a glass vacuum system. The most reactive is xenon hexafluoride as typically illustrated by the above hydrolytic properties.

In many cases the results of a reaction are products of fluorination of the attacking species, as with hydrogen,

$$XeF_2 + H_2 \xrightarrow{400\ °C} Xe + 2HF,$$

$$XeF_4 + 2H_2 \xrightarrow{130\ °C} Xe + 4HF,$$

$$XeF_6 + 3H_2 \xrightarrow[\text{violent}]{20\ °C} Xe + 6HF.$$

The first two systems were used in the initial studies on the xenon fluorides as a means of analysis; for xenon hexafluoride it was found more convenient to use mercury as the reducing agent

$$XeF_6 + 3Hg \longrightarrow Xe + 3HgF_2.$$

Xenon tetrafluoride can be estimated titrimetrically by using potassium iodide solution

$$XeF_4 + 4KI \longrightarrow Xe + 2I_2 + 4KF,$$

the liberated iodine being determined in the usual manner with standard thiosulphate solution.

It is possible that the xenon fluorides could become useful as fluorinating reagents because, unlike most other such reagents, they give rise to no messy, involatile by-products; the xenon gas produced can be pumped away leaving the required products behind:

$$XeF_6 + 3\,C_5F_8 \longrightarrow 3\,C_5F_{10} + Xe,$$

(cyclic structures: ring with F_2, F_2, F_2 and C=C bearing F, F → ring with F_2 at each of five positions)

$$XeF_2\ (\text{or } XeF_4 \text{ or } XeF_6) + \rangle C{=}C\langle \longrightarrow -\overset{|}{\underset{F}{C}}-\overset{|}{\underset{F}{C}}-,$$

$$XeF_6 + 8NH_3 \longrightarrow Xe + 6NH_4F + N_2,$$

$$XeF_6 + 6HCl \longrightarrow Xe + 6HF + 3Cl_2,$$

$$XeF_4 + 2SF_4 \longrightarrow Xe + 2SF_6.$$

Sometimes, however, the reactant is a more powerful fluorinating agent as in the case of dioxygen difluoride, which produces xenon hexafluoride when treated with xenon tetrafluoride

$$XeF_4 + O_2F_2 \xrightarrow{-80\,^{\circ}C} XeF_6 + O_2.$$

Liquid hydrogen fluoride is a solvent which readily supports ionization of dissolved species and, since it is incapable of further fluorination, it was an automatic choice as a possible solvent for the xenon fluorides: the difluoride and tetrafluoride are slightly soluble but the solutions do not conduct electricity, leading to the conclusion that no ions are formed. On the other hand, xenon hexafluoride dissolves to give a strongly conducting solution, probably due to an ionization process such as

$$XeF_6 \xrightarrow{HF(l)} XeF_5^+ + F^-.$$

Therefore XeF_6 acts as a base in liquid hydrogen fluoride by giving anions characteristic of the solvent's self-ionization

$$2HF \rightleftharpoons H_2F^+ + F^-.$$

A few derivatives of the XeF_5^+ cation are known, such as $XeF_5^+BF_4^-$ and $XeF_5^+Pt^{V}F_6^-$.

11.4.6 *Bonding in the xenon fluorides*

The simplest explanation of the bonding in xenon difluoride and xenon tetrafluoride is to assume that the xenon atom is sp^3d hybridized in xenon difluoride (11.5) and sp^3d^2 hybridized in xenon tetrafluoride (11.6). Although the 5d orbitals

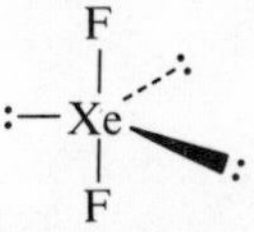

(11.5) linear molecule
(three lone pairs)

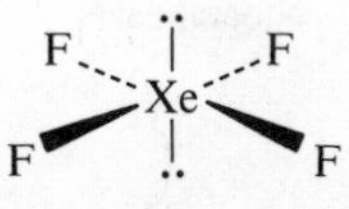

(11.6) planar molecule
(two lone pairs)

of an isolated xenon atom in the $5p^55d^1$ state have their radial electron density maximum at about 4·9 Å from the nucleus, apparently making them far too large to contribute effectively to the bonding in xenon difluoride where the Xe—F bond length is only 2·00 Å, calculations show that they contract when other atoms surround the xenon atom; it is possible that they contract sufficiently to participate in bonding to the fluorine atoms. However, experimental verification of this is lacking and at the present time many chemists are of the opinion a better approximation to the bonding in xenon difluoride considers that the xenon uses only *one*

Table 42 Structures of Xenon–Fluorine and Xenon–Oxygen Compounds

Compound	*Shape*	*Number of lone pairs*	*Comments*
Xenon difluoride XeF_2	F—Xe—F (180°) linear	3	Krypton difluoride KrF_2 and xenon dichloride $XeCl_2$ thought to be similar
Xenon tetrafluoride XeF_4	F, F—Xe—F, F square planar	2	Earlier reports of a similar tetrafluoride of krypton are now known to be incorrect
Xenon hexafluoride XeF_6	slightly distorted octahedral	1	Tetramer in solid state
Xenon trioxide XeO_3	Xe, O O O pyramidal	1	
Perxenate ion XeO_6^{4-}	O, O O, Xe, O O, O octahedral	0	
Xenon tetroxide XeO_4	O, Xe, O O, O tetrahedral	0	
Xenon oxide tetrafluoride $XeOF_4$	O, F F, Xe, F F square pyramidal	1	
Xenon dioxide difluoride XeO_2F_2	O, Xe, O F, F tetrahedral	0	

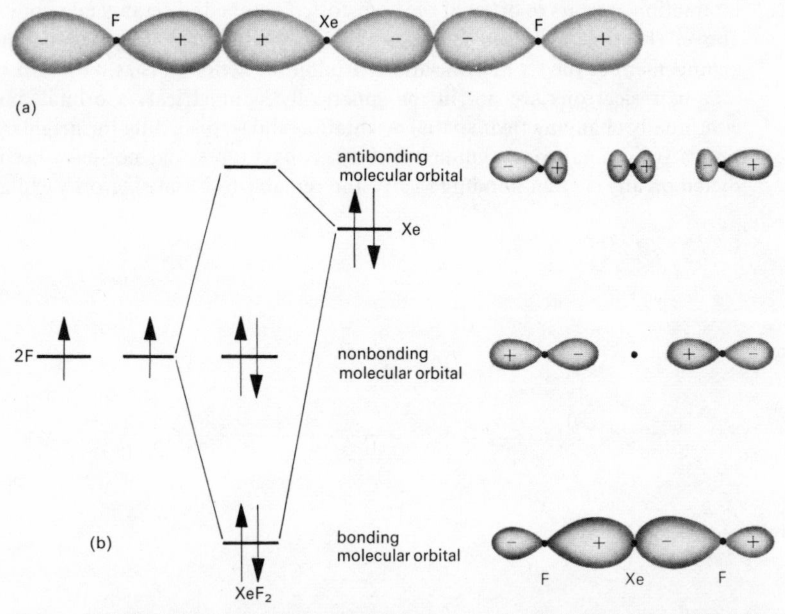

Figure 103 The formation of three-centre, four-electron bond in xenon difluoride. (a) The atomic orbitals used in the interaction. (For simplicity, no nodes are shown in the 5p orbital of the xenon atom.) (b) Energy-level diagram and approximate shape of the three-centre molecular orbitals

of its 5p orbitals to bond to the *two* fluorine atoms by forming a three-centre molecular orbital system of the type described in Figure 103.

By assuming that in xenon tetrafluoride the xenon atom bonds to the four fluorine atoms by using two of its p-orbitals and forming two three-centre molecular-orbital systems at right angles to each, the planar shape of xenon tetrafluoride is apparent (11.6a). This bonding scheme, when extrapolated to xenon hexafluoride, would predict a regular octahedral shape for this molecule – three three-centre molecular-orbital systems mutually at right angles (11.7).

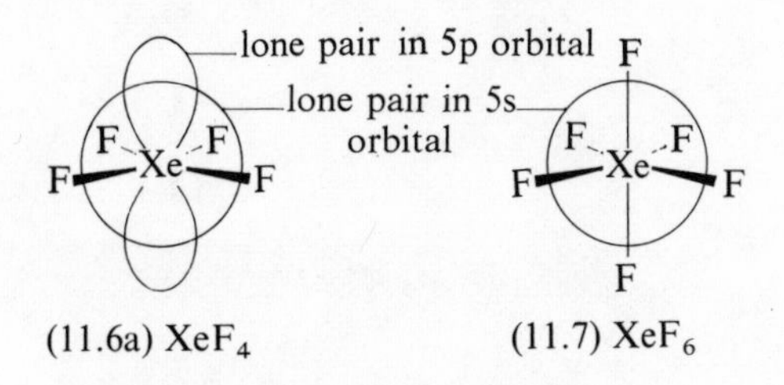

(11.6a) XeF_4 (11.7) XeF_6

However, xenon hexafluoride is a rather peculiar molecule in that its shape fluctuates rapidly with time. Although the equilibrium position for the Xe—6F interaction appears to occur at the undistorted octahedral arrangement predicted above, the time-averaged shape is that of a somewhat distorted octahedral arrangement of the six fluorine atoms around the xenon (it is as if the two xenon lone-pair electrons are not in the spherically symmetrical 5s orbital, but are continually changing their spatial orientation and so distorting the neighbouring bonds from octahedral symmetry). Such behaviour would not have been predicted on any current bonding theory and remains to be satisfactorily explained.

Chapter 12
Non-Aqueous Solvents

12.1 Introduction

For centuries alchemists and philosophers searched for the perfect solvent which would dissolve everything – presumably giving little thought as to the type of vessel they would require to hold the liquid! Ironically, as it turned out, the most universal solvent yet known to man was all around them in huge quantities: water. Consequently, over the past century chemists have spent a good deal of their time studying reactions in water and have demonstrated that many salts ionize extensively in the medium and that such solutions conduct electricity. Conversely, covalent compounds, which do not ionize in solution, were found to be mainly insoluble.

It is only comparatively recently that a serious search has been made for other solvents which might support ionization of salts; indeed many eminent chemists who were responsible for the growth of the ionic theory believed water to be unique in this respect. The wheel has now turned full circle and currently there are few liquids which have not received some attention regarding their possible solvent power, examples ranging from liquid hydrogen chloride which boils at $-83{\cdot}7$ °C to metals and salts which are liquid only above 1000 °C.

Solvents are exceedingly useful to the chemist because they permit intimate contact between reactants which otherwise may not react. They also allow control of reactions which might otherwise be violent and, finally, separation and purification of reaction products are much more easily achieved using solvents than by other means. To have as wide a range of solvents as possible is highly desirable because of the great diversity of reactants, ranging from ionic salts to covalent, non-polar molecules, all of which require different physical characteristics on the part of a solvent to render them soluble. Furthermore, a solute may be found to be unstable towards solvolysis in one solvent but to be freely soluble in another; for example, water reacts vigorously with many non-metallic halides, but these can often be studied quite conveniently in either liquid sulphur dioxide or phosphorus oxychloride $OPCl_3$.

Obviously, physical constants such as melting point and boiling point govern the usefulness of a particular solvent and dictate the methods which can be used for handling it. The solvents which boil below room temperature are normally studied in glass vacuum systems at low temperatures; they can, however, be handled in pressure vessels at ambient temperature, but such apparatus is difficult to use and is necessarily opaque, which makes direct observation of a reaction difficult.

Solvents normally liquid at about 20 °C are the most easily handled, since their manipulation often requires few changes from those used for aqueous solutions. In most cases, however, stringent precautions must be taken to exclude the entry of air and water from the solvent, since even traces of these can often change the characteristics of a solvent.

Molten metals and salts are normally studied in sealed vessels heated in ovens or furnaces. Here attack of the containing vessel is a major problem, especially at temperatures in excess of 1000 °C. The materials used in construction, as with any solvent, depend on the characteristics of the particular liquid under study; glass, silica, metals and graphite have found most use above about 200 °C.

The force of attraction between two charges $+e_1$ and $-e_2$ (for example, two ions) is described by the relation

$$F = \frac{e_1 e_2}{\varepsilon r^2},$$

where ε is the dielectric constant of the medium in which the ions are situated and r is the distance between them. Obviously those liquids with the largest dielectric constant will be the best solvents for metallic salts, since such liquids will be able to overcome the lattice energy of the salts and separate the ions. In water ($\varepsilon = 78{\cdot}5$ at 25 °C) salts have been demonstrated not to exhibit complete ionization, some ion pairs being present in solution

$$\begin{array}{ccc} \underset{\text{(solid)}}{A^+B^-} & \rightleftharpoons & A^+ + B^- \\ & \searrow\!\!\!\nwarrow \quad \nearrow\!\!\!\swarrow & \\ & \underset{\text{(solution)}}{A^+B^-} & \end{array}$$

Pure sulphuric acid has a larger dielectric constant (100 at 25 °C) than water and in this solvent all ions are thought to be completely 'free'; that is, there is no ion–ion interaction, although, of course, there is bound to be some ion–solvent interaction. Unfortunately, few liquids have a dielectric constant greater than water and the chemist has often to be satisfied with the study of, for example, glacial acetic acid, which has a dielectric constant of only 6·2 at 25 °C. In this solvent few free ions exist in solution, most solutes being present as ion pairs, or even ion triplets and ion quadruplets:

$A^+B^-A^+$ $B^-A^+B^-$
ion triplets

$\begin{array}{l} A^+B^- \\ B^-A^+ \end{array}$ $A^+B^-A^+B^-$
ion quadruplets

However, even in glacial acetic acid, ionic precipitation reactions (metatheses) occur very readily,

$$KBr + AgNO_3 \longrightarrow AgBr\downarrow + KNO_3.$$
$$(Br^- + Ag^+ \longrightarrow AgBr\downarrow)$$

This is to be contrasted with solutions in pure sulphuric acid, which do not undergo rapid metatheses even though 'free' ions are present. Here the failure is due to the high viscosity of pure sulphuric acid (about thirty times more viscous than water), which does not allow rapid mixing of the reactants. Similarly, migration of ions under an electric field is a very slow process in sulphuric acid.

12.2 Solubility

Solubilities in any solvent are almost impossible to predict from first principles, but it is possible to make one or two generalizations. The first is so common that many chemists use it without even pausing to consider the matter: polar solutes dissolve in polar solvents and non-polar solutes dissolve in non-polar solvents; or 'like dissolves like'. For example, few people attempt the purification of a typically organic compound by recrystallization from water – they use an 'organic' solvent such as benzene which is non-polar and has a low dielectric constant.

The polarity of a solvent molecule is normally reflected in the bulk dielectric constant ε of the liquid, so that one can use ε as a rough guide to the polarity. Polar molecules form associated liquids, since the molecules tend to attract each other by dipole–dipole interactions. When the polar molecule contains hydrogen, this dipole–dipole interaction is often augmented by the much stronger phenomenon of hydrogen bonding, which gives the liquid a semi-rigid, polymeric structure at temperatures close to the melting point. This, of course, accounts for the high viscosity of pure sulphuric acid.

When a solute attempts to enter the solvent it must break open this polymeric structure in some way. A covalent, non-polar solute has no mechanism for doing this and is therefore insoluble. On the other hand, a salt is composed of the ultimate in dipoles – ions – and can enter into ion–dipole interactions, or in some instances can even form strong hydrogen bonds, with the solvent molecules. Such compounds then have a method of opening up the solvent so as to enter into solution.

A phenomenon closely allied to the above ion–solvent interactions is that of solvate formation. In some salts either, or both, the cations and anions have such strong interaction with the solvent that solid (or sometimes liquid) complexes can be isolated. Hydrates, such as $CuSO_4.5H_2O$, are well known and have their counterpart in most of the other non-aqueous solvents; see Table 43. The formation of a solvate by a salt usually implies solubility in that particular solvent because certain ions making up the salt lattice are surrounded by co-ordinated solvent which has the effect of reducing the lattice energy of the crystal. Another way of looking at solvation is to regard it as the formation of a very concentrated solution of the salt – when further solvent is added the only effect is one of dilution as the ions are separated by more and more solvent.

Table 43 The Structures of Several Solvates

Solvent	*Solvate*	*Structure*
$POCl_3$	$BCl_3.POCl_3$	$Cl_3B \leftarrow OPCl_3$
	$AsCl_3.POCl_3$	held by dipole–dipole forces only
	$(TiCl_4.POCl_3)_2$	Cl_3PO, Cl, Cl, Cl around Ti; Cl bridges Ti–Ti; Cl, Cl, Cl, $OPCl_3$ (dimer with two bridging Cl)
HF	$H_2O.HF$	$H_3O^+F^-$
	$KF.HF$	$K^+[F—H—F]^-$
	$KF.2HF$	$K^+[F—H—F—H—F]^-$ (H–F–H angle 130°)
BrF_3	$KF.BrF_3$	$K^+BrF_4^-$
H_2SO_4	$K_2SO_4.H_2SO_4$	$K^+HSO_4^-$
	$Li_2SO_4.9H_2SO_4$	$LiHSO_4.4H_2SO_4$ (i.e. $[Li(H_2SO_4)_4]^+HSO_4^-$)

12.3 Self-ionization of solvents

Many solvents, including water, possess a definite electrical conductivity which can be explained by assuming a slight self-ionization

$$2H_2O \rightleftharpoons H_3O^+ + OH^-.$$

Typical self-ionization processes which have been postulated for several non-aqueous solvents are shown in Table 44.

If, as in many instances, the self-ionization is minute, how is one to check the validity of such postulated mechanisms? Direct evidence is very hard to come by, but two main methods have been used: (a) study of solvates, (b) isotopic exchange reactions.

It is often found that solvates contain ions characteristic of the solvent's self-ionization and therefore provide some proof for the latter. For example, potassium fluoride dissolves in liquid bromine trifluoride and a solvate $KF.BrF_3$ may be isolated from the solution. X-ray crystallography of such bromine trifluoride solvates has shown them to contain the planar tetrafluorobromate-(III) ion BrF_4^-, the anion postulated in the self-ionization of bromine trifluoride

$$2BrF_3 \rightleftharpoons BrF_2^+ + BrF_4^-.$$

Table 44 Proposed Self-Ionization Reactions in Some Non-Aqueous Solvents

Solvent	*Self-ionization schemes*	*Remarks*
NH_3	$2NH_3 \rightleftharpoons NH_4^+ + NH_2^-$	
HF	$2HF \rightleftharpoons H_2F^+ + F^-$	complicated by solvation of F^- ion to give $F(HF)_n^-$
BrF_3	$2BrF_3 \rightleftharpoons BrF_2^+ + BrF_4^-$	
ICl	$2ICl \rightleftharpoons I^+ + ICl_2^-$	
N_2O_4	$N_2O_4 \rightleftharpoons NO_2^+ + NO_2^-$ $\updownarrow$ $NO^+ + NO_3^-$	further complicated by the dissociation $N_2O_4 \rightleftharpoons 2NO_2$
H_2SO_4	$2H_2SO_4 \rightleftharpoons H_3SO_4^+ + HSO_4^-$ and $2H_2SO_4 \rightleftharpoons H_3O^+ + HS_2O_7^-$	see text
$POCl_3$	$POCl_3 \rightleftharpoons POCl_2^+ + Cl^-$ or $2POCl_3 \rightleftharpoons POCl_2^+ + POCl_4^-$	both apparently somewhat doubtful.

Other fluorides, such as antimony pentafluoride, form solvates which are thought to contain the difluorobromonium ion BrF_2^+

$$SbF_5 + BrF_3 \longrightarrow BrF_2^+ + SbF_6^-.$$

This method fails in the case of phosphorus oxychloride, since all its solvates are covalent and therefore contain no species which could arise from a self-ionization such as

$$2POCl_3 \rightleftharpoons POCl_2^+ + POCl_4^-.$$

Experiments using radioactive chlorine-36 show that rapid exchange of chlorine atoms occurs between phosphorus oxychloride and labelled tetraethylammonium chloride. Though this exchange is consistent with a process involving the self-ionization of phosphorus oxychloride, it is thought to take place via the mechanism

$$NEt_4Cl^* \longrightarrow NEt_4^+ + Cl^{*-},$$
$$Cl^{*-} + POCl_3 \longrightarrow [POCl_4^*]^- \longrightarrow Cl^- + POCl_3^*.$$

At present, the weight of evidence from these and other experiments seems to point to the fact that phosphorus oxychloride undergoes no self-ionization.

Early investigators into the solvent chemistry of sulphur dioxide postulated the self-ionization of the pure liquid to be

$$2SO_2 \rightleftharpoons SO^{2+} + SO_3^{2-}.$$

However, thionyl chloride $SOCl_2$ does not exchange either sulphur-35 or oxygen-18 with the solvent, and sulphur trioxide only undergoes exchange of sulphur in the presence of *added* sulphite ions. Therefore the concentration of either thionyl or sulphite ions derived from the solvent must be vanishingly small and it is now considered that the solvent takes no part in ionic reactions other than to act as an ion carrier.

The high dielectric constant of pure sulphuric acid supports an extensive self-ionization

$$2H_2SO_4 \rightleftharpoons H_3SO_4^+ + HSO_4^-,$$

and many metallic salts undergo solvolysis to give hydrogen sulphates containing the HSO_4^- ion when dissolved in the liquid. A sulphur trioxide exchange reaction also occurs simultaneously to give the hydrogen disulphate ion $HS_2O_7^-$

$$2H_2SO_4 \rightleftharpoons H_3O^+ + HS_2O_7^-.$$

The total concentration of ionic species produced by these two dissociation reactions is 0·043 molar, which is sufficient to make the melting point of sulphuric acid occur some 0·26 °C lower than if no self-ionization had occurred.

12.4 Acid–base behaviour in non-aqueous solvents

Essentially, the classification of substances into acids and bases gives us two types of active compounds such that members of one type have a greater tendency to react with members of the other type than among themselves. Such a classification should therefore allow the correlation of known facts and ideas and also assist in the planning of new research.

In protonic solvents, of which water is typical, acids can be defined as proton donors:

$$\underset{\text{acid}}{HCl} + H_2O \longrightarrow H_3O^+ + Cl^-,$$

$$\underset{\text{acid}}{H_3O^+} + Cl^- \longrightarrow HCl + H_2O,$$

and bases as proton acceptors:

$$\underset{\text{base}}{H_2O} + HCl \longrightarrow H_3O^+ + Cl^-,$$

$$\underset{\text{base}}{Cl^-} + H_3O^+ \longrightarrow HCl + H_2O.$$

In non-protonic solvents such a system of nomenclature is of course impossible. Considering the self-ionization of water for a moment,

$$2H_2O \rightleftharpoons H_3O^+ + OH^-,$$

it can be seen that acids, for example hydrochloric acid, produce cations characteristic of the water self-ionization. This type of behaviour prompted another definition of acids and bases (the solvent-system theory) in which acids are considered to be compounds which produce cations characteristic of the particular solvent either by direct dissociation or by reaction with its solvent

$$NH_4Cl \xrightarrow{NH_3(l)} NH_4^+ + Cl^- \qquad \text{(direct dissociation),}$$
$$SbF_5 \xrightarrow{BrF_3(l)} BrF_2^+ + SbF_6^- \qquad \text{(reaction with solvent),}$$

whilst a base is a compound which produces anions characteristic of the solvent

$$NaOH \longrightarrow Na^+ + OH^- \qquad \text{(in water),}$$
$$NaNH_2 \longrightarrow Na^+ + NH_2^- \qquad \text{(in liquid ammonia),}$$
$$KF \longrightarrow K^+ + BrF_4^- \qquad \text{(in liquid bromine trifluoride).}$$

As we have seen, certain solvents undergo no self-ionization (e.g. sulphur dioxide and phosphorus oxychloride) and therefore this acid–base theory is of no use when correlating results in these solvents.

In liquid hydrogen fluoride, which is a strongly acidic solvent (i.e. it readily donates protons), no protonic acid is known since even the strongest acids from the aqueous system are protonated by this solvent and therefore behave as bases,

$$\text{e.g.} \quad \underset{\text{base}}{HNO_3} + \underset{\text{acid}}{HF} \rightleftharpoons \underset{\text{acid}}{H_2NO_3^+} + \underset{\text{base}}{F^-}.$$

(transfer of H^+ from HF to HNO_3)

Other compounds, of which antimony pentafluoride is typical, can however behave as acids in liquid hydrogen fluoride because they increase the concentration of H_2F^+ ions (i.e. solvated protons) by a fluoride-ion transfer mechanism

$$\underset{\text{acid}}{SbF_5} + \underset{\text{base}}{2HF} \rightleftharpoons \underset{\text{acid}}{H_2F^+} + \underset{\text{base}}{SbF_6^-}$$

(transfer of F^- from 2HF to SbF_5; transfer of F^- from SbF_6^- to H_2F^+)

This allows the possibility of another definition that acids are anion acceptors or cation donors whereas bases are anion donors or cation acceptors. Using this concept it is possible to discuss acid–base behaviour in all polar, non-aqueous solvents. However, like the acid–base theory of Lewis which this new concept so greatly resembles, the definition of acids and bases is now so broad and general that it begins to lose its usefulness. Normally workers use the solvent-system theory, wherever it is applicable, because of its specificity.

Acid–base neutralization reactions are possible in non-aqueous solvents just as they are in water, two typical examples being

$$KBrF_4 + BrF_2SbF_6 \xrightarrow{BrF_3(l)} KSbF_6 + 2BrF_3,$$
$$\underset{\text{base}}{KNH_2} + \underset{\text{acid}}{NH_4Cl} \xrightarrow{NH_3(l)} \underset{\text{salt}}{KCl} + \underset{\text{solvent}}{2NH_3}.$$

In most cases these reactions can be followed by titration, the end points being detected either by coloured indicators or, more often, by potentiometric and conductimetric techniques.

12.5 Levelling action of a protonic solvent

In solvents which contain ionizable protons, the solvated proton is responsible for the acidity, whereas the base changes with each solvent. When a strong proton donor HA (i.e. a strong acid) dissolves in a solvent HB which is a proton acceptor (i.e. is basic), complete proton exchange between solute and solvent occurs

$$HA + HB \longrightarrow H_2B^+ + A^-,$$

and cations characteristic of the solvent's self-ionization are formed. The acid strength of HA is therefore equal to that of the cation H_2B^+ and it follows that all strong acids will display this same acidity – i.e. they are all 'levelled' to the acid strength of the H_2B^+ ion. To take examples, the mineral acids hydrochloric, nitric and sulphuric, are all levelled to the acidity of the oxonium ion H_3O^+ in water

$$HCl + H_2O \longrightarrow H_3O^+ + Cl^-,$$
$$H_2SO_4 + 2H_2O \longrightarrow 2H_3O^+ + SO_4^{2-},$$
$$HNO_3 + H_2O \longrightarrow H_3O^+ + NO_3^-,$$

and to the acidity of the ammonium ion NH_4^+ in liquid ammonia,

e.g. $HCl + NH_3 \longrightarrow NH_4^+ + Cl^-.$

If, on the other hand, the solvent HB is also a proton donor, an equilibrium will be set up when the acid HA dissolves,

$$HA + HB \rightleftharpoons H_2B^+ + A^-,$$

and fewer H_2B^+ ions will be formed; the acid strength of the solution will depend on the number of such ions and will be a function of the proton–donor power of HA. Thus acids, when dissolved in an acidic medium, are differentiated according to their acid strength. Conversely bases are differentiated in basic solvents and are levelled in acidic solvents.

This levelling action has several far-reaching consequences on the usefulness of a solvent for particular reactions. For example, it is obviously impossible to use a comparatively acidic solvent such as water for dissolving very strong bases like the hydride and ethoxide ions, since these will immediately be levelled by protonation to the basicity of the hydroxide ion

$$H^- + H_2O \longrightarrow H_2 + OH^-,$$
$$OEt^- + H_2O \longrightarrow EtOH + OH^-.$$

Such compounds must be handled in a basic solvent.

In a basic solvent all acids, including those classed as weak in the aqueous system, will be levelled to the acid strength of the solvent cation. Analysts have exploited this fact by titrating weak acids such as carboxylic acids and phenols in anhydrous primary amine solvents, for example butylamine and ethylenediamine

$$R'COOH + RNH_2 \longrightarrow RNH_3^+ + R'COO^-,$$
levelling reaction

$$RNH_3^+ + OR''^- \longrightarrow RNH_2 + R''OH.$$
standard base (e.g. ethoxide) — neutralization reaction

The end-point of such a titration is normally detected potentiometrically by placing an indicator electrode in the solution and comparing the potential between this and some standard electrode; at the end-point there is an abrupt change in the observed potential (see Figure 104).

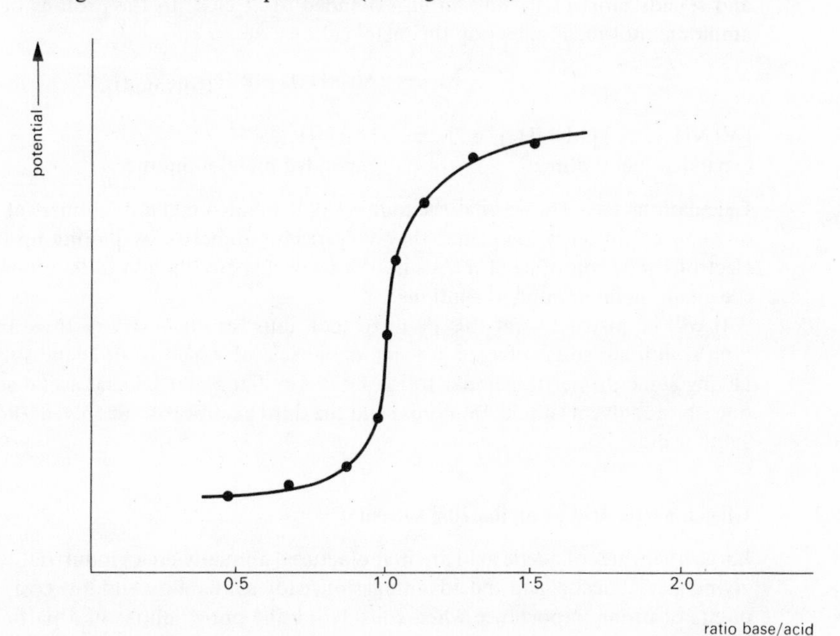

Figure 104 A typical potential curve obtained for an acid–base potentiometric titration in an ionizing solvent

12.6 Metal solutions in non-aqueous solvents

The fact that several metals dissolve in liquid ammonia to form deep-blue or bronze coloured solutions has been known for many years and the great reducing power of such solutions has been exploited in both preparative inorganic and organic chemistry. The phenomenon has since been shown not to be restricted to liquid ammonia, and coloured solutions of metals are now known in ethers, amines and molten metallic salts.

Although the alkali-metal–ammonia solutions have received concentrated study using every possible physical technique, their nature is not yet completely understood. The current theory suggests that when the alkali metal dissolves, its outermost electron leaves the metal atomic orbital – the fate of this electron apparently then depends upon the concentration of the solution under study. In dilute solutions ($<0{\cdot}05$ molar) the electron is completely 'free' from the solvated metal cation and the solutions are paramagnetic; here the electron is considered to be solvated; that is, it exists in a cavity in the solution, the nearest-neighbour ammonia molecules being orientated with their hydrogen atoms nearest to the electron. As the concentration increases, the electron becomes more localized and spends most of its time in an expanded orbit close to the protons of the ammonia molecules solvating the metal cation

$$M \longrightarrow M(NH_3)_x^+ + e^- \text{ (solvated)}$$

$$\upharpoonleft\downharpoonright$$

$$[M(NH_3)_x^+e^-][M(NH_3)_x^+e^-] \rightleftharpoons [M(NH_3)_x^+e^-]$$

expanded metal dimer — expanded metal monomer

Calculations have shown that the number of expanded metal monomers at any one concentration is low, since they very readily dimerize by pairing up their electrons. The solutions of metals in other solvents are thought to be similar to the alkali-metal ammonia solutions.

It will be instructive at this point to look into the chemistry of three fairly typical non-aqueous solvents in some detail one of which is an acidic solvent having some properties similar to the more familiar water (glacial acetic acid), one a basic solvent (liquid ammonia) and the third a non-protonic solvent (liquid sulphur dioxide).

12.7 Glacial acetic acid as an ionizing solvent

Large quantities of acetic acid are manufactured annually on an industrial scale, giving glacial acetic acid the advantages of ready availability and low cost, two points of prime importance when considering the potentialities of a particular solvent.

The 99·7 per cent commercial acetic acid has water as its main impurity; this can be removed by refluxing the acid with a small quantity of acetic anhydride for about three hours

$$(CH_3CO)_2O + H_2O \longrightarrow 2CH_3COOH.$$

A fractional distillation from chromic oxide CrO_3 (added to destroy any oxidizable organic impurities) gives the anhydrous solvent, the purity of which may be checked by its melting point.

As can be seen from Table 45, the melting and boiling points allow glacial acetic acid to be handled in ordinary vessels at room temperature and pressure, although precautions must be taken to stop the entry of moisture, for example,

Table 45 Physical Properties of Glacial Acetic Acid

Melting point	16·6 °C
Boiling point	118·1 °C
Dielectric constant (25 °C)	6·2
Specific conductivity (25 °C)	$2{\cdot}4-3{\cdot}0\times10^{-9}\ \Omega^{-1}\,cm^{-1}$
Viscosity (25 °C)	1·15 centipoise

by working in a dry-box. If traces of water are of no consequence for a particular experiment then it is possible to work with acetic acid in a fume cupboard providing the reaction vessels are normally kept closed with corks, rubber bungs, or ground-glass stoppers and are only opened to the atmosphere for brief periods during manipulations. When corks and bungs are used they must be protected with thin polythene sheet otherwise moisture in them will be extracted into the solvent.

From the low dielectric constant, one would predict metal salts to be largely insoluble in glacial acetic acid, but in practice it is found that many salts containing large ions (i.e. those salts having comparatively low lattice energies) are soluble. These include nitrates, thiocyanates, cyanides and iodides; acetates, the base analogues in glacial acetic acid, are also often soluble.

On the other hand, the low dielectric constant suggests that the solvent molecules are only slightly polar and probably have little interaction with each other – it is therefore not surprising that many covalent compounds, such as stannic chloride, arsenic trichloride, germanium tetrachloride and hydrogen sulphide, are soluble in acetic acid. Water, ammonia and the common (anhydrous) acids dissolve by interacting with the solvent either by forming hydrogen bonds, as in the case of water, or by protonation reactions such as

$$NH_3+CH_3COOH \longrightarrow NH_4^+ + CH_3COO^- \quad \text{(protonation by the solvent),}$$

$$HClO_4+CH_3COOH \longrightarrow CH_3C(OH)_2^+ + ClO_4^- \quad \text{(protonation of the solvent).}$$

A useful solubility rule which has been found to hold true is that compounds insoluble in water are also insoluble in acetic acid. Realizing this, it is possible to predict many ionic precipitation reactions from a knowledge of aqueous solution chemistry:

$$AgNO_3+KI \longrightarrow AgI\downarrow + KNO_3,$$

$$AgNO_3+KSCN \longrightarrow AgSCN\downarrow + KNO_3,$$

$$ZnI_2+\underset{\text{(sodium oxalate)}}{Na_2C_2O_4} \longrightarrow ZnC_2O_4\downarrow + 2NaI.$$

Although ionic precipitation reactions take place readily in glacial acetic acid, electrical conductivity measurements show that dissolved salts are only very slightly dissociated into single ions, mainly being present as ion multiplets as we noted in the introduction. For example the ion-pair dissociation constant for the reaction

$$K^+Br^- \rightleftharpoons K^+ + Br^-$$

is about 10^{-7}. This is, of course, again a natural outcome of the low dielectric constant of acetic acid.

The specific conductivity of pure glacial acetic acid, though being very low, has been assumed to be due to a slight self-ionization

$$2CH_3COOH \rightleftharpoons CH_3C(OH)_2^+ + CH_3COO^-.$$

Such a self-ionization is compatible with the rapid exchange noted between potassium acetate labelled with radioactive carbon-14 and acetic acid. However, such an exchange could also take place by direct proton transfer between an acetate ion and a solvent molecule.

Glacial acetic acid is an acidic solvent, and reaction can therefore be expected between it and solutes capable of attacking protons. For example, sodium carbonate evolves carbon dioxide quite readily when added to the solvent

$$Na_2CO_3 + 2CH_3COOH \longrightarrow 2NaOOCCH_3 + CO_2 + H_2O$$

and sodium hydroxide dissolves to give sodium acetate

$$NaOH + CH_3COOH \longrightarrow NaOOCCH_3 + H_2O.$$

Electropositive metals such as sodium react to give hydrogen

$$2Na + 2CH_3COOH \longrightarrow 2NaOOCCH_3 + H_2.$$

In the case of zinc metal only slight reaction to give hydrogen and a precipitate of zinc acetate occurs in the pure solvent, but when sodium acetate (which behaves as a base in acetic acid) is added, the zinc dissolves readily due to the formation of a soluble complex, sodium tetra-acetatozincate,

$$2NaOOCCH_3 + Zn(OOCCH_3)_2 \longrightarrow Na_2Zn(OOCCH_3)_4.$$

This reaction demonstrates the amphoteric nature of zinc and is analogous to the dissolution of zinc hydroxide in aqueous sodium hydroxide which gives sodium zincate. Neodymium triacetate is similar to zinc acetate in being soluble in the presence of potassium acetate to give a bluish-coloured complex; under the influence of an electric field the colour migrates towards the anode as expected for an acetatoneodymiate(III) ion,

$$Nd(OOCCH_3)_3 + nKOOCCH_3 \longrightarrow K_nNd(OOCCH_3)_{3+n}.$$

Using the solvent-system theory of acids and bases, acetates can be recognized as bases in acetic acid whilst compounds producing solvated protons will act as acids. The number of solvated protons formed by an acid HA will be proportional to the strength of the acid, due to the differentiating effect of an acidic solvent on dissolved acids (see section 12.5).

$$HA + CH_3COOH \rightleftharpoons CH_3C(OH)_2^+ + A^-.$$

It is therefore possible to compare the strengths of a series of acids by observing the magnitude of their electrical conductivities in acetic acid, since these will give a measure of the number of ions present in each solution. Such experiments have shown that the strengths of several inorganic acids are in the order

$$HClO_4 > HBr > H_2SO_4 > HCl > HNO_3.$$

Conversely, when bases which are classed as 'weak' in the aqueous system are dissolved in glacial acetic acid they will be levelled to the basic strength of the acetate ion by protonation,

e.g. $$NH_3 + CH_3COOH \longrightarrow NH_4^+ + CH_3COO^-,$$

and, since the acetate ion can be titrated readily against a strong acid like perchloric acid, these so-called weak bases can be easily estimated in glacial acetic acid.

The acid–base neutralization

$$CH_3COO^- + CH_3(OH)_2^+ \longrightarrow 2CH_3COOH$$

can be followed by noting changes in the electrical conductivity of the solution as base is added to acid (or vice versa) or by noting the change in potential between an indicator electrode placed in the solution and a standard reference electrode. At the end-point of the neutralization, when equivalent amounts of acid and base have been mixed, there is an abrupt change in both the conductivity and the potential (see Figures 104 and 105).

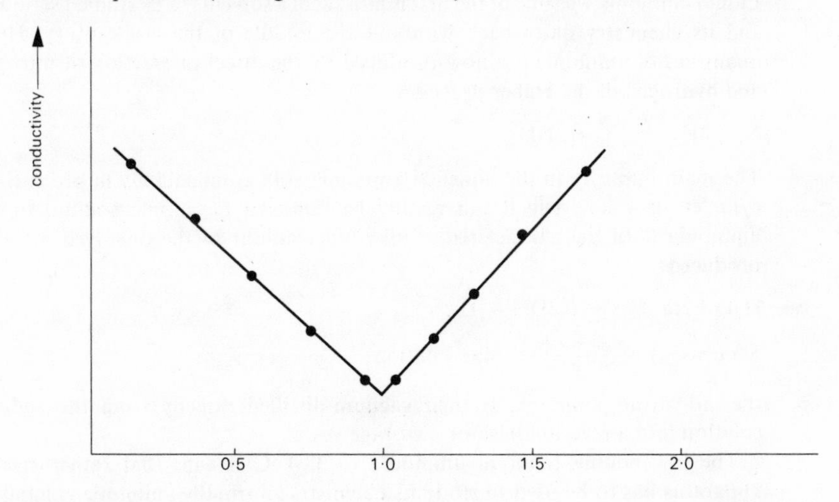

Figure 105 A typical conductivity curve obtained for an acid–base conductimetric titration in an ionizing solvent

By a suitable procedure the constituents of a mixture of primary, secondary and tertiary aliphatic amines (all of which are weak bases) can be determined by a potentiometric titration against perchloric acid in acetic acid. The mixture is divided into three aliquots; one aliquot is titrated directly against the perchloric acid to give the total amine content. The second aliquot is treated with acetic anhydride for three hours at room temperature before being titrated; in this way the primary and secondary amines are acylated to non-basic products and the titration therefore gives the tertiary amine content. Salicylaldehyde is added to the final aliquot and forms a Schiff's base with the primary amine only; the titration with perchloric acid then gives the combined secondary and tertiary amine content. From the three titrations it is possible to calculate the concentration of each amine in the original mixture.

From these few reactions it is possible to obtain a general idea of the chemistry which can occur in glacial acetic acid. As was pointed out, this solvent shows certain similarities to water, which makes its study slightly less strange to those unfamiliar with the non-aqueous solvents. Most solvents, however, bear little or no resemblance to water; they may be exceedingly reactive like bromine trifluoride, in which even asbestos dissolves with incandescence, or, as in the case of salt melts, the liquid structure may be composed completely of ions. Although salt melts have been used in industrial processes for many years (e.g. liquid sodium chloride is used in the extraction of metallic sodium), it is only recently that a concentrated effort had been made to study such systems in detail.

12.8 Liquid ammonia as a solvent

Liquid ammonia was one of the first non-aqueous solvents to be studied seriously and its chemistry dates back to about the middle of the last century. Huge quantities of ammonia are now produced by the direct interaction of nitrogen and hydrogen in the Haber process

$$N_2 + 3H_2 \xrightarrow[\text{pressure}]{\text{catalyst}} 2NH_3.$$

The main impurity in the liquefied ammonia sold commercially in pressurized cylinders is water, which can readily be removed by adding sodium to the ammonia until the characteristic 'inky-blue' colour of the dissolved metal is produced:

$$H_2O + Na \longrightarrow NaOH + \tfrac{1}{2}H_2$$

$$\text{Na (excess)} \xrightarrow[\text{NH}_3\text{ is dry}]{\text{dissolves when}} \text{blue solution;}$$

the anhydrous ammonia is then vacuum-distilled directly from the sodium solution into a reaction flask or a storage vessel.

The low boiling point of ammonia (−33·4 °C) means that rather special apparatus has to be used to study its chemistry; normally ammonia is handled in a glass vacuum apparatus and the temperature of the liquid held below −33 °C by a suitable refrigerant (e.g. the readily obtainable 'dry ice', solid carbon dioxide, gives a temperature of about -78 °C).

Table 46 Physical Properties of Liquid Ammonia

Melting point	$-77{\cdot}7$ °C
Boiling point	$-33{\cdot}4$ °C
Dielectric constant (-33 °C)	22
Specific conductivity (-15 °C)	$4 \times 10^{-10}\ \Omega^{-1}\ \mathrm{cm}^{-1}$
Viscosity (-33 °C)	0·254 centipoise

The relatively low dielectric constant of ammonia (Table 46) leads us to expect that fewer salts will be soluble in liquid ammonia than in water; indeed few salts of divalent or trivalent cations are soluble (high lattice energies) unless the anion is large and polarizable, as are, for example, thiocyanate and iodide. Fluorides and chlorides, even those of monovalent ions, have lattice energies which are too large to allow appreciable solubility in ammonia; this is often used to advantage in synthetic chemistry by arranging that a metal chloride is precipitated from solution in 'double decomposition' (metathetic) reactions.

12.8.1 *Acid–base reactions*

The slight electrical conductivity of ammonia is explained by assuming a self-ionization reaction

$$2NH_3 \rightleftharpoons NH_4^+ + NH_2^-,$$

so that ammonium salts behave as acids, and amides as bases, in this solvent. Therefore, when a reactive metal is added to an ammonium salt in liquid ammonia, hydrogen is evolved,

$$NH_4Cl + Na \longrightarrow \tfrac{1}{2}H_2 + NaCl\downarrow,$$

and the reaction between a metal amide and an ammonium salt is an acid–base neutralization. In agreement with this it has been found calorimetrically that the heats of such reactions are constant at about 109 kJ mol^{-1}, irrespective of the amide or ammonium salt used because the same reaction is occurring in all cases,

$$NH_4^+ + NH_2^- \longrightarrow 2NH_3.$$

Titrations can be carried out in ammonia, as in water, but the end-points are usually detected either conductimetrically or potentiometrically, because indicators do not function satisfactorily in liquid ammonia; indicators which are weak acids in water are fully ionized in ammonia due to the levelling action of the solvent, whereas indicators which are weak bases in water will not be dissociated at all in ammonia.

Solutions of alkali-metal amides are convenient reagents for synthesizing other amides,

$$IrBr_3 + 3KNH_2 \longrightarrow 3KBr + Ir(NH_2)_3,$$
$$AgCl + NaNH_2 \longrightarrow NaCl + AgNH_2,$$
$$ZnI_2 + 2KNH_2 \longrightarrow 2KI + Zn(NH_2)_2,$$

and in several cases it has been observed that the precipitated amides dissolve on the addition of an excess of amide ions, due to the formation of a soluble complex anion,

e.g. $Zn(NH_2)_2 + 2KNH_2 \longrightarrow K_2Zn(NH_2)_4$
potassium tetra-amidozincate

This amphoteric nature of zinc amide in liquid ammonia is entirely analogous to the behaviour of zinc hydroxide in aqueous bases,

$$Zn(OH)_2 + 2KOH \longrightarrow K_2Zn(OH)_4.$$

12.8.2 *Solvate formation*

The lone-pair electrons on the ammonia molecule make it an excellent donor and solvates are formed with many metallic salts including those of the alkali metals (e.g. $NaI.4NH_3$) and the transition metals (normally giving octahedrally coordinated $M[NH_3]_6^{n+}$ cations).

12.8.3 *Solvolysis reactions*

Many covalent halides are decomposed by liquid ammonia,

e.g. $BCl_3 + 6NH_3 \longrightarrow B(NH_2)_3 + 3NH_4Cl,$

$$GeCl_4 + 8NH_3 \longrightarrow Ge(NH_2)_4 + 4NH_4Cl$$
$$\downarrow$$
$$Ge(NH)_2 + 2NH_3$$
$$ZrCl_4 + 2NH_3 \longrightarrow Cl_3ZrNH_2 + NH_4Cl.$$

The initial step in these reactions probably involves the coordination of an ammonia molecule to the central atom, giving a highly unstable solvate which rapidly loses hydrogen halide,

$$BCl_3 + NH_3 \longrightarrow [Cl_3B \leftarrow NH_3] \longrightarrow HCl + Cl_2BNH_2 \xrightarrow{NH_3} \text{etc.},$$
$$HCl + NH_3 \longrightarrow NH_4Cl.$$

12.8.4 *Reductions in liquid ammonia using alkali-metal solutions*

The deep-blue solutions which the alkali metals form in liquid ammonia make excellent reducing agents which are widely used in both inorganic and organic chemistry; only a few examples will be mentioned here.

(a) *Reduction of salts.* Simple salts are normally reduced to the finely divided metal,

$$CuBr + K \longrightarrow KBr + Cu\downarrow,$$

but the reductions are often complicated by a side reaction in which the precipitated metal catalyses the solvolysis of the alkali metal,

$$K + NH_3 \xrightarrow[\text{catalyst}]{\text{metal}} KNH_2 + \tfrac{1}{2}H_2 \text{ (rapid)}.$$

The alkali-metal solutions are thermodynamically unstable with respect to amide formation but, unless a catalyst is present, the kinetic barrier to decomposition is sufficiently high for the solutions to be kept for many months at −78 °C.

In more complex salts the anions are often attacked and, for example, iodates are reduced to iodides and permanganates are reduced to manganates.

(b) *Production of low oxidation states of metals.* Potassium tetracyanonickelate(II) $K_2Ni(CN)_4$ is reduced first to the red tetracyanonickelate(I) and then to the yellow tetracyanonickelate(0) on treatment with potassium in liquid ammonia. Similar cyanide complexes of metals in the zero oxidation state which can be prepared in this manner are $Co^0(CN)_4^{4-}$, $Mn^0(CN)_6^{6-}$ and $Cr^0(CN)_6^{6-}$. Complexes containing metals in the oxidation state −1 can sometimes be made when a metal carbonyl (in which the metal's oxidation state is already zero) is reduced.

$$2K + Mn_2(CO)_{10} \longrightarrow 2KMn^{-1}(CO)_5,$$
$$2Na + Co_2(CO)_8 \longrightarrow 2NaCo^{-1}(CO)_4.$$

(c) *Fission of hydrogen atoms.* The hydrides of group IV (except carbon) and group V react with alkali-metal solutions evolving hydrogen,

$$GeH_4 + K \longrightarrow KGeH_3 + \tfrac{1}{2}H_2,$$
$$R_3GeH + K \longrightarrow KGeR_3 + \tfrac{1}{2}H_2,$$
$$SnH_4 + Na \longrightarrow \tfrac{1}{2}H_2 + NaSnH_3 \xrightarrow{\text{Na}} Na_2SnH_2 + \tfrac{1}{2}H_2,$$
$$PH_3 + K \longrightarrow KPH_2 + \tfrac{1}{2}H_2.$$

These derivatives are useful to the preparative chemist for facilitating the synthesis of compounds containing metal–metal bonds,

$$R_3GeK \xrightarrow{R'_3GeCl} R'_3Ge—GeR_3 + KCl\downarrow$$

$$R_3GeK \xrightarrow{R'_3SnCl} R'_3Sn—GeR_3 + KCl\downarrow$$

$$R_3GeK \xrightarrow{ClMn(CO)_5} R_3Ge—Mn(CO)_5 + KCl\downarrow.$$

12.8.5 *Metathetic reactions*

Normally these reactions are designed so that an alkali-metal chloride is precipitated, leaving the desired compound in solution; filtration followed by crystallization then allows the product to be isolated in a pure state prior to analysis. Among the many reactions of this type which have been studied, perhaps the most interesting are those involving tetrahydroborates (borohydrides). Potassium tetrahydroborate KBH_4 is hydrolysed in aqueous solution, but dissolves unchanged in liquid ammonia, and its solutions can thus be used to prepare new tetrahydroborates,

e.g. $NH_4Cl + KBH_4 \longrightarrow KCl\downarrow + NH_4BH_4.$

(Ammonium tetrahydroborate is stable only at low temperatures and on warming it slowly loses hydrogen.)

$$[Cr(NH_3)_6]Cl_3 + 3KBH_4 \longrightarrow [Cr(NH_3)_6][BH_4]_3 + 3KCl\downarrow,$$
$$[Mg(NH_3)_6]Cl_2 + 2KBH_4 \longrightarrow [Mg(NH_3)_6][BH_4]_2 + 2KCl\downarrow.$$

From this short discussion we note that liquid ammonia is readily available and easily purified (two essential points if a solvent is to have wide applicability), that many compounds which are hydrolytically unstable in aqueous media can be studied in ammonia solution (which opens up new areas of synthetic chemistry) and that, since ammonia is 'basic', the effect of levelling action (e.g. on weak acids) is opposite to that of an acidic medium like glacial acetic acid (which levels weak bases).

12.9 **Liquid sulphur dioxide as a solvent**

Commercially available sulphur dioxide is made by roasting sulphur or metal sulphides in air and contains small quantities of sulphur trioxide and water as impurities. These may be removed by passing the gas first through concentrated sulphuric acid and then over phosphorus pentoxide. The low boiling point of sulphur dioxide (Table 47) requires that it has to be studied in closed pressure vessels, in a vacuum apparatus at low temperatures or, less satisfactorily, in Dewar flasks provided with drying tubes and bungs.

The low dielectric constant leads to low solubility of many metal salts and when solution does occur it will be accompanied by extensive ion-pair formation;

Table 47 Physical Properties of Liquid Sulphur Dioxide

Melting point	−75·5 °C
Boiling point	−10·2 °C
Dielectric constant	17·3 (−16 °C) 24·6 (−69 °C)
Viscosity	0·39 centipoise (0 °C) 0·43 centipoise (−10 °C)

however, notwithstanding the extensive ion association, it is found that ionic precipitation reactions occur rapidly in sulphur dioxide. Many covalent compounds, especially the halogens and halides of non-metals, are very soluble, making sulphur dioxide a valuable addition to the chemist's list of solvents, because such compounds either react with water and many other protonic solvents or are, at best, only slightly soluble. As was noted on p. 284, it is now thought that liquid sulphur dioxide does not undergo any self-ionization and behaves merely as an essentially inert reaction medium.

12.9.1 *Solvate formation*

Sulphur dioxide readily forms solvates with many metallic salts, including those of the alkali metals: $LiI.2SO_2$, $NaI.2SO_2$, $NaI.4SO_2$, $KNCS.SO_2$. The fluorides of the alkali metals form 1:1 'solvates' $MF.SO_2$, which are isomorphous with the corresponding chlorates and have been found to contain the fluorosulphinate ion SO_2F^-. These are efficient fluorinating agents and many of their reactions can be carried out in sulphur dioxide solution,

e.g. $AsCl_3 + 3MSO_2F \longrightarrow AsF_3 + 3MCl + 3SO_2$.

Many organic compounds, such as tertiary amines, pyridine, dioxan, ethylene oxide and dialkylsulphides, also form 1:1 solvates with sulphur dioxide.

12.9.2 *Solvolysis reactions*

Several covalent halides of elements in high oxidation states react with sulphur dioxide to give oxy-halides,

$$PX_5 + SO_2 \longrightarrow OPX_3 + SOX_2 \quad (X = Cl, Br),$$
$$WCl_6 + SO_2 \longrightarrow OWCl_4 + SOCl_2.$$

12.9.3 *Complex formation*

Although phosphorus pentahalides react with sulphur dioxide, antimony pentachloride dissolves unchanged and liquid sulphur dioxide is a particularly useful solvent in which to make the water-sensitive hexachloroantimonates,

$$R_4NCl + SbCl_5 \longrightarrow [R_4N][SbCl_6],$$
$$NOCl + SbCl_5 \longrightarrow [NO][SbCl_6],$$
$$[NO][SbCl_6] + [R_4N][PF_6] \longrightarrow [NO][PF_6]\downarrow + [R_4N][SbCl_6].$$

Nitryl chloride NO_2Cl and antimony pentachloride react together in liquid chlorine to give nitronium hexachloroantimonate $NO_2^+SbCl_6^-$, which is soluble in liquid sulphur dioxide and can be used to prepare other nitronium salts,

e.g. $[NO_2][SbCl_6] + [Me_4N][BF_4] \longrightarrow [NO_2][BF_4] + [Me_4N][SbCl_6]$.

Nitronium tetrafluoroborate is an efficient nitrating agent for aromatic compounds and was used in the initial experiments designed to prove that the nitronium ion was the active species responsible for the nitrating action of sulphuric acid–nitric acid mixtures. (Nitric acid dissolves in pure sulphuric acid to give the nitronium ion

$HNO_3 + 2H_2SO_4 \longrightarrow NO_2^+ + H_3O^+ + 2HSO_4^-$.)

The solubility of iodine in sulphur dioxide is increased by the addition of up to one mole of a heavy alkali-metal iodide due to the formation of the tri-iodide ion

$MI + I_2 \longrightarrow MI_3$ (M = K, Rb, Cs).

Similarly, the solubility of mercuric iodide is increased by the addition of potassium iodide, which gives the HgI_4^{2-} ion. Both these reactions also occur in aqueous solution.

Quaternary ammonium sulphites precipitate the white aluminium sulphite when added to solutions of aluminium chloride

$$3[Me_4N]_2[SO_3] + 2AlCl_3 \longrightarrow 6Me_4NCl + Al_2(SO_3)_3\downarrow.$$

Further addition of sulphite ions causes the precipitated aluminium sulphite to dissolve due to the formation of a complex anion

$$Al_2(SO_3)_3 + 3[Me_4N]_2[SO_3] \longrightarrow 2[Me_4N]_3[Al(SO_3)_3].$$

Gallium trichloride and stannic chloride behave in a similar fashion.

12.9.4 *Metathetic reactions*

As in liquid ammonia, metal chlorides are insoluble in sulphur dioxide, and hence metathetic reactions are generally designed so that a chloride is precipitated from solution. A relatively simple example of this is the reaction between thionyl chloride $SOCl_2$ and caesium sulphite

$$Cs_2SO_3 + SOCl_2 \longrightarrow 2CsCl\downarrow + 2SO_2.$$

This reaction can be extended to prepare other thionyl derivatives from the chloride,

e.g. $2AgOOCCH_3 + SOCl_2 \longrightarrow 2AgCl\downarrow + SO(OOCCH_3)_2.$
thionyl acetate

Thionyl acetate is stable in solution but it decomposes to give acetic anhydride and sulphur dioxide if the solvent is removed; several other thionyl carboxylates made in the same way are stable enough to be isolated at room temperature (e.g. $SO(OOCCH_2C_6H_5)_2$ and $SO(OOCCH_2Cl)_2$).

Potassium thiocyanate reacts smoothly with many covalent halides in liquid sulphur dioxide, e.g.

$$BCl_3 + 3KSCN \longrightarrow 3KCl\downarrow + B(NCS)_3,$$
$$SiCl_4 + 4KSCN \longrightarrow 4KCl\downarrow + Si(NCS)_4.$$

Infrared spectra of these two products suggest that they have the isothiocyanato structure in which the nitrogen atom is coupled to the boron or silicon atom.

Thus with sulphur dioxide we find that the lack of self-ionization does not detract from the chemical usefulness of the solvent and the chemistry carried out in it is very similar to that observed in, say, water or liquid ammonia except for the lack of acid–base equilibria (as defined by the solvent-system theory). Probably it is safe to say that the lack of self-ionization may increase a solvent's usefulness because extensive self-ionization (as with sulphuric acid for example) often leads to very troublesome solvolysis of a large percentage of solutes.

Formula Index

Al 126
Al_2Br_6 129
$Al_2(CH_3)_6$ 141
$AlCl_3$ 128
Al_2Cl_6 128
$(AlCl_4)_2Cd_2$ 113
$(AlCl_4)Ga$ 140
$(AlH_3)_n$ 133
AlH_4Li 81, 82, 133
$AlM(SO_4)_2.12H_2O$ 72, 138
AlO_2^- 135

Ar 265
Ar_2^+ 267
$ArNe^+$ 267
$ArXe^+$ 267

As 176
AsF_5 180
AsF_3Cl_2 206
AsH_3 52, 187
AsH_4Br 177
AsH_4I 177
As_2O_3 196
$(AsR)_5$ 181
$(AsR)_6$ 181
As_4S_6 206

At 241
At_2 250
At_2^+ 242
$AtBr_2^-$ 250
$AtCl_2^-$ 250

B 126
$B(C_5H_5N)_4Br_3$ 128
$B(C_6H_5)_4Na$ 81
B_2Cl_4 139
B_4Cl_4 139
B_8Cl_8 139
B_9Cl_9 139
BF_3 130
BF_4^- 130
B_3F_5 139
BH_3 54, 133
B_2H_6 54, 55, 133, 142
B_4H_{10} 56
B_5H_9 56
B_5H_{11} 56
B_9H_{15} 57
$B_{10}H_{14}$ 56, 57, 58
$B_{10}H_{16}$ 57
$B_{12}H_{12}^{2-}$ 58
$B_{10}H_{10}C_2$ 58
BH_3CO 160
BH_4Na 81
BN 135, 136
$B_3N_3H_6$ 137
$B(OH)_3$ 134
BX_3 130
B_2X_4 139

Ba 93
$BaCO_3$ 104
BaF_2 97
$Ba(OH)_2$ 103

Be 93
$Be(BH_4)_2$ 102
BeC_2 107
Be_2C 107
$Be(CH_3)_2$ 109
$Be[C(CH_3)_3]_2$ 102, 110
$BeCl_2$ 97
$BeCl_2(NC_5H_5)_2$ 105
$BeCl_2[O(C_2H_5)_2]_2$ 98

BeH_2 102
$Be(H_2O)_4^{2+}$ 93, 95, 98, 226
$Be(NH_3)_4Cl_2$ 98
$Be(NO_3)_2$ 104
$Be(NO_3)_2.2N_2O_4$ 104
$Be_4O(NO_3)_6$ 104
$Be_4O(OOCR)_6$ 99
$BeSO_4$ 104

Bi 176
Bi_2 187
Bi_9^{5+} 176
BiH_3 177
$Bi(NO_3)_3.5H_2O$ 210
BiO^+ 176
$BiO(NO_3)$ 210

Br 241
Br^+ 242
Br_2^+ 242
BrF_5 249
BrF_4Cs 88
BrO^- 260
BrO_2 257
BrO_3^- 261
Br_2O 256
(BrO)H 260
$(BrO_4)H$ 233, 261

C 144, 153
C^{4-} 172
C_2^{2-} 88, 106, 172
C_3^{4-} 107, 172
C_2D_2 42
$(CF)_n$ 156
C_6F_6 170, 263
C_6F_{12} 263
C_6F_5Cl 264
CF_3OF 259
CH_5^+ 35
CH_3COOH 288
C_8K 88, 157, 158
$C_{24}K$ 88, 157, 158
$C_{36}K$ 88, 157, 158
$C_{48}K$ 88, 157, 158
$C_{60}K$ 88, 157, 158
C_2Li_2 88
C_3Mg_2 107
CNCl 251
$CN.N_3$ 251

CO 159, 163
CO_2 37, 164
CO_3^{2-} 165
$C_2O_6^{2-}$ 231
C_3O_2 166
$COBH_3$ 160
$(CO)_9Fe_2$ 161
$(CO)_2Fe_2(NO)_2$ 193
$(CO_3H)Na$ 66, 166
$(CO)_4Ni$ 160
CX_2 145
CX_4 145

Ca 93
CaC_2 106, 107
$Ca(CN_2)$ 107
$CaCO_3$ 104
CaF_2 105
$Ca(H_2O)_8^{2+}$ 226
$Ca(H_2O)_8O_2$ 103, 226
$Ca(NH_3)_n$ 79
CaO_2 103
$Ca(OH)_2$ 103
$CaSO_4$ 104

Cd 111
Cd_2^{2+} 113
$Cd_2(AlCl_4)_2$ 113
$Cd(C_5H_5N)_2Cl_2$ 122
$(CdCl_6)K_4$ 121
CdF_2 120
$Cd(H_2O)_6^{2+}$ 123
CdI_2 115
$Cd(NH_3)_6^{2+}$ 123
$Cd(NH_3)_4Cl_2$ 122
CdX_3^- 121
CdX_4^{2-} 121
CdX_6^{4-} 121

Cl 241
Cl^+ 242
ClCN 251
ClF_3 246, 249
Cl_2I^+ 247
Cl_2I^- 248
Cl_4I^- 248
ClNO 193
ClO_2 257
ClO_3 257
ClO_3^- 261

ClO_4^- 39, 261
Cl_2O 256
Cl_2O_7 258
$(ClO)H$ 260
$(ClO_2)H$ 258
$(ClO_3)H$ 258
$(ClO_4)H$ 258
Cl_2Xe 272
$(ClXeO_3)Cs$ 273

$Cr(H_2O)_6^{3+}$ 226

Cs 67
Cs_2 70
$CsBrF_4$ 88
CsCl 86, 87
$CsClXeO_3$ 273
Cs_2S_6 84
Cs_2ZnBr_4 122

Cu 90
Cu^+ 30, 91
Cu^{2+} 30, 91

D_2 42
D_2C_2 42
D_2O 42
D_2SO_4 42

F 241
F_2 27, 242
F_3B 130
F_5Br 249
$(FC)_n$ 156
FH 54, 65, 255
F_2H^- 65, 85, 255, 282
$F(HF)_2^-$ 85, 255, 282
F_3H_2P 189
F_4HP 189
FI 250
F_4I^+ 247
F_5I 246, 250
F_6I^+ 248
F_6I^- 248
F_7I 247
F_2Kr 271
F_2Li_2 71
F_3N 207
F_4N_2 208
F_2Na 68

F_2O 256
F_2O_2 257
F_2O_4 256
$FOCF_3$ 259
F_2O_2Xe 272
F_4OXe 272
F_6PtXe 266, 269
F_2S_2 237
F_4S 212, 239
$F_{10}S_2$ 237
F_2Xe 34, 270, 271, 274, 275, 277
F_4Xe 34, 270, 271, 274, 275
F_6Xe 34, 270, 274, 277
$F_5Xe^+PtF_6^-$ 269

$Fe_2(CO)_9$ 161
$Fe(CO)_2(NO)_2$ 193
$Fe(NO)_4$ 193
$[Fe(NO)_2I]_2$ 193

Ga 126
Ga_2^{4+} 127
$GaAlCl_4$ 140
Ga_2Br_6 138
$GaBr_3N(CH_3)_3$ 129
Ga_2Cl_6 138
$GaGaCl_4$ 139
GaX_4^- 127, 128

Ge 144, 153, 173
$GeCl_2$ 171
GeF_2 147
$GeF_4N(CH_3)_3$ 148, 171
Ge_5H_{12} 50
GeH_3CH_3 51
$GeHCl_3$ 171
$GeH_3Mn(CO)_5$ 51
GeH_3SiH_3 51, 159
$(GeR_2)_4$ 173
GeR_nCl_{4-n} 172

H 14, 21, 40, 46
H^+ 47
H^- 22, 48
$H(H_2O)_{10}^+$ 225
H_2O 61, 224
H_2O_2 53, 229
H_3O^+ 47, 48, 228
$H_9O_4^+$ 47, 228

He 265
He^{+} 22
He_2^{+} 267
HeF_2 24

Hg 111
Hg_2^{2+}–Hg^{2+} 112, 124
Hg_3^{2+} 113
$Hg(DMSO)_6(ClO_4)_2$ 123
HgI_4^{2-} 121
$Hg(NH_3)_2^{2+}$ 122
$Hg(NH_3)_2Cl_2$ 122
$Hg(NH_3)_4(NO_3)_2$ 125
$Hg_2(NO_3)_2.2H_2O$ 125

I 241
I^{+} 241
I_2^{+} 242
I_3^{+} 242
I_3^{-} 246
I_5^{+} 242
$I(C_5H_5N)_2^{+}$ 241
ICl_2^{+} 247
ICl_2^{-} 248
ICl_4^{-} 248
I_2Cl_6 249
$IClO_4$ 242
IF 249, 250
IF_4^{+} 247
IF_5 246, 250
IF_6^{+} 248
IF_6^{-} 34
IF_7 247
IH 54, 255
I_3K 88
$I_2N(CH_3)_3$ 253
IO_4^{-} 262
I_2O_5 258
$(IO_4)H$ 262
$(IO_6)H_5$ 262
$I(OH)_6^{+}$ 262
$I(OOCCH_3)_3$ 242
IR_2^{+} 243
$I[SC(NH_2)_2]_2^{+}$ 241

In 175

K 67
K_2 70
KC_8 88, 157, 158
KC_{24} 88, 157, 158
KC_{36} 88, 157, 158
KC_{48} 88, 157, 158
KC_{60} 88, 157, 158
K_4CdCl_6 121
$KGeH_3$ 295
K_2HgI_4 122
KI_3 88

Kr 266
KrF_2 271

Li 67
Li_2 70
$LiAlH_4$ 81, 82, 133
Li_2C_2 88
$(LiCH_3)_4$ 70
Li_2Cl_2 71
Li_2F_2 71
LiH 81
Li_3N 84
Li_2O 82
LiR 89
$Li_2Zn(CH_3)_4$ 122

Mg 93
MgC_2 107
Mg_2C_3 107
$MgC_2H_5Br(OEt_2)_2$ 107
$MgC_6H_5Br(OEt_2)_2$ 107
$MgCH_3Br(THF)_3$ 107
MgC_6F_5Cl 264
$MgCO_3$ 104
$MgCl$ 95
$MgCl_2.6C_2H_5OH$ 106
$Mg(H_2O)_6^{2+}$ 95, 226
$Mg(NH_3)_6Cl_2$ 98
$Mg(OH)_2$ 103
MgR_2 108
$MgRX$ 107, 108
$MgSO_4$ 104

N 176
N_2 179, 184, 185
N_3^{-} 189, 190
NCl_3 185
NF_3 207
N_2F_2 207
N_2F_4 208

NH_2^- 293
NH_3 61, 187, 292
NH_4^+ 293
N_2H_2 52, 188
N_2H_4 52, 188
N_3H 189
NH_4F 65
NH_2OH 190
NO 191, 192
NO^+ 193
NO_2 194, 195
NO_2^+ 195, 200, 237
NO_2^- 198
NO_3^- 199
N_2O 191
N_2O_3 194
N_2O_4 194
N_2O_5 195
NOCl 193
$(NO)_4Fe$ 193
$(NO)_2Fe(CO)_2$ 193
$[(NO)_2FeI]_2$ 193
$(NO_2)H$ 198
$(NO_3)H$ 198
$N_2O_2H_2$ 197
$N_3P_3Cl_6$ 182
$N_4P_4Cl_8$ 182
$N_2Ru(NH_3)_5^{2+}$ 183
N_4S_4 204
NS_7H 205, 214
$N_4S_4H_4$ 204

Na 67
Na^- 45
Na_2 70
$NaB(C_6H_5)_4$ 77
$NaBH_4$ 81
NaCl 68, 69, 86
NaF_2 68
$NaHCO_3$ 66, 84
$NaI.4NH_3$ 85
NaN_3 84

$Ni(CO)_4$ 160
$Ni(PCl_3)_4$ 208
$Ni(PF_3)_4$ 208

O 211
^{17}O 226
O_2 213
O_2^+ 211, 213
O_2^- 214
O_2^{2-} 214, 230
O_3 220
OH_2 61, 63, 64, 224
O_2H_2 53, 229

P 176, 186
P_2 178
P_4 178, 186
PCl_3 207
PCl_5 180
P_2Cl_4 208
$(PCl_3)_4Ni$ 208
$PCl_4^+PCl_6^-$ 209
PF_5 35, 209
PF_6^- 180
PH_3 52, 187
PH_5 33
PHF_4 189
PH_2F_3 189
$P_3N_3Cl_6$ 182
$P_4N_4Cl_8$ 182
P_4O_6 196
P_4O_{10} 196
$POCl_3$ 38, 209
PO_2H_3 200
$(PO_3)H$ 202
$(PO_3)H_3$ 201
$(PO_4)H_3$ 201
$(P_2O_6)H_4$ 203
$(P_2O_7)H_4$ 202
POX_3 209
$(PR)_n (n = 4, 5, 6)$ 210
PR_3 210
P_2R_4 181, 210
P_4S_3 205
P_4S_5 205
P_4S_7 205
P_4S_{10} 205

Pb 144
Pb_{aq}^{2+} 148
$Pb(C_2H_5)_4$ 172
$PbCl_4$ 146
PbF_2 147
PbH_4 159
$Pb[Pb(C_6H_5)_3]_4$ 49
PbR_4 172

Po 211, 220
Po_2 220

Rb 67
Rb_2 70

Rn 266, 268
RnO_3 272
RnO_4H^- 272

S 211
S_8 214, 222
S_n^{2-} 214
SF_4 212, 239
SF_6 212, 239
S_2F_2 237
S_2F_{10} 238
SH_2 224
SH_6 33
S_2H_2 224
S_3H_2–S_8H_2 53, 224
$S_{17}H_2$ 53
S_nH_2 214, 224
S_4N_4 204
$S_4N_4H_4$ 204
$S_6(NH)_2$ 205
S_7NH 205, 214
SO_2 296
SO_3 233
S_2O 232
$S_2O_7^{2-}$ 39, 234
$S_3O_{10}^{2-}$ 39, 234
$(SO_3)H_2$ 236
$(SO_4)H_2$ 234, 236
$(SO_5)H_2$ 231
$(S_2O_7)H_2$ 234
$(S_2O_8)H_2$ 231
SR_2 239
S_2R_2 240
S_2R_3 240

Sb 176
$Sb(CH_3)_5$ 180
$Sb(CH_3)_6^-$ 180
$SbCl_6^-$ 180
SbF_5 206
SbH_3 52, 187
SbO^+ 176
Sb_2O_3 196
$Sb(OH)_6M$ 204
$Sb_2(SO_4)_3$ 210

Se 211, 221
Se_4^{2+} 211
Se_8 214, 222
SeH_2 224
SeO_2 233
SeO_4^{2-} 237
$(SeO_3)H_2$ 237
SeX_6^{2-} 35

Si 144, 153, 173
$Si(CH_3)_nCl_{4-n}$ 172
$SiCl_4.2P(CH_3)_3$ 149
SiF_2 146, 170
$(SiF_2)_n$ 170
SiF_5^- 148
SiF_6^{2-} 149
Si_nF_{2n+2} 170
$SiF_4.2L$ 149, 171
SiH_4 50
Si_2H_6 51
Si_3H_8 51
SiH_3GeH_3 51, 159
SiO_2 37, 166
SiO_4^{4-} 168
SiR_4 172
$(SiR_2O)_n$ 38, 151
$SiR_3(SiR_2)_nSiR_3$ 173

Sn 144, 153
$SnCl_2$ 148
$SnCl_5^-$ 148
$SnCl_2.2H_2O$ 148
SnF_2 147
SnH_4 50
SnS 148

Sr 93
$Sr(OH)_2$ 103
$SrSO_4$ 104

Te 211
TeH_2 223
TeO_3H_2 237
$Te(OH)_6$ 237
TeX_6^{2-} 35

Tl 126
Tl^+ 143
TlF 143

TlI_3 138
TlI_4^- 138
Tl_2O 134
Tl_2O_3 134

Xe 266
Xe_2^+ 267
$XeAr^+$ 267
$XeCH_3^+$ 268
$XeCl_2$ 272
$XeCl_4$ 272
Xe^+F^- 266
XeF_2 34, 270, 271, 274, 275, 277
XeF_4 34, 270, 271, 274, 275
XeF_6 34, 270, 274, 277
$XeF_5^+PtF_6^-$ 269
$Xe_2F_{11}^+PtF_6^-$ 270
XeO_3 272
XeO_4 273
XeO_3ClCs 273
$XeOF_4$ 272
XeO_2F_2 272
$XeO_4HNa.1{\cdot}5H_2O$ 273
$XeO_6Na_4.6H_2O$ 273
$XeO_6Na_4.8H_2O$ 273

Zn 111
$(ZnBr_4)Cs_2$ 122
$[Zn(CH_3)_4]Li_2$ 122
$Zn(C_5H_5N)_2Cl_2$ 114, 122
$ZnCl_2$.terpyridyl 123
ZnF_2 120
ZnH_2 119
$Zn(H_2O)_6^{2+}$ 114, 123
$Zn(H_2O)_2Cl_4^{2-}$ 114
$Zn(NH_3)_6^{2+}$ 123
$Zn(NH_3)_4Cl_2$ 114, 122
$Zn_4O(OOCCH_3)_6$ 117
ZnS 116, 117

Subject Index

Alkali metals 67
- bicarbonates 66, 84
- carbonates 84
- comparison to copper 90
- compounds with carbon 88
- cyclic polyether complexes 73
- extraction 80
- flame test 76
- gaseous molecules of 72
- gravimetric determination of 77
- halides 85
- hydrides 81
- isonitrosoacetophenone complexes 72
- nitrates 85
- nitrides 84
- nitrites 85
- occurrence 79
- oxides 82
- properties 67
- solubilities in liquid ammonia 78
- solutions of 78
- structure 74
- sulphides 84

Alkali-metal dimers 70

Alkaline-earth metals 93
- carbides 106, 107
- carbonates 104
- complexes 98, 105
- extraction 100
- gaseous halides 97
- halides 104
- hydrates of salts 94
- hydrides 102
- hydroxides 103
- nitrates 103
- nitrides 103
- occurrence 99
- organometallic compounds 107
- oxides 102
- properties 93
- similarities of beryllium and aluminium 110
- sulphates 104
- univalent species of 95

Aluminium 126
- alums 72, 138
- similarities to beryllium 110
- trichloride 128, 138
- *see also* Group III elements

Alums 72, 138

Ammonium fluoride, hydrogen bonding in 65

Anti-fluorite structure 105, 232

Antimony 176
- *see also* Group V elements

Argon 265
- *see also* Group 0 elements

Arsenic 176
- *see also* Group V elements

Astatine 241
- *see also* Group VII elements

Aufbau principle 23

Azide ion 181, 189, 190

Beryllium 93
- atom 24
- basic acetate 99
- basic nitrate 104
- carboxylates 99
- comparison with zinc 115
- similarities to aluminium 110
- *see also* Alkaline-earth metals

Bicarbonate, hydrogen bonding in 66

Bismuth 176
  see also Group V elements
Body-centred cubic structure 74
Bond
  $d_{\pi}-p_{\pi}$ 38, 180, 212
  hydrogen bonding 60
  *n*d orbitals and 32
  $p_{\pi}-p_{\pi}$ 35
  three centre, in diborane 55
  three-centre, four-electron 33, 34, 277
Boranes 54
Borazine 135
Borazon 135
Born–Haber cycle 68, 96, 266
Boron 126
  atom 24
  boric acid 134
  hydroboration 142
  icosahedra in elemental boron 132
  monomeric halides 54, 129
  see also Group III elements
Bromine 241
  see also Group VII elements

Cadmium 111
  see also Group IIb elements
Caesium 67
  see also Alkali metals
Caesium chloride structure 87
Carbon 144
  dioxide 164
  monoxide 159
  suboxide 166
  see also Group IV elements
Carboranes 58
Caro's acid 235
Chlorine 241
  see also Group VII elements
Copper 30, 90

Decaborane 56, 57
Deuterium 40, 42
Diamond synthesis 154
Diborane 54
Disilane 51

Element 112 32
Element 114 32

First long period 29
First short period 24
Flame photometer 76
Fluorine 241
  atom 25
  dissociation energy of F_2 28
  see also Group VII elements
Fluorite structure 105
Francium 67
  isotopes of 80
  see also Alkali metals
Frasch process 219

Gallium 126
  see also Group III elements
Germanium 144
  see also Group IV elements
Glacial acetic acid
  properties of 289
  as solvent 288
Graphite 136, 152, 153
  lamellar compounds of 156
Grignard reagents 108
Group I elements 67
  see also Alkali metals
Group IIa elements 93
  see also Alkaline-earth metals
Group IIb elements: zinc, cadmium and mercury 111
  carbonates 120
  complexes 114, 121, 125
  extraction 118
  halides 120
  hydrides 119
  hydroxides 120
  mercury(I) compounds 112, 124
  nitrates 120
  occurrence 117
  organometallic compounds 123
  oxides 120
  properties 112
Group III elements 126
  extraction 131
  hydrides 132
  hydrolysis of salts 128
  hydroxides 134
  lower halides 138
  nitrides 135
  occurrence 130

Group III Elements – *continued*
organometallic compounds 141
oxides 134
properties 126
salts of oxy-acids 137
structure of 132
sulphides 135
trihalides 138
Group IV elements 144
bond energies of 147
carbides 171
extraction 152
halides 169
hydrides 49, 158
occurrence 152
organometallic compounds 172
oxides 159
properties 145
semiconductor properties of 173
structures 153
Group V elements 176
extraction 184
halides 206
hydrides 51, 187
occurrence 184
organometallic compounds 210
oxides 191, 195
oxy-acids 197, 200, 203
oxy-halides 209
properties 177
salts of oxy-acids 210
structure 185
sulphides 204
Group VI elements 211
extraction 218
halides 237
hydration numbers 228
hydration of ions 226
hydrides 52, 223
hydrogen peroxide 229
metallic oxides 231
occurrence 218
organometallic compounds 239
oxides 232
oxy-acids of sulphur, selenium, tellurium 234
peroxides 230
peroxy-salts 230
properties 211
semi-conductor properties of selenium 221
Group VII elements 241
extraction 252
fluorocarbons 263
hydrides 54, 254
mixed halogen compounds 245
occurrence 251
oxidation–reduction potentials of 259
oxides 255
oxy-acids 255
properties 241
pseudohalides 250
pseudohalogens 250
Group 0 elements 265
bonding in xenon fluorides 275
chlorides of xenon 272
fluorides 269
occurrence 268
oxidation potentials 273
oxides of xenon 272
properties 265
reactions of xenon fluorides 274
structures of xenon oxides and fluorides 276

Haber process 292
Haemoglobin 217
Heavy elements 31
Helium 265
atom 22
see also Group 0 elements
Hexagonal close-packed structure 75
Higher boron hydrides 56
Hydration numbers of cations 226
Hydroboration 142
Hydrogen 40
atomic 21, 46
covalent hydrides 49
diazine (di-imide) 52, 188
Group IV hydrides 49, 158
Group V hydrides 51, 187
Group VI hydrides 52, 223
Group VII hydrides 54, 254
hydrazine 52, 188
hydride ion 22, 48
hydride-ion radius 22, 49
hydrides of boron 54, 56
hydrogen bonding 60

isotopes of 40
liquid 43
metallic hydrides 59
ortho 43
para 43
in periodic table 44
Hydrolysis of aquated ions 95, 128, 229

Icosahedron 132
Indium 126
see also Group III elements
Iodine 241
see also Group VII elements
Iron enneacarbonyl 161

Krypton 265
see also Group 0 elements

Lead 144
see also Group IV elements
Lewis acid–base theory 285
Liquid ammonia as a solvent 292
Liquid sulphur dioxide as a solvent 296
Lithium 67
aluminium hydride
(tetrahydroaluminate) 50
atom 24
see also Alkali metals
London forces 63

Mercury 111
mercury(I) derivatives 124
mercury(I)–mercury(II) equilibrium 112
see also Group IIb elements
Metal solutions in liquid ammonia 78
Monogermane 51
Monosilane 50, 51
Morse curve 41

Neon 265
atom 25
see also Group 0 elements
Nickel carbonyl 160
Nitrogen 176
atom 25
fixation 183
hydrazine 188
hydrazoic acid 189
hydroxylamine 190
oxides 191
oxy-acids 197
see also Group V elements
Nitrophenols, hydrogen bonding in 66
Non-aqueous solvents 279
acid–base behaviour in 284
glacial acetic acid 288
levelling action of solvent 286
liquid ammonia 292
liquid sulphur dioxide 296
self-ionization of 282
solution of metals in 287
Non-stoichiometric hydrides 59

Orbitals 13, 19, 20
d 18
and bonding 32
p 18
s 16
three-centre molecular 55
Organolithium derivatives 70, 89
Oxonium ion 47, 228
Oxy-anions of phosphorus, sulphur, chlorine 212
Oxygen 211
atom 25
hydrogen peroxide 229
oxyhaemoglobin 217
ozone 214, 233
peroxides 229
peroxy-salts 230
water 42, 64, 224
see also Group VI elements

Pauli exclusion principle 13, 24
Penetration 18, 24, 26, 91
Pentagermanes 50
Periodic table 13
possible extensions to 32
Peroxy-acids 230
Phosphazenes 182
Phosphorus 176
allotropes 185
oxy-acids 200
phosphazenes 182
see also Group V elements
Polonium 211
see also Group VI elements
Polytetrafluoroethylene 170

Potassium 67
atom 29
radioactive isotope 80
see also Alkali metals
Protium 40

Quantum numbers 13
Quartz 152, 166

Radius ratios of alkali-metal halides 87
Radon 265
see also Group 0 elements
Rare gases 265
see also Group 0 elements
Rubidium 67
radioactive isotopes 80
see also Alkali metals
Rutile structure 114

Schrödinger wave equation 13, 16, 21
Second long period 31
Second short period 29
Selenium 211
allotropes 220, 222
semi-conductor properties of 221
see also Group VI elements
Self-ionization 282
of liquid ammonia 283, 293
of solvents 282
of water 228
Silica 37, 166
Silicates 168
Silicon 144
silicones 151
see also Group IV elements
Size contraction 27
Sodium 67
see also Alkali metals
Sodium chloride structure 86
Sulphanes 53, 214
Sulphur 211
acids 234
allotropes 220, 222
behaviour of liquid 214
imides 205, 214
see also Group VI elements
Sulphuric acid as a solvent 282, 284

Tellurium 211
see also Group VI elements
Thallium 126
thallium(I) 143
halides 143
see also Group III elements
Three-centre, four-electron bond 34, 277
Tin 144
see also Group IV elements
Trisilane 51
Tritium 40

Water 42, 64, 224
hydration numbers of cations 226
hydrogen bonding in 64
self-ionization of 228
Wave function 16, 18, 20
Wurtzite 117

Xenon 265
see also Group 0 elements

Zeolites 168
Zinc 111
blende 116, 117
comparison with beryllium 115
see also Group IIb elements
Zone refining 173

The Periodic Table

emphasizing the differing electron configurations among the non-transition elements

Group I	Group II							
							1 **H** 1·008 $1s^1$	2 **He** 4·003 $1s^2$
3 **Li** 6·941 [He] $2s^1$	4 **Be** 9·012 [He] $2s^2$							
11 **Na** 22·990 [Ne] $3s^1$	12 **Mg** 24·305 [Ne] $3s^2$	← d transition elements —						
19 **K** 39·102 [Ar] $4s^1$	20 **Ca** 40·08 [Ar] $4s^2$	21 **Sc** 44·956 [Ar] $3d^1\ 4s^2$	22 **Ti** 47·90 [Ar] $3d^2\ 4s^2$	23 **V** 50·941 [Ar] $3d^3\ 4s^2$	24 **Cr** 51·996 [Ar] $3d^5\ 4s^1$	25 **Mn** 54·938 [Ar] $3d^5\ 4s^2$	26 **Fe** 55·847 [Ar] $3d^6\ 4s^2$	27 **Co** 58·933 [Ar] $3d^7\ 4s^2$
37 **Rb** 85·47 [Kr] $5s^1$	38 **Sr** 87·62 [Kr] $5s^2$	39 **Y** 88·906 [Kr] $4d^1\ 5s^2$	40 **Zr** 91·22 [Kr] $4d^2\ 5s^2$	41 **Nb** 92·906 [Kr] $4d^3\ 5s^2$	42 **Mo** 95·94 [Kr] $4d^5\ 5s^1$	43 **Tc** (99) [Kr] $4d^5\ 5s^2$	44 **Ru** 101·07 [Kr] $4d^6\ 5s^2$	45 **Rh** 102·91 [Kr] $4d^7\ 5s^2$
55 **Cs** 132·91 [Xe] $6s^1$	56 **Ba** 137·34 [Xe] $6s^2$	57 **La** 138·91 [Xe] $5d^1$ $6s^2$	72 **Hf** 178·49 [Xe] $4f^{14}$ $5d^2 6s^2$	73 **Ta** 180·95 [Xe] $4f^{14}$ $5d^3 6s^2$	74 **W** 183·85 [Xe] $4f^{14}$ $5d^4 6s^2$	75 **Re** 186·2 [Xe] $4f^{14}$ $5d^5 6s^2$	76 **Os** 190·2 [Xe] $4f^{14}$ $5d^6 6s^2$	77 **Ir** 192·22 [Xe] $4f^{14}$ $5d^7 6s^2$
87 **Fr** (223) [Rn] $7s^1$	88 **Ra** 226·025 [Rn] $7s^2$	89 **Ac** 227·0 [Rn] $6d^1$ $7s^2$	104 **Eka-Hf**	105 **Eka-Ta**	106 **Eka-W**			
			continuation of 6d transition series					
			58 **Ce** 140·12	59 **Pr** 140·91	60 **Nd** 144·24	61 **Pm** (147)	62 **Sm** 150·35	63 **Eu** 151·96
			90 **Th** 232·04	91 **Pa** (231)	92 **U** 238·03	93 **Np** (237)	94 **Pu** (242)	95 **Am** (243)